METHODS IN MOLECULAR BIOLOGY

Series Editor
John M. Walker
School of Life and Medical Sciences
University of Hertfordshire
Hatfield, Hertfordshire, UK

For further volumes:
http://www.springer.com/series/7651

For over 35 years, biological scientists have come to rely on the research protocols and methodologies in the critically acclaimed Methods in Molecular Biology series. The series was the first to introduce the step-by-step protocols approach that has become the standard in all biomedical protocol publishing. Each protocol is provided in readily-reproducible step-by step fashion, opening with an introductory overview, a list of the materials and reagents needed to complete the experiment, and followed by a detailed procedure that is supported with a helpful notes section offering tips and tricks of the trade as well as troubleshooting advice. These hallmark features were introduced by series editor Dr. John Walker and constitute the key ingredient in each and every volume of the Methods in Molecular Biology series. Tested and trusted, comprehensive and reliable, all protocols from the series are indexed in PubMed.

Manuscript Instructions https://drive.google.com/file/d/1K4Az5vpAFVzYadJ3jKx2-QSo7Dqa_SNA/view?usp=sharing

Muscle Stem Cells

Methods and Protocols

Edited by

Nathalie Didier

Institut Mondor de Recherche Biomédicale, INSERM U955, Créteil, France

Caroline E. Brun

Pathophysiology and Genetics of Neuron and Muscle (PGNM), NeuroMyoGene Institute, CNRS UMR5261, Inserm U1315, Université Claude Bernard Lyon 1, Lyon, France

Humana Press

Editors

Nathalie Didier
Institut Mondor de Recherche Biomédicale
INSERM U955
Créteil, France

Caroline E. Brun
Pathophysiology and Genetics of Neuron and Muscle (PGNM)
NeuroMyoGene Institute
CNRS UMR5261, Inserm U1315
Université Claude Bernard Lyon 1
Lyon, France

ISSN 1064-3745 ISSN 1940-6029 (electronic)
Methods in Molecular Biology
ISBN 978-1-0716-5642-6 ISBN 978-1-0716-5643-3 (eBook)
https://doi.org/10.1007/978-1-0716-5643-3

This Humana imprint is published by the registered company Springer Science+Business Media, LLC, part of Springer Nature.
The registered company address is: 1 New York Plaza, New York, NY 10004, U.S.A.

Preface

Adult muscle stem cells, also known as satellite cells, play a central role in skeletal muscle growth, maintenance, and regeneration, and they remain a major focus of investigation in both basic and translational research. Our understanding of the fine-tuned regulation of muscle stem cells continues to expand, thanks to methodological advances and the development of innovative protocols that enable the dissection of the molecular and cellular changes these cells undergo throughout life. This volume of *Methods in Molecular Biology* series provides an experimental roadmap for interrogating muscle stem cell biology, from tissue dissociation and cell isolation to ex vivo analysis, in vitro modeling, functional assays, and multiomics technologies.

One strength of this protocol collection is its coverage of methods that capture or preserve the physiological state of muscle stem cells in situ and ex vivo. Several chapters focus on the analysis of quiescent cells, freshly isolated populations, and intact myofiber-associated stem cells, allowing investigators to study muscle stem cells within their native niche or under minimally perturbing conditions. Other contributions describe approaches for profiling gene expression, chromatin accessibility, chromatin architecture, and proteomic states using low-input or highly optimized protocols, reflecting the technical innovations that now make it possible to study rare cell populations with increasing precision. Underscoring the importance of environmental context for muscle stem cell biology, we also provide methods for in situ imaging, computational analysis of immunofluorescence, cross-grafting, and coculture with niche-associated cells. Altogether, these protocols provide valuable tools for investigating how intrinsic properties and extrinsic cues cooperate to regulate stem cell fate decisions, including quiescence, activation, proliferation, differentiation, and self-renewal. Finally, this *Muscle Stem Cells* volume brings together protocols applicable to human muscle stem cells and skeletal muscle tissue, as well as approaches to generate human muscle progenitors and model human muscle stem cell fate in a dish. These chapters reflect the growing need to bridge mechanistic insights from animal models with human-relevant systems, and to support studies increasingly oriented toward disease modeling, regenerative medicine, and therapeutic translation.

We are deeply grateful to all contributing authors for sharing their expertise and for preparing clear, practical, and carefully validated protocols. We hope that this volume will serve as a useful resource for both new and experienced investigators working in the field of muscle stem cell biology, and that it will facilitate rigorous experimentation, methodological innovation, and further discoveries in skeletal muscle regeneration.

Créteil, France — *Nathalie Didier*
Lyon, France — *Caroline E. Brun*

Contents

Contributors

LUCIA ACAMPORA • *Institut Pasteur, Université Paris Cité, UMR CNRS 3738, Stem Cells and Development Unit, Paris, France; Department of Clinical Medicine and Surgery, University of Naples Federico II, Naples, Italy*

LUCIE ANCEL • *Pathophysiology and Genetics of Neuron and Muscle (PGNM), NeuroMyoGene Institute, CNRS UMR5261, Inserm U1315, Université Claude Bernard Lyon 1, Lyon, France*

GHISLAIN BANOS • *Univ Paris Est Creteil, INSERM, EFS, IMRB, Creteil, France*

JESUS R. BARAJAS • *Center for Cardiovascular and Muscular Diseases, Sanford Burnham Prebys Medical Discovery Institute, La Jolla, CA, USA*

MONA BENSALAH • *Sorbonne Université, Inserm, Institut de Myologie, Centre de Recherche en Myologie, Paris, France*

JULIETTE BERTHIER • *Univ Paris Est Creteil, IMRB, INSERM U955, Creteil, France*

ANNE BIGOT • *Sorbonne Université, Inserm, Institut de Myologie, Centre de Recherche en Myologie, Paris, France*

DARREN M. BLACKBURN • *Department of Biochemistry, Microbiology and Immunology, University of Ottawa, Ottawa, ON, Canada*

ANAÏS BLEUZEN • *Univ Paris Est Creteil, INSERM, EFS, IMRB, Creteil, France*

CAROLINE E. BRUN • *Pathophysiology and Genetics of Neuron and Muscle (PGNM), NeuroMyoGene Institute, CNRS UMR5261, Inserm U1315, Université Claude Bernard Lyon 1, Lyon, France*

MARTINA CANATELLI-MALLAT • *Laboratory Lille Neuroendocrinology, Lille Neuroscience and Cognition, CHU Lille University of Lille, Inserm, Lille, France*

LUCA CAPUTO • *Center for Cardiovascular and Muscular Diseases, Sanford Burnham Prebys Medical Discovery Institute, La Jolla, CA, USA*

INDIGO T. C. CHAN • *Division of Life Science, The Hong Kong University of Science and Technology, Hong Kong, China*

TOM H. CHEUNG • *Division of Life Science, The Hong Kong University of Science and Technology, Hong Kong, China; Hong Kong Center for Neurodegenerative Diseases, Hong Kong, China; Center for Stem Cell Research, State Key Laboratory of Nervous System Disorders, Daniel and Mayce Yu Molecular Neuroscience Center, SIAT-HKUST Joint Laboratory for Brain Science, The Hong Kong University of Science and Technology, Hong Kong, China; Guangdong Provincial Key Laboratory of Brain Science, Disease and Drug Development, Shenzhen-Hong Kong Institute of Brain Science, HKUST Shenzhen Research Institute, Shenzhen, China*

EMILIE CHRISTIN • *Pathophysiology and Genetics of Neuron and Muscle (PGNM), NeuroMyoGene Institute, CNRS UMR5261, Inserm U1315, Université Claude Bernard Lyon 1, Lyon, France*

DELIA CICCIARELLO • *Pathophysiology and Genetics of Neuron and Muscle (PGNM), NeuroMyoGene Institute, CNRS UMR5261, Inserm U1315, Université Claude Bernard Lyon 1, Lyon, France*

COLIN CRIST • *Department of Human Genetics, McGill University, Montreal, QC, Canada; Lady Davis Institute for Medical Research, Jewish General Hospital, Montreal, QC, Canada*

CHIARA D'ERCOLE • *Sorbonne Université, INSERM UMRS 974, Association Institut de Myologie, Centre de Recherche en Myologie, Paris, France*

REBECCA DESAULNIERS • *CHU Sainte-Justine Azrieli Research Center, Montreal, QC, Canada; Faculty of Medicine, Department of Pharmacology and Physiology, Université de Montréal, Montréal, QC, Canada*

NATHALIE DIDIER • *Univ Paris Est Creteil, INSERM, EFS, IMRB, Creteil, France*

CHRISTELLE DUBOIS • *Univ Paris Est Creteil, IMRB, INSERM U955, Creteil, France*

NICOLAS A. DUMONT • *CHU Sainte-Justine Azrieli Research Center, Montreal, QC, Canada; Faculty of Medicine, School of Rehabilitation, Université de Montréal, Montréal, QC, Canada*

JOANA ESTEVES DE LIMA • *University Paris Est Creteil, INSERM, EnvA, EFS, AP-HP, IMRB, Creteil, France*

BRENDAN EVANO • *Stem Cells and Development, Department of Developmental and Stem Cell Biology, Institut Pasteur, Université Paris Cité, Paris, France; CNRS UMR 3738, Institut Pasteur, Paris, France*

EMELINE FARAÜS • *Univ Paris Est Creteil, INSERM, EFS, IMRB, Creteil, France*

VINCENT GACHE • *Pathophysiology and Genetics of Neuron and Muscle (PGNM), NeuroMyoGene Institute, CNRS UMR5261, Inserm U1315, Université Claude Bernard Lyon 1, Lyon, France*

PAULINE GARCIA • *CHU Sainte-Justine Azrieli Research Center, Montreal, QC, Canada; Faculty of Medicine, Department of Pharmacology and Physiology, Université de Montréal, Montréal, QC, Canada*

BARBARA GAYRAUD-MOREL • *Institut Pasteur, Université Paris Cité, UMR CNRS 3738, Stem Cells and Development Unit, Paris, France*

MARIANNE GERVAIS • *Univ Paris Est Creteil, IMRB, INSERM U955, Creteil, France*

LORENZO GIORDANI • *Sorbonne Université, INSERM UMRS 974, Association Institut de Myologie, Centre de Recherche en Myologie, Paris, France*

STÉPHANIE GOBERT • *Pathophysiology and Genetics of Neuron and Muscle (PGNM), NeuroMyoGene Institute, CNRS UMR5261, Inserm U1315, Université Claude Bernard Lyon 1, Lyon, France*

NIMISHA GUPTA • *Department of Microbiology, Immunology and Molecular Genetics, University of California, Los Angeles, CA, USA*

ZACHARY GURLEKIAN • *CHU Sainte-Justine Azrieli Research Center, Montreal, QC, Canada; Faculty of Medicine, Department of Pharmacology and Physiology, Université de Montréal, Montréal, QC, Canada*

GARY J. HE • *Division of Life Science, The Hong Kong University of Science and Technology, Hong Kong, China; Hong Kong Center for Neurodegenerative Diseases, Hong Kong, China*

MICHAEL R HICKS • *Sue and Bill Gross Stem Cell Research Center, University of California, Irvine, CA, USA; Department of Physiology and Biophysics, School of Medicine, University of California, Irvine, CA, USA*

DIEU-HUONG HOANG • *Department of Microbiology, Immunology and Molecular Genetics, University of California, Los Angeles, CA, USA*

FAYEZ ISSA • *University Paris Est Creteil, INSERM, EnvA, EFS, AP-HP, IMRB, Creteil, France*

GAËTAN JUBAN • *Pathophysiology and Genetics of Neuron and Muscle (PGNM), NeuroMyoGene Institute, CNRS UMR5261, Inserm U1315, Université Claude Bernard Lyon 1, Lyon, France*

PIERRE KLEIN • *Sorbonne Université, Inserm, Institut de Myologie, Centre de Recherche en Myologie, Paris, France*

ANITA KNEPPERS • *Pathophysiology and Genetics of Neuron and Muscle (PGNM), NeuroMyoGene Institute, CNRS UMR5261, Inserm U1315, Université Claude Bernard Lyon 1, Lyon, France*

ROBERT S. KRAUSS • *Department of Stem Cell Biology and Regenerative Medicine, Icahn School of Medicine at Mount Sinai, New York, NY, USA*

RÉMI KURDIAN • *Laboratory Lille Neuroendocrinology, Lille Neuroscience and Cognition, CHU Lille University of Lille, Inserm, Lille, France; Department of Human Genetics, McGill University, Montreal, QC, Canada; Lady Davis Institute for Medical Research, Jewish General Hospital, Montreal, QC, Canada*

THOMAS LAUMONIER • *Department of Orthopedic Surgery, Geneva University Hospitals and Faculty of Medicine, Geneva, Switzerland; Department of Cell Physiology and Metabolism, Faculty of Medicine, Geneva, Switzerland*

LOLA E. R. LESSARD • *Pathophysiology and Genetics of Neuron and Muscle (PGNM), NeuroMyoGene Institute, CNRS UMR5261, Inserm U1315, Université Claude Bernard Lyon 1, Lyon, France*

HSIAO-FAN LO • *Department of Stem Cell Biology and Regenerative Medicine, Icahn School of Medicine at Mount Sinai, New York, NY, USA*

HA MY LY • *Pathophysiology and Genetics of Neuron and Muscle (PGNM), NeuroMyoGene Institute, CNRS UMR5261, Inserm U1315, Université Claude Bernard Lyon 1, Lyon, France*

JIMMY MASSENET • *Center for Cardiovascular and Muscular Diseases, Sanford Burnham Prebys Medical Discovery Institute, La Jolla, CA, USA*

ALICIA MAYEUF-LOUCHART • *Laboratory Lille Neuroendocrinology, Lille Neuroscience and Cognition, CHU Lille University of Lille, Inserm, Lille, France*

ORANE MERCIER • *Pathophysiology and Genetics of Neuron and Muscle (PGNM), NeuroMyoGene Institute, CNRS UMR5261, Inserm U1315, Université Claude Bernard Lyon 1, Lyon, France*

DIEGO MICHEL • *Department of Orthopedic Surgery, Geneva University Hospitals and Faculty of Medicine, Geneva, Switzerland; Department of Cell Physiology and Metabolism, Faculty of Medicine, Geneva, Switzerland*

JULIEN MIGNOT • *Univ Paris Est Creteil, INSERM, EFS, IMRB, Creteil, France*

RÉMI MOUNIER • *Pathophysiology and Genetics of Neuron and Muscle (PGNM), NeuroMyoGene Institute, CNRS UMR5261, Inserm U1315, Université Claude Bernard Lyon 1, Lyon, France*

LAURA MURAINE • *Sorbonne Université, Inserm, Institut de Myologie, Centre de Recherche en Myologie, Paris, France*

MELANIA MUROLO • *Institut Pasteur, Université Paris Cité, UMR CNRS 3738, Stem Cells and Development Unit, Paris, France*

ELISA NEGRONI • *Sorbonne Université, Inserm, Institut de Myologie, Centre de Recherche en Myologie, Paris, France*

Monica Nicolau • *Center for Cardiovascular and Muscular Diseases, Sanford Burnham Prebys Medical Discovery Institute, La Jolla, CA, USA; Graduate School of Biomedical Sciences, Sanford Burnham Prebys Medical Discovery Institute, La Jolla, CA, USA*

Chiara Nicoletti • *Center for Cardiovascular and Muscular Diseases, Sanford Burnham Prebys Medical Discovery Institute, La Jolla, CA, USA*

Teoman Ozturk • *Univ Paris Est Creteil, INSERM, EFS, IMRB, Creteil, France*

Eléonore Parisel • *Department of Orthopedic Surgery, Geneva University Hospitals and Faculty of Medicine, Geneva, Switzerland; Department of Cell Physiology and Metabolism, Faculty of Medicine, Geneva, Switzerland; Department of Pathology and Immunology, Faculty of Medicine, Geneva, Switzerland*

Alexandre Pintart • *Pathophysiology and Genetics of Neuron and Muscle (PGNM), NeuroMyoGene Institute, CNRS UMR5261, Inserm U1315, Université Claude Bernard Lyon 1, Lyon, France*

Pier Lorenzo Puri • *Center for Cardiovascular and Muscular Diseases, Sanford Burnham Prebys Medical Discovery Institute, La Jolla, CA, USA*

April D Pyle • *Department of Microbiology, Immunology and Molecular Genetics, University of California, Los Angeles, CA, USA; Eli and Edythe Broad Center of Regenerative Medicine and Stem Cell Research, University of California, Los Angeles, CA, USA*

Marie Quétin • *Univ Paris Est Creteil, IMRB, INSERM U955, Creteil, France*

Anna Rausch De Traubenberg • *Pathophysiology and Genetics of Neuron and Muscle (PGNM), NeuroMyoGene Institute, CNRS UMR5261, Inserm U1315, Université Claude Bernard Lyon 1, Lyon, France*

Ingo Riederer • *Laboratory on Thymus Research, Oswaldo Cruz Institute, Oswaldo Cruz Foundation, Rio de Janeiro, Brazil; National Institute of Science and Technology on Neuroimmunomodulation (INCT-NIM); Oswaldo Cruz Institute, Oswaldo Cruz Foundation, Rio de Janeiro, Brazil; Rio de Janeiro Research Network on Neuroinflammation, Oswaldo Cruz Institute, Oswaldo Cruz Foundation, Rio de Janeiro, Brazil*

Ludivine Rotard • *Pathophysiology and Genetics of Neuron and Muscle (PGNM), NeuroMyoGene Institute, CNRS UMR5261, Inserm U1315, Université Claude Bernard Lyon 1, Lyon, France*

Korin Sahinyan • *Lady Davis Institute for Medical Research, Jewish General Hospital, Montréal, QC, Canada*

Liza Sarde • *Stem Cells and Development, Department of Developmental and Stem Cell Biology, Institut Pasteur, Université Paris Cité, Paris, France; CNRS UMR 3738, Institut Pasteur, Paris, France*

Audrey Saugues • *Pathophysiology and Genetics of Neuron and Muscle (PGNM), NeuroMyoGene Institute, CNRS UMR5261, Inserm U1315, Université Claude Bernard Lyon 1, Lyon, France*

Isabella Scionti • *Pathophysiology and Genetics of Neuron and Muscle (PGNM), NeuroMyoGene Institute, CNRS UMR5261, Inserm U1315, Université Claude Bernard Lyon 1, Lyon, France*

Muriel Sébastien • *Pathophysiology and Genetics of Neuron and Muscle (PGNM), NeuroMyoGene Institute, CNRS UMR5261, Inserm U1315, Université Claude Bernard Lyon 1, Lyon, France*

MARIE-CLAUDE SINCENNES • *Centre Armand-Frappier Santé Biotechnologie, Institut National de la Recherche Scientifique, Laval, QC, Canada; Unité Mixte de Recherche INRS-UQAC en Santé Durable, Chicoutimi, QC, Canada*

VAHAB D. SOLEIMANI • *Department of Biochemistry, Microbiology and Immunology, University of Ottawa, Ottawa, ON, Canada; Regenerative Medicine Program, Ottawa Hospital Research Institute, Ottawa, ON, Canada; Department of Human Genetics, McGill University, Montréal, QC, Canada*

DANIEL STOCKHOLM • *Généthon, Evry, France; École Pratique des Hautes Études, PSL University, Paris, France*

SHAHRAGIM TAJBAKHSH • *Institut Pasteur, Université Paris Cité, UMR CNRS 3738, Stem Cells and Development Unit, Paris, France; Stem Cells and Development, Department of Developmental and Stem Cell Biology, Institut Pasteur, Université Paris Cité, Paris, France; CNRS UMR 3738, Institut Pasteur, Paris, France*

NING-SUM TO • *Division of Life Science, The Hong Kong University of Science and Technology, Hong Kong, China; Hong Kong Center for Neurodegenerative Diseases, Hong Kong, China*

SAM TORRES • *Department of Microbiology, Immunology and Molecular Genetics, University of California, Los Angeles, CA, USA*

LAURA VIRTANEN • *Sorbonne Université, INSERM UMRS 974, Association Institut de Myologie, Centre de Recherche en Myologie, Paris, France*

WENXIN ZHANG • *Division of Life Science, The Hong Kong University of Science and Technology, Hong Kong, China*

Part I

Tissue Preparation for Muscle-Resident Cell Isolation

Chapter 1

Optimized Dissociation of Skeletal Muscle Tissue to Recover Muscle Stem Cells

Teoman Ozturk, Julien Mignot and Nathalie Didier

Abstract

Adult skeletal muscle regenerative potential mainly relies on resident muscle stem cells (MuSC), tightly anchored in a specialized niche. These cells are widely used for mechanistic *in vitro* assays and are promising candidates for cell-based therapies. However, muscle tissue enzymatic dissociation required to isolate MuSC remains a critical step that can deeply alter their viability and functionality. Here, we provide a step-by-step method to dissociate skeletal muscle tissue, optimized to preserve the viability of MuSC and improve their purification yield.

Keywords Skeletal muscle tissue, Dissociation, Muscle stem cells

1 Introduction

Adult skeletal muscle consists of multinucleated muscle fibers, surrounded and maintained together by successive and intricate networks of extracellular matrix (ECM), largely composed of collagens, types I and III being the most abundant [1]. In addition, the regenerative capacity of this tissue is sustained by resident muscle stem cells (MuSC). In homeostatic muscle, MuSC are quiescent and express the transcription factor Pax7 [2]. Quiescent MuSC harbor long cytoplasmic extensions and are tightly anchored between the basal lamina and the myofiber sarcolemma, within a specialized niche actively maintaining their quiescent state and functional integrity [3–6]. Indeed, quiescent MuSC establish polarized adhesive interactions with the myofiber *via* cadherins on their apical side and with the basal lamina through laminin–integrin interactions on the basal side [7–10]. Moreover, the MuSC niche contains type V and VI collagen, which also contribute to quiescence maintenance [11,12].

Upon injury or stress stimuli, MuSC exit quiescence, activate, and then divide to subsequently differentiate and fuse to form new myofibers, or self-renew to restore the pool of quiescent MuSC

Nathalie Didier and Caroline E. Brun (eds.), *Muscle Stem Cells*, Methods in Molecular Biology, vol. 3093,
https://doi.org/10.1007/978-1-0716-5643-3_1,

[13,14]. Interestingly, this progression of MuSC in the myogenic lineage program can be recapitulated by *in vitro* assays, providing a simplified and tractable system to study myogenesis. Moreover, the robust engraftment potential of MuSC following *in vivo* transplantation makes them promising candidates for cell-based therapies [15,16].

However, the intricate anatomical location of MuSC renders their isolation a time-consuming and laborious process, which remains a limiting factor for these applications. Indeed, releasing MuSC from their niche requires extensive enzymatic digestion of the muscle tissue leading to a massive activation of stress signaling pathways, quiescence break, rapid decline of Pax7 and reduced viability [17–19]. In agreement, inhibition of stress signaling pathways (*e.g.*, P38 MAPK and JNK) by pharmacological treatments during the dissociation process, significantly improved MuSC purification yield, viability, and engraftment efficacy [19]. Therefore, muscle enzymatic dissociation is a critical step to purify MuSC with preserved myogenic properties.

Here, we provide a protocol for muscle tissue dissociation combining enzymatic digestion of collagens and basal lamina components by Collagenase A (from *Clostridium histolyticum*) and Dispase II (from *Bacillus polymyxa*), respectively [20,21], and mechanical dissociation. This protocol does not require costly or dedicated equipment and has been optimized to reduce the number of steps and digestion time to preserve MuSC viability and improve purification yield. We describe in detail the different steps and critical points from muscle harvest to the recovery of muscle tissue mononucleated cells.

2 Materials

2.1 Equipment and Materials

1. Sterile 60 mm Petri dishes
2. Curved Iris scissors
3. Microdissecting forceps
4. Water bath (37°C)
5. Centrifuge for 50 mL tubes
6. 15 and 50 mL plastic tubes
7. 5 and 10 mL plastic pipettes
8. Sterile glass Pasteur pipettes
9. 2 mL pipette latex bulbs
10. Bunsen burner
11. 40 μm cell strainers
12. 0.22 μm filters

2.2 Buffers and Media

1. Sterile Hanks' Balanced salt solution 1× (HBSS 1×)
2. Digestion buffer (**DB**): 3 U/mL type II Dispase, 0.4–0.5 U/mL Collagenase A, 8 mM $CaCl_2$, 5 mM $MgCl_2$, 20 U/mL DNAse I (*see* **Notes 1 and 2**)
3. Washing buffer (**WB**): HBSS 1× added with 0.2% Bovine Serum Albumin (BSA), 0.5% antibiotics
4. Recovery buffer (**RB**): HBSS 1× added with 0.2% Bovine Serum albumin (BSA), 0.5% antibiotics, 10% fetal calf serum (FCS)

3 Methods

The method and the volumes indicated below are given for the digestion of 1 g of adult mouse muscles. Mice were handled according to French and European Community guidelines (APAFIS#51690-2024091111597155 v11). To ensure effective digestion and recover high-quality MuSC, it is essential to strictly follow the indicated times (Fig. 1a–d). This protocol can also be applied to the dissociation of rat and human muscles.

3.1 Preparation of Buffers and Materials (Timing: 30 min)

1. Prepare 40 mL of digestion buffer (DB) per gram of muscle to digest (*see* **Note 3**). Resuspend Dispase II in 40 mL of HBSS1×, and then add the other components. Filter the solution through a 0.22 μm filter.
2. Prepare 500 mL of washing buffer (WB, *see* **Note 4**).
3. Prepare a glass Pasteur pipette by heating its tip over a Bunsen burner to smooth the edges and reduce the opening. Fill a 15 mL tube with 8 mL of DB. Coat the glass Pasteur pipette topped with a latex bulb with the DB by performing several up-and-downs and leave it aside filled with DB for the recovery step. This is important to avoid muscle pieces sticking into the pipette afterward.
4. Add 10 mL of recovery buffer (RB) in two 50 mL tubes and keep on ice.

3.2 Muscle Harvesting and Mincing (Timing: 15 min)

1. Euthanize the mouse and clean the skin with 70% ethanol.
2. Incise the skin and dissect the muscles using sterile dissection tools and immediately place them in a sterile Petri dish containing 1 mL of WB (Fig. 2a and *see* **Note 5**).
3. Wash off the hairs and remove tendons and fat tissue if needed and transfer the muscles in a new Petri dish.
4. Add 500 μL of DB and start mincing the muscles using sharp, sterile curved Iris scissors until getting a homogeneous slurry paste (Fig. 2b and *see* **Note 6**).

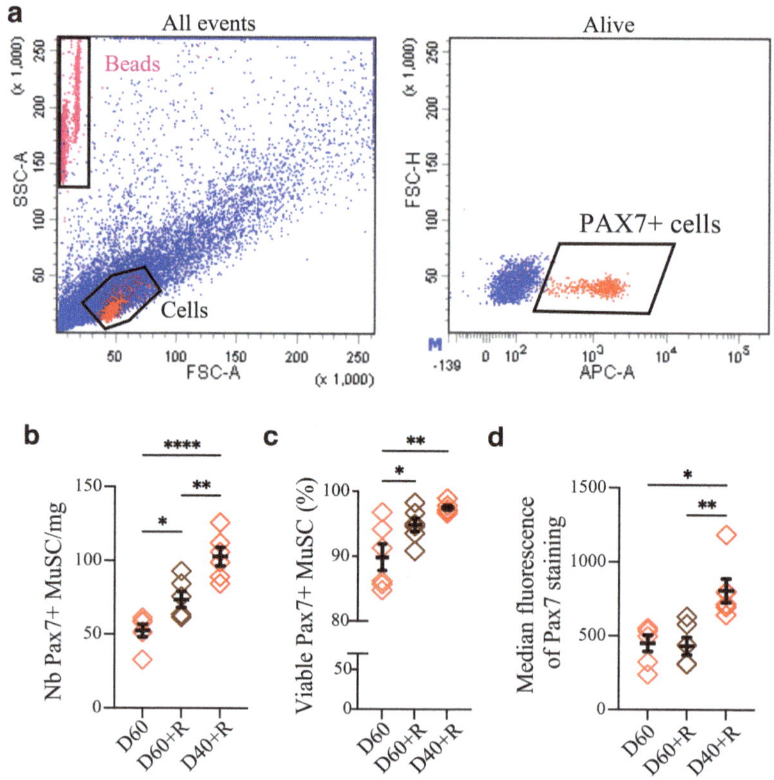

Fig. 1 Digestion time is a crucial factor in the isolation of MuSC. **a,** Representative dot plots showing the strategy used to quantify Pax7+ MuSC on mononucleated cells from single Tibialis Anterior (TA) muscle by flow cytometry, using Trucount tubes (BD). Doublets and dead cells were excluded, and acquisitions were performed on 20,000 beads. **b,** Quantification of the number of Pax7+ MuSC normalized per mg of muscle, depending on different protocols of digestion: D60: continuous digestion during 60 min; D60+R: 60 min of digestion with intermediate recoveries every 20 min; D40+R: 40 min of digestion with intermediate recovery at 20 min. **c,** Percentage of viable Pax7+ MuSC. **c,** Median of Pax7 staining fluorescence intensity. Data are presented as the mean ± sem of six independent experiments ($n = 6$). One Way ANOVA, with $*p < 0.05$, $**p < 0.01$, and $****p < 0.0001$

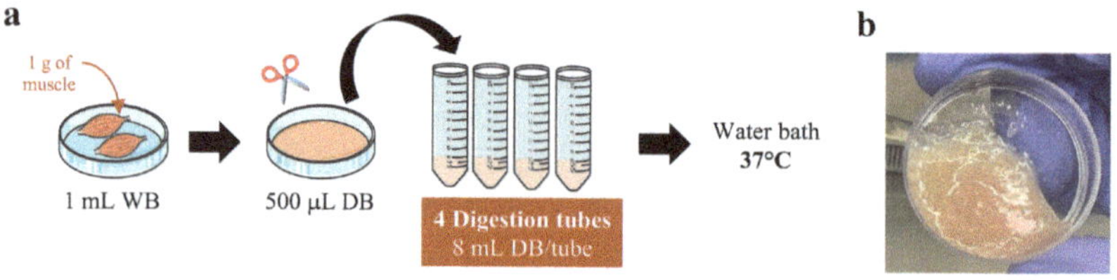

Fig. 2 Muscle harvesting and mincing. **a,** Schematic representation of the key steps from muscle harvest to the enzymatic dissociation. WB: washing buffer, DB: digestion buffer. **b,** Photo showing the suitable consistence of the muscle paste obtained after mincing

3.3 Enzymatic Dissociation of the Muscles (Timing: 50 min)

1. Transfer the muscle paste in four 50 mL tubes containing 8 mL of DB each (referred to as Digestion tubes) using a 5 mL pipette previously coated with DB.
2. Cap the Digestion tubes and place them in a water bath at 37°C to start the digestion. Carefully agitate the Digestion tubes every 5–10 min to avoid muscle paste aggregation (*see* **Note 7**).
3. After 20 min of digestion, recover the Digestion tubes and proceed to eight to ten up-and-downs with the glass Pasteur pipette topped with a latex bulb (previously coated with DB) and wait 2 min for the non-digested pieces to fall down (Fig. 3a, b).
4. Recover 2 mL of the upper phase of each Digestion tubes using the Pasteur pipette, transfer them in the Recovery tubes and keep on ice.
5. Add 2 mL of fresh DB in the Digestion tubes, put them back in the water bath at 37°C and agitate every 5–10 min as explained in Step 2.
6. After 20 min, removed the Digestion tubes from the water and bath and proceed to up-and-downs with the glass Pasteur pipette as in Step 3 (*see* **Note 8**).
7. Transfer the contents of the four Digestion tubes into the two Recovery tubes and add WB up to 40 mL.

3.4 Filtration and Washes (Timing: 20 min)

1. Split the two Recovery tubes in four tubes and fill up to 30 mL with WB.

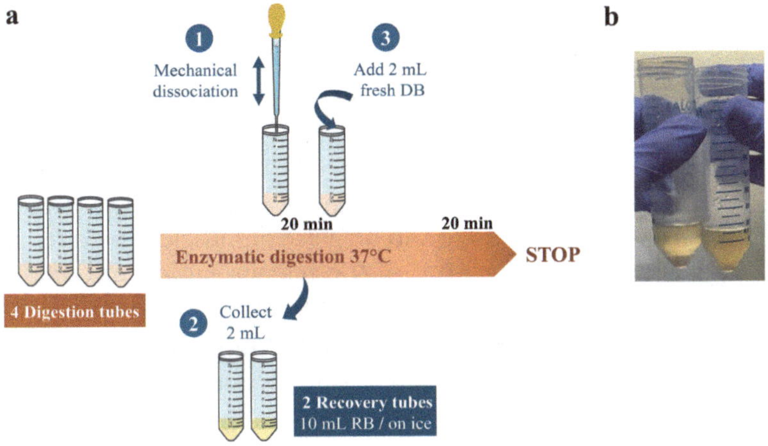

Fig. 3 Dissociation of the muscle tissue. **a,** Schematic representation of the recovery step at 20 min of enzymatic digestion. DB: digestion buffer; RB: recovery buffer. **b,** Photo of Digestion tubes after 20 min of digestion, prior (left) and after mechanical dissociation (right)

2. Filter the cell suspensions through 40 μm cell strainers using a 10 mL pipette, in four Filtration tubes (*see* **Note 9**). Wash the strainers with 5 mL of WB.
3. Centrifuge the Filtration tubes for 10 min at 300 g.
4. Carefully aspire the supernatant, resuspend the pellets in 500 μL of WB and pool them in a single tube. At this step, the cell suspension obtained contains the mononucleated cells of the muscle tissue, including the MuSC (*see* **Note 10**).

4 Notes

1. Stock solutions of Collagenase A (100 mg/mL) and DNAse I (10 mg/mL) can be prepared in advance in HBSS 1×, filtered through 0.22 μm filters, aliquoted and frozen at −20°C for further use. However, we recommend preparing Dispase II solution extemporaneously for optimal digestion.
2. Digestion efficiency and cell viability can significantly vary depending on Dispase II batches. We recommend testing new batch and adjust the enzymatic activity if needed prior conducting crucial experiment.
3. As a general guideline, 40 mL of digestion buffer is required for dissociating the muscles of two hindlimbs of an adult mouse (800 mg to 1 g), and 80 mL for the muscles of the four limbs.
4. Washing Buffer can be stored at 4°C for up to 1-month.
5. From this step, work under a laminar flow hood if cells are dedicated to *in vitro* culture.
6. Avoid leaving big muscles pieces. This step usually takes 6–8 min of active mincing.
7. If shaking the Digestion tube is not sufficient to eliminate the aggregates, use a 5 mL pipette coated with DB to disperse them. This is important for effective digestion.
8. At this step (40 min of digestion), the digestion should be completed, and no piece of muscle should remain. Otherwise, transfer 7 mL of the cell suspension into the recovery tubes, and put the digestion tubes back in the water bath for 10 min.
9. Start by collecting and filtrating the cell suspension from the surface to prevent obstruction of the strainer.
10. At the end of the digestion process, mononucleated cells can be utilized for the intended applications or frozen in FCS/DMSO (90/10) for further use.

Acknowledgments

N.D. receives financial support from the Agence Nationale de la Recherche (ANR-23-CE17-0035 and ANR-25-CE14-2467-01) and the Association Française contre les Myopathies (AFM Téléthon, Translamuscle II Pole). Some images of the figures were adapted from Servier Medical Art, licensed under CC BY 4.0.

Competing Interests The authors declare no competing interests.

References

1. McKee TJ, Perlman G, Morris M, Komarova SV (2019) Extracellular matrix composition of connective tissues: a systematic review and meta-analysis. Sci Rep 9: 10542
2. Seale P, Sabourin LA, Girgis-Gabardo A et al. (2000) Pax7 is required for the specification of myogenic satellite cells. Cell 102:777–786
3. Hung M, Lo H-F, Jones GEL, Krauss RS (2023) The muscle stem cell niche at a glance. J Cell Sci 136:jcs261200
4. Relaix F, Bencze M, Borok MJ et al. (2021) Perspectives on skeletal muscle stem cells. Nat Commun 12: 692
5. Kann AP, Hung M, Wang W et al. (2022) An injury-responsive Rac-to-Rho GTPase switch drives activation of muscle stem cells through rapid cytoskeletal remodeling. Cell Stem Cell 29:933–947.e6
6. Guillon E, Bacar H, Gilquin L et al. (2025) Unraveling the matrisome signatures of quiescent and activated muscle stem cells. Stem Cell Rep 20:102635
7. Hung M, Lo H-F, Jones GEL, Krauss RS (2023) The muscle stem cell niche at a glance. J Cell Sci 136:jcs261200
8. Hung M, Lo H-F, Beckmann AG et al. (2024) Cadherin-dependent adhesion is required for muscle stem cell niche anchorage and maintenance. Development 151:dev202387
9. Goel AJ, Rieder M-K, Arnold H-H et al. (2017) Niche cadherins control the quiescence-to-activation transition in muscle stem cells. Cell Rep 21:2236–2250
10. Rozo M, Li L, Fan C-M (2016) Targeting β 1-integrin signaling enhances regeneration in aged and dystrophic muscle in mice. Nat Med 22:889–896
11. Baghdadi MB, Castel D, Machado L et al. (2018) Reciprocal signalling by Notch–Collagen V–CALCR retains muscle stem cells in their niche. Nature 557:714–718
12. Urciuolo A, Quarta M, Morbidoni V et al. (2013) Collagen VI regulates satellite cell self-renewal and muscle regeneration. Nat Commun 4: 1964
13. Chargé SBP, Rudnicki MA (2004) Cellular and molecular regulation of muscle regeneration. Physiol Rev 84:209–238
14. Zammit PS, Golding JP, Nagata Y et al. (2004) Muscle satellite cells adopt divergent fates: a mechanism for self-renewal?. J Cell Biol 166:347–357
15. Sacco A, Doyonnas R, Kraft P et al. (2008) Self-renewal and expansion of single transplanted muscle stem cells. Nature 456:502–506
16. Montarras D, Morgan J, Collins C et al. (2005) Direct isolation of satellite cells for skeletal muscle regeneration. Science 309:2064–2067
17. Machado L, Esteves de Lima J, Fabre O et al. (2017) In situ fixation redefines quiescence and early activation of skeletal muscle stem cells. Cell Rep 21:1982–1993
18. Machado L, Geara P, Camps J et al. (2021) Tissue damage induces a conserved stress response that initiates quiescent muscle stem cell activation. Cell Stem Cell 28:1125–1135.e7
19. Ozturk T, Mignot J, Gattazzo F et al. (2024) Dual inhibition of P38 MAPK and JNK pathways preserves stemness markers and alleviates premature activation of muscle stem cells during isolation. Stem Cell Res Ther 15:179
20. Lim LS, Riau A, Poh R et al. (2009) Effect of dispase denudation on amniotic membrane. Mol Vis 15:1962–1970
21. Stenn KS, Link R, Moellmann G et al. (1989) Dispase, a neutral protease from *Bacillus polymyxa*, is a powerful fibronectinase and type IV collagenase. J Invest Dermatol 93:287–290

Check for updates

Chapter 2

Isolation of Mouse, Rat, and Human Muscle Stem Cells by Fluorescence-Activated Cell Sorting Based on α7-Integrin

Anaïs Bleuzen, Ghislain Banos, Teoman Ozturk, Emeline Faraüs, Julien Mignot and Nathalie Didier

Abstract

Muscle stem cells (MuSC) are necessary for skeletal muscle tissue homeostasis and repair following injury. These cells are therefore extensively studied and are considered promising candidates for cell-based therapies, rendering their purification a recurring technical requirement. For this purpose, adhesion molecules expressed on MuSC surface represent ideal positive selection markers, for magnetic-activated cell sorting (MACS) or fluorescence-activated cell sorting (FACS). Here, we describe a method for efficiently isolating MuSC from postnatal and adult muscles of mice, rats, and humans by FACS, using α7-integrin (ITGA7) as common selection marker. We provide a detailed protocol for the immunostaining of mononucleated cells derived from digested skeletal muscle and the gating strategies to obtain highly pure MuSC from the muscles of these three species.

Keywords Skeletal muscle, Muscle stem cells, Fluorescence-activated cell sorting, α7-integrin

1 Introduction

Adult skeletal muscle regenerative capacity relies on resident muscle stem cells (MuSC) expressing Pax7 transcription factor and residing in a specialized niche between the myofiber sarcolemma and the surrounding basal lamina [1–4]. Within this niche, adhesion molecules play a crucial role both for their physical anchoring, and for the regulation of the signaling cues that maintain their quiescent state or trigger their activation in response to muscle injury. Indeed, the basal surface of MuSC is enriched in integrins, notably α7β1-integrins, ensuring anchoring to the basal lamina, whereas on their apical side, interaction with the myofiber is mediated by cadherins, with M-Cadherin being the most abundant [5–8].

Owing to their low abundance, intricate anatomical localization, and the presence of others non-myogenic interstitial cell types (i.e., fibro-adipogenic progenitors, immune cells, endothelial cells),

Nathalie Didier and Caroline E. Brun (eds.), *Muscle Stem Cells*, Methods in Molecular Biology, vol. 3093,
https://doi.org/10.1007/978-1-0716-5643-3_2,

purification of MuSC remains a central technical requirement for in vitro mechanistic studies, cell-transplantation experiments, or cell-based therapeutic approaches. Therefore, aside from their functional roles, surface adhesion molecules have been advantageously used as positive selection markers for isolating MuSC by magnetic-activated cell sorting (MACS) and fluorescence-activated cell sorting (FACS) in the last two decades. Antibody-based recognition of α7-integrin (ITGA7) was the first method described for isolating MuSC-derived myoblasts from mouse muscle [9]. Next, the suitability of ITGA7 for isolating highly pure murine MuSC was further confirmed, and other positive selection markers were identified, including CD34, CXCR4, and CD106 (VCAM-1) [10–13]. Similarly, several purification strategies have been described for human MuSC based on single or combined selection markers, among which CD56 (NCAM), CD29 (β1-integrin), CXCR4, ITGA7, and CD82 [14–19]. However, likely due to the lack of commercially available conjugated antibodies, only CD106 has been reported as positive surface marker for the isolation of MuSC from rat muscle [20, 21].

In this chapter, we describe the immunostaining protocol for mononucleated cells derived from digested skeletal muscle, and three gating strategies for isolating highly pure MuSC (≥98% Pax7+ cells) from postnatal and adult mouse, rat, and human muscles by FACS. These strategies rely on ITGA7 as positive selection marker using the same antibody for the three species, combined with species-specific markers for excluding nonmyogenic lineages.

2 Materials

2.1 Equipment and Materials

1. 5 and 10 mL plastic pipettes
2. 40 μm cell strainers
3. Flow cytometry cell sorter with 405, 488, and 640 nm lasers and filters appropriate for each fluorophore used (see list of conjugated antibodies below)
4. Sorting tubes compatible with the cell sorter
5. Centrifuge for sorting tubes
6. Compensation beads

2.2 Buffers and Media

All buffers should be prepared under sterile conditions and kept on ice or at 4°C.

1. Washing buffer (**WB**): Hanks' balanced salt solution (HBSS) 1× added with 0.2% bovine serum albumin (BSA), 0.5% antibiotics.

2. Immunostaining and sorting buffer (**ISB**): WB added with 10 µg/mL DNAse I.
3. Red Blood cell lysis buffer.
4. Viability dye compatible with FACS (*e.g.*, 7AAD).
5. Fetal calf serum (FCS).
6. Blocking solution: 10% BSA.

2.3 Conjugated Antibodies (see Note 1)

1. **Antibodies for mouse MuSC sorting:** anti-mouse CD45-PE-Cy7 1:120 (#552848, BD Biosciences), anti-mouse TER-119-PE-Cy7 1:120 (#557853, BD Biosciences), anti-mouse CD31-PE-Cy7 1:120 (#561410, BD Biosciences), anti-mouse CD34-bV421 1:100 (#562608, BD Biosciences), anti-mouse Ly-6A/E-PE 1:100 (#553336, BD Biosciences), and anti-mouse ITGA7-FITC 1:80 (#LS-C179570, LSBio).
2. **Antibodies for rat MuSC sorting:** anti-rat CD45 PE-Cy7 1:100 (#561588, BD Biosciences), anti-rat CD31 PE-Cy7 1:100 (#50-112-3278, eBioscience), anti-rat CD29 Pacific Blue 1:150 (#102224, Biolegend), and anti-mouse ITGA7 FITC 1:80 (#LS-C179570, LSBio).
3. **Antibodies for human MuSC sorting:** anti-human CD45-PE 1:100 (#555482, BD Biosciences), anti-human CD235a-PE 1:150 (#555570, BD Biosciences), anti-human CD11b-PE 1:100 (#555388, BD Biosciences), anti-human CD31-PE 1:100 (#555466, BD Biosciences), anti-human CD34-bv421 1:100 (#562577, BD Biosciences), anti-human CD56-PE-Cy7 1:100 (#557747, BD Biosciences), anti-human CD29-AF700 1:150 (#303019, Biolegend), and anti-mouse ITGA7-FITC 1:80 (#LS-C179570, LSBio).

3 Methods

3.1 Immunostaining of Digested Muscle Mononucleated Cells

This protocol is given for the immunostaining of mononucleated cells obtained from the digestion of 1 g of muscle tissue (*see* **Notes 2 and 3**).

1. At the end of the enzymatic digestion process, centrifuge the cells for 10 min at 300 × *g*.
2. Aspirate the supernatant and leave around 300 µL volume.
3. Add 1 mL of red blood cell lysis buffer, resuspend the cells, and incubate for 15–20 min at room temperature (RT).
4. Add 2 mL of WB and centrifuge for 5 min at 300 × *g*.

5. Aspire, leave around 300 µL and resuspend the pellet.
6. Transfer the cells in two Sorting tubes (150 µL/tube).
7. Block with 150 µL of 10% BSA for 10 min at RT.
8. Add the mix of conjugated antibodies and incubate for 1 h on ice (*see* **Note 4**).
9. During antibody incubation, prepare the tubes that will be used for collecting the sorted cells (Collection tubes). Fill the tubes with 10% FCS, incubate for 1 h at RT, and then aspirate leaving a volume of 500 µL in the tubes (*see* **Note 5**).
10. After 1 h of incubation with antibodies, wash the cells with 1 mL of WB and centrifuge for 5 min at 300 × *g*.
11. Aspire the supernatant, resuspend the pellet with 300 µL of ISB and keep on ice until sorting.

3.2 Gating Strategies and MuSC Sorting

1. Just before sorting, filter the cells through 40 µm strainer and wash the strainer with 200 µL of ISB.
2. Add the viability dye according to manufacturer recommendation and incubate for 5 min.
3. Make an acquisition on a small number of events (50,000 total events) to properly position the gates, as illustrated in Fig. 1 (mouse muscle), Fig. 2 (rat muscle), and Fig. 3 (human muscle).

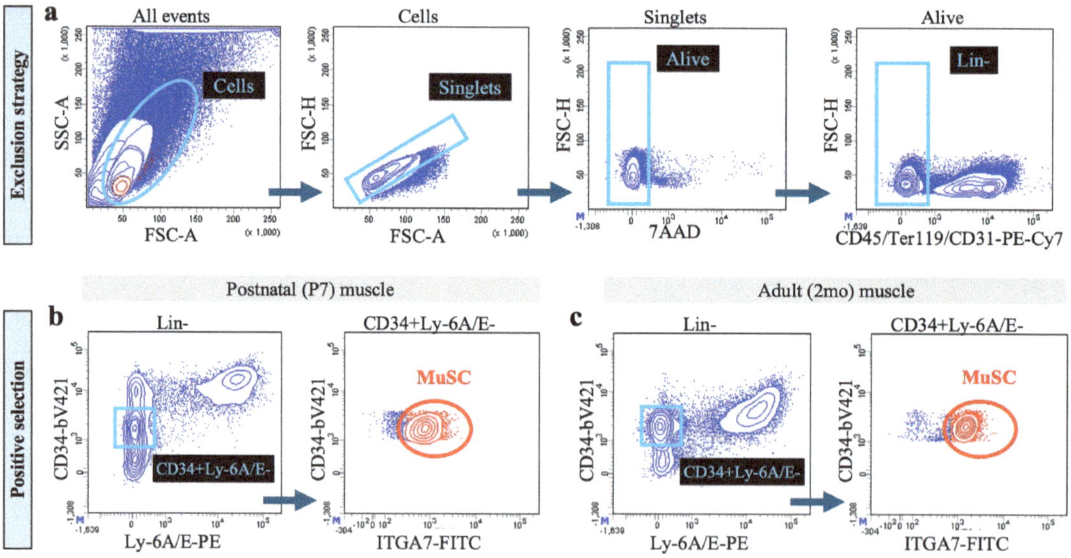

Fig. 1 Gating strategy to purify MuSC from postnatal or adult mouse muscles. Representative contour plots showing the gating strategy used to purify CD34+ITGA7+ MuSC from mononucleated cells derived from digested postnatal (P7) and adult (2-month-old) mouse muscles. **a,** Exclusion strategy: Debris, doublets, dead cells, and CD45+/Ter119+/CD31+ (Lin+) cells are excluded. **b** and **c**, Representative plots showing the positive selection strategy of CD34+ITGA7+ MuSC from postnatal (**b**), and adult (**c**) mouse muscles. Data were generated using BD FACSAria™ Fusion flow cytometer and BD FACSDiva™ Software

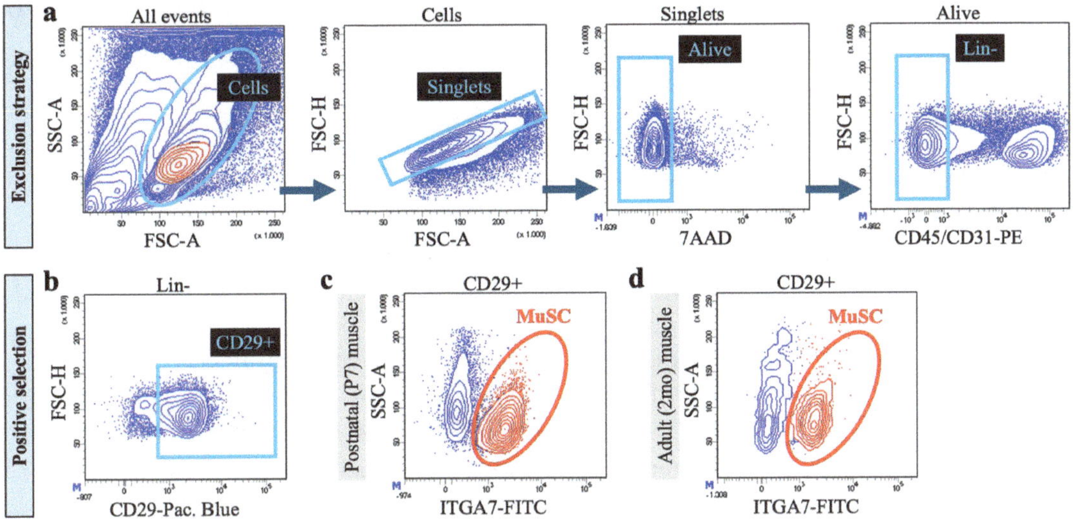

Fig. 2 Gating strategy to purify MuSC from postnatal or adult rat muscles. Representative contour plots showing the gating strategy used to purify CD29+ITGA7+ MuSC from mononucleated cells derived from digested postnatal (P7) and adult (2-month-old) rat muscles. **a,** Exclusion strategy: Debris, doublets, dead cells and CD45+/CD31+ (Lin+) cells are excluded. **b–d,** Representative plots showing the positive selection strategy of CD29+ITGA7+ MuSC from postnatal (**c**), and adult (**d**) rat muscles. Data were generated using BD FACSAria™ Fusion flow cytometer and BD FACSDiva™ Software

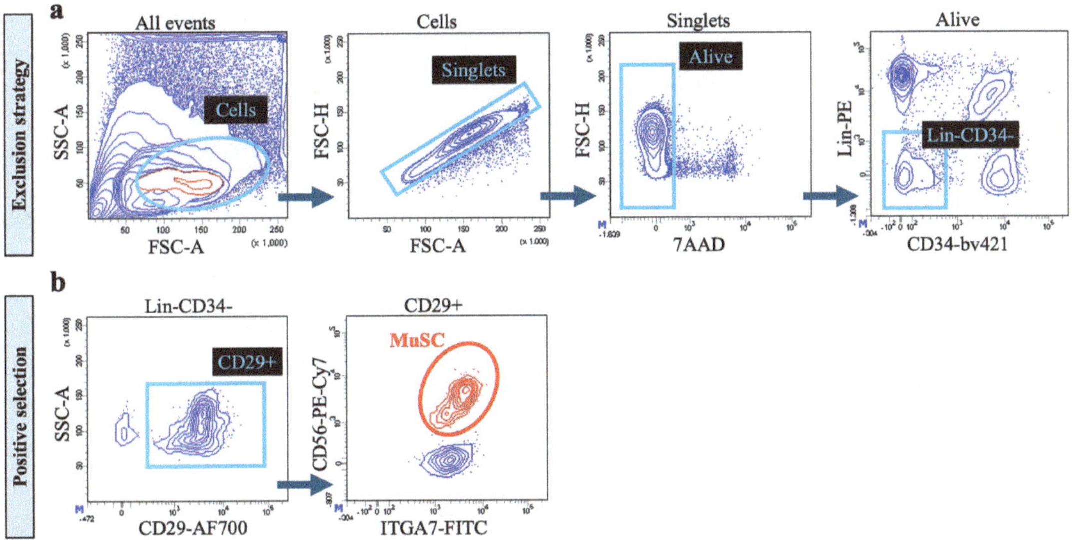

Fig. 3 Gating strategy to purify MuSC from adult human muscle. Representative contour plots showing the gating strategy used to purify CD29+ITGA7+CD56+ MuSC from mononucleated cells derived from digested human muscles. **a,** Exclusion strategy: Debris, doublets, dead cells, and Lin+ (CD45+/CD31+/CD235a+/CD11b+) and CD34+ cells are excluded. **b,** Representative plots showing the positive selection strategy of CD29+ITGA7+CD56+ MuSC from adult human muscle. The muscle biopsy used to generate these data was obtained from brain-dead organ donor with informed consent of the relatives and approval of the French Biomedicine Agency (PFS21-008), using BD FACSAria™ Fusion flow cytometer and BD FACSDiva™ Software

4. Proceed to the sorting in coated Collection tube (*see* **Note 6**).
5. After sorting, fill the Collection tube with WB and centrifuge for 5 min at 300 × *g* to pellet the cells.

4 Notes

1. Specificity of the staining obtained with all the listed conjugated antibodies was verified using the corresponding isotype controls, purchased from the same manufacturer. However, it is recommended to repeat all these controls, when transposing the method to new experimental conditions and equipment.
2. As a general guideline, mononucleated cells obtained from the muscles of two hindlimbs of an adult mouse have to be divided in two tubes while those obtained from the four limbs have to be divided in four tubes for optimal immunostaining.
3. If possible, turn-on the cell sorter, proceed to the startup procedure and prepare the template during the antibody incubation step. Sorting of MuSC can be performed both with 0.70 μm or 0.85 μm nozzle.
4. For the very first experiment, mono-stained cell tubes (i.e., cells immunostained with each antibody, separately) and compensation beads must be prepared to adjust the PMT voltages of the cell sorter and create the compensation matrix. In addition, although the sorting strategies have been validated, the cell profiles may slightly vary depending on the cell preparation and sorting equipment. It is therefore strongly recommended to perform Fluorescence Minus One (FMO) controls (*i.e.*, cells immunostained with all antibodies except one) to properly position the gates.
5. It is important to perform the coating of the Collection tubes to prevent the cells from sticking to the walls of the tubes during sorting.
6. MuSC can be directly sorted in lysis buffer for RNA extraction if needed.

Acknowledgments

N.D. receives financial support from the Agence Nationale de la Recherche (ANR-23-CE17-0035 and ANR-25-CE14-2467-01) and the Association Française contre les Myopathies (AFM Téléthon, Translamuscle Pole).

Competing Interests The authors declare no competing interests.

References

1. Seale P, Sabourin LA, Girgis-Gabardo A et al (2000) Pax7 is required for the specification of myogenic satellite cells. Cell 102:777–786
2. Chargé SBP, Rudnicki MA (2004) Cellular and molecular regulation of muscle regeneration. Physiol Rev 84:209–238
3. Relaix F, Zammit PS (2012) Satellite cells are essential for skeletal muscle regeneration: the cell on the edge returns centre stage. Development 139:2845–2856
4. Relaix F, Bencze M, Borok MJ et al (2021) Perspectives on skeletal muscle stem cells. Nat Commun 12:692
5. Hung M, Lo H-F, Jones GEL, Krauss RS (2023) The muscle stem cell niche at a glance. J Cell Sci136:jcs261200
6. Hung M, Lo H-F, Beckmann AG et al (2024) Cadherin-dependent adhesion is required for muscle stem cell niche anchorage and maintenance. Development 151:dev202387
7. Goel AJ, Rieder M-K, Arnold H-H et al (2017) Niche cadherins control the quiescence-to-activation transition in muscle stem cells. Cell Rep 21:2236–2250
8. Rozo M, Li L, Fan C-M (2016) Targeting β1-integrin signaling enhances regeneration in aged and dystrophic muscle in mice. Nat Med 22:889–896
9. Blanco-Bose WE, Yao CC, Kramer RH, Blau HM (2001) Purification of mouse primary myoblasts based on alpha 7 integrin expression. Exp Cell Res 265:212–220
10. Sacco A, Doyonnas R, Kraft P et al (2008) Self-renewal and expansion of single transplanted muscle stem cells. Nature 456:502–506
11. Maesner CC, Almada AE, Wagers AJ (2016) Established cell surface markers efficiently isolate highly overlapping populations of skeletal muscle satellite cells by fluorescence-activated cell sorting. Skelet Muscle 6:35
12. Gattazzo F, Laurent B, Relaix F et al (2020) Distinct phases of postnatal skeletal muscle growth govern the progressive establishment of muscle stem cell quiescence. Stem Cell Rep 15:597–611
13. Liu L, Cheung TH, Charville GW, Rando TA (2015) Isolation of skeletal muscle stem cells by fluorescence-activated cell sorting. Nat Protoc 10:1612–1624
14. Renault V, Rolland E, Thornell L-E et al (2002) Distribution of satellite cells in the human vastus lateralis muscle during aging. Exp Gerontol 37:1513–1514
15. Garcia SM, Tamaki S, Lee S et al (2018) High-yield purification, preservation, and serial transplantation of human satellite cells. Stem Cell Rep 10:1160–1174
16. Bareja A, Holt JA, Luo G et al (2014) Human and mouse skeletal muscle stem cells: convergent and divergent mechanisms of myogenesis. PLoS One 9:e90398
17. Castiglioni A, Hettmer S, Lynes MD et al (2014) Isolation of progenitors that exhibit myogenic/osteogenic bipotency in vitro by fluorescence-activated cell sorting from human fetal muscle. Stem Cell Rep 2:92–106
18. Ozturk T, Mignot J, Gattazzo F et al (2024) Dual inhibition of P38 MAPK and JNK pathways preserves stemness markers and alleviates premature activation of muscle stem cells during isolation. Stem Cell Res Ther 15:179
19. Alexander MS, Rozkalne A, Colletta A et al (2016) CD82 is a marker for prospective isolation of human muscle satellite cells and is linked to muscular dystrophies. Cell Stem Cell 19:800–807
20. Boscolo Sesillo F, Wong M, Cortez A, Alperin M (2020) Isolation of muscle stem cells from rat skeletal muscles. Stem Cell Res 43:101684
21. Te LJI, Doherty C, Correa J, Batt J (2021) Identification, isolation, and characterization of fibro-adipogenic progenitors (FAPs) and myogenic progenitors (MPs) in skeletal muscle in the rat. J Vis Exp 9:(172)

Check for updates

Chapter 3

Co-cultures of Macrophages with Muscle Stem Cells or Fibroadipogenic Precursor Cells from Skeletal Muscle

Stéphanie Gobert, Orane Mercier and Gaëtan Juban

Abstract

Skeletal muscle is capable of complete regeneration after an acute injury thanks to the properties of muscle stem cells (MuSCs), which can activate, proliferate, differentiate, and fuse to replace the damaged myofibers. However, this regeneration process involves the interaction of several cell types, including fibroadipogenic precursor cells (FAPs) that reorganize the extracellular matrix and macrophages (MPs), which modulate MuSC and FAP behavior. Here, we present a protocol to isolate MuSCs, FAPs, and MPs from skeletal muscle and assess the regulation of MuSC and FAP function by MPs through a short-term co-culture system, which allows to preserve key in vivo characteristics of the isolated cells.

Keywords Skeletal muscle, Macrophage, Muscle stem cell, Fibroadipogenic precursor cell, Co-cultures, Proliferation, Differentiation

1 Introduction

Skeletal muscle is a heterogeneous tissue composed of different cell types, including multinucleated myofibers, muscle stem cells (MuSCs) known as satellite cells, endothelial cells, and mesenchymal progenitors termed fibroadipogenic precursors (FAPs). Following muscle injury, MuSCs are rapidly activated and initiate the adult myogenic program, leading to the formation of new myofibers. Muscle damage also induces a strong inflammatory response including neutrophils, macrophages (MPs), and regulatory T cells that are present at a very low levels in resting muscle. Together, these cell populations act in a coordinated manner to ensure efficient muscle regeneration and functional recovery [1].

Among immune cells, MPs play a crucial role during the regeneration process. After an injury, circulating monocytes quickly infiltrate the damaged muscle and differentiate into pro-inflammatory macrophages expressing the surface marker Ly6C [2]. These pro-inflammatory MPs clear cellular debris by efferocytosis, promote MuSC

Nathalie Didier and Caroline E. Brun (eds.), *Muscle Stem Cells*, Methods in Molecular Biology, vol. 3093,
https://doi.org/10.1007/978-1-0716-5643-3_3,

proliferation while inducing apoptosis of FAPs to prevent excessive extracellular matrix deposition [2–5]. As efferocytosis progresses, macrophages undergo a phenotypic transition toward a restorative state. Restorative MPs support MuSC differentiation and fusion, thereby contributing to myofiber growth [2, 3]. In addition, they participate in extracellular matrix remodeling by stimulating collagen production by FAPs [4, 5]. Dysregulation of macrophage subset functions leads to impaired muscle regeneration [6–8] and contributes to the development of fibrosis in degenerative myopathies [5].

For these reasons, interactions between macrophages and other muscle-resident cell types are of particular interest in both physiological and pathological contexts. Studying these interactions can provide insights into the molecular pathways regulating muscle regeneration. One approach consists of isolating the relevant cell populations and analyzing their behavior in co-culture systems under controlled conditions. Here, we describe a protocol for the isolation of primary MuSCs, MPs, and FAPs using magnetic-activated cell sorting (MACS) or preplating, together with short-term co-culture methods designed to preserve key in vivo characteristics of the isolated cells. Cellular parameters such as proliferation and differentiation of MuSCs and FAPs are assessed by immunofluorescence.

2 Materials

2.1 Mouse Model

For the purpose of this study, wild-type C57BL/6J mice from 8- to 12-week-old were used.

2.2 Cell Isolation

1. Material for mouse dissection (thin forceps, thin scissors, thin sharp scissors).
2. 70% Ethanol.
3. Parafilm.
4. Heat-inactivated (56°C for 30 min) fetal bovine serum (FBS).
5. Heat-inactivated (56°C for 30 min) horse serum (HS).
6. 50 mL Polypropylene conical tubes.
7. 0.22 μm filters.
8. 30 μm cell strainers.
9. 70 μm cell strainers.
10. 100 μm cell strainers.
11. MS Columns (Miltenyi Biotec) or LS Columns (Miltenyi Biotec).
12. 75 cm^2 flasks for culture.
13. 100 mm Petri dishes.

14. 30 mL syringes.
15. 20-gauge needles.
16. FcR-blocking reagent (Miltenyi Biotec).
17. Primary antibodies: Anti-CD64-AF647 antibody (Biolegend #139322), Anti-Cy5/Anti-Alexa Fluor 647 Microbeads (Miltenyi Biotec #130-091-395).
18. Satellite Cell Isolation Kit (Miltenyi Biotec, #130-104-268).
19. MuSC digestion medium I: Ham's F10 medium (Gibco), 800 U/mL Collagenase II (Gibco #17101-105), 10% HS, 1% Penicillin/Streptomycin (P/S).
20. MuSC digestion medium II: Phosphate-Buffered Saline (PBS), 3000 U/mL Collagenase II (Gibco), 33 U/mL Dispase (Gibco #17105-041), 1% P/S.
21. MuSC washing medium: Ham's F-10, 10% Horse serum (HS), 1% P/S.
22. ACK lysing buffer (Gibco).
23. ACK STOP solution: PBS, 2% FBS.
24. FAP digestion medium: DMEM high glucose with GlutaMAX and pyruvate (Gibco), 0.36 U/mL Collagenase B (Roche #11 088 831 001), 2.4 U/mL Dispase II (Roche #04 942 078 001), 1% P/S.
25. MP digestion medium: DMEM high glucose with GlutaMAX and pyruvate (Gibco), 2.4 U/mL Collagenase B (Roche), 2.4 U/mL Dispase II (Roche), 1% P/S.
26. MP washing medium: DMEM high glucose with GlutaMAX and pyruvate (Gibco), 10% FBS, 1% P/S.
27. MACS buffer: PBS, 0.5% BSA, 2 mM EDTA filtered through 0.22 μm.

2.3 Cell Culture

1. 24-well plates
2. 96-well plates
3. Lab-Tek 8-chamber slides (Thermo Scientific)
4. Culture inserts pore size 0.4 μm (NUNC #056408)
5. Matrigel coating solution: DMEM/F12 medium (Gibco), 10% Matrigel Growth Factor Reduced (Corning #356231)
6. MuSC proliferation medium: DMEM/F12 medium (Gibco), 10% FBS, 2% Ultroser G (Pall #15950-017), 1% P/S
7. MuSC differentiation medium: DMEM/F12 medium (Gibco), 2% HS, 1% P/S
8. FAP proliferation medium: DMEM high glucose with GlutaMAX and pyruvate (Gibco), 10% FBS, 1% P/S

9. FAP adipogenic medium: DMEM high glucose with GlutaMAX and pyruvate (Gibco), 10% FBS, 0.5 mM IBMX (Sigma-Aldrich), 0.25 mM Dexamethasone (Sigma-Aldrich), 10 mg/mL Insulin (Sigma-Aldrich), 5 mM Rosiglitazone (Sigma-Aldrich), 1% P/S
10. DMEM low serum medium: DMEM high glucose with GlutaMAX and pyruvate (Gibco), 2.5% FBS, 1% P/S

2.4 Immunostaining

1. PBS
2. Fixation solution: 4% Paraformaldehyde in PBS
3. Permeabilization solution: 0.5% Triton X-100 in PBS
4. Blocking solution: 4% BSA in PBS
5. Antibodies (see Fig. 2)
6. EdU Click-iT for imaging AF488 (Invitrogen)
7. EdU revealing solution (for 100 μL): 87.5 μL PBS; 2 μL $CuSO_4$; 0.5 μL AF488 Fluorescent azide; 10 μL 1× Solution G
8. Nuclei staining solution: 2 μM Hoechst in PBS
9. Fluoromount mounting medium (Invitrogen)

3 Methods

A schematic overview of the workflow for the isolation and co-culture of MuSCs, FAPs, and MPs is described in Fig. 1.

3.1 Muscle Dissection

1. Before dissection, prepare the appropriate digestion medium under a sterile hood (see Sect. 3.2).
2. Switch on the water bath at 37°C.
3. Spray dissection tools with ethanol 70% and put them under the dissection hood.
4. Euthanize the mouse following approved guidelines of your laboratory (e.g., isoflurane and cervical dislocation).
5. Spray the mouse (hind-limbs and belly) with 70% ethanol.
6. At the basis of the mouse leg, use big scissors to make a small sock-like cut (be careful not to cut the tendons).
7. With the scissors, make a cut in the skin along the leg, up to the top of the thigh.
8. Use tweezers to remove the skin up to the top of the leg (take care to pull the skin over the front of the leg to avoid taking away muscles at the back of the leg).
9. Apply 70% ethanol again to remove the hair.

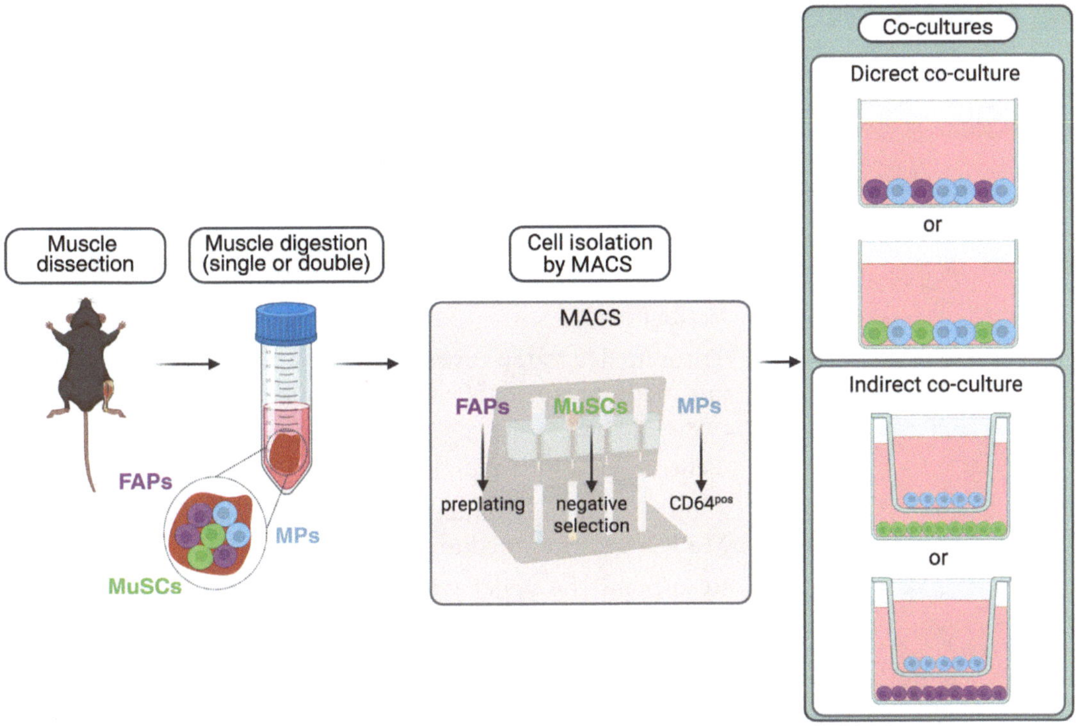

Fig. 1 Experimental workflow for the isolation and co-culture of FAPs, MuSCs, and MPs. Schematic representation of the sequential steps involved in the experimental procedure, including skeletal muscle dissection, enzymatic digestion, isolation of specific cell populations and subsequent co-culture of FAPs and MuSCs with MPs

10. Cut the tendons at the bottom with thin scissors, pull the leg muscle forward and backward: recover first the muscles (gastrocnemius, plantaris, soleus at the back; tibialis anterior, extensor digitorum longus at the front) by passing the forceps between the muscle and the tendon, cut the tendon and recover the muscle by cutting at its basis (*see* **Note 1**).
11. Collect all hind limbs muscles and place them one by one in a Petri dish (100 mm), which is open and placed on ice.
12. Do the same with the upper leg muscles (quadriceps and hamstrings).
13. Repeat the same process on the other leg.
14. When all the muscles are placed on ice in the Petri dish, mince the muscles with a scalpel blade or thin scissors into very small pieces to obtain a pulp (proceed fast).
15. Place the pulp in a 50 mL tube and proceed to digestion.

3.2 Muscle Digestion

Three different protocols are presented here depending on the cell type that is isolated. These correspond to the protocols which, in our hands, give the best yield for the cell type considered.

3.2.1 MuSCs

1. Add 10 mL of MuSC digestion medium I (*see* **Note 2**).
2. Seal the tube with Parafilm and incubate horizontally in the water bath at 37°C with gentle shaking (60–70 rpm) for 45–60 min (*see* **Note 3**).
3. Clean the tube with alcohol and from now on, work under a sterile hood.
4. Triturate five to ten times with a pipette and add MuSC washing medium up to 50 mL, and gently invert a few times to mix.
5. Centrifuge at 400*g* for 10 min at 4°C.
6. Aspirate the supernatant down to 20 mL and add 1 mL of MuSC digestion medium II.
7. Resuspend the pellet with a 10 mL serological pipette, triturate 10–15 times, and add MuSC washing medium up to 30 mL.
8. Seal the tube with Parafilm and incubate in the water bath at 37°C with gentle shaking for 30 min.
9. Clean the tube with alcohol and from now on, work under a sterile hood.
10. Using a 30 mL syringe with a 20-gauge needle, go back and forth ten times to triturate the muscle suspension (*see* **Note 4**). Filter the cell suspension through a cell strainer (70 μm).
11. Rinse the cell strainer with another 10 mL of MuSC washing medium and place the tube on ice. From now on, keep the samples at 4°C.
12. Centrifuge at 400 g for 10 min at 4°C.
13. Discard the supernatant and lyse red blood cells with 0.5 mL of ACK Lysing Buffer. Vortex until the solution gets a pink/red color (3–5 s) and stop the lysis with 5 mL of ACK STOP solution.
14. Centrifuge at 400*g* for 10 min at 4°C.
15. Remove the supernatant and resuspend the pellet in 5 mL of MACS buffer.
16. Count the cells and evaluate their viability (e.g., with Trypan blue dye).

3.2.2 FAPs

1. Add 10 mL of FAP digestion medium (*see* **Note 2**).
2. Seal the tube with Parafilm and incubate horizontally in the water bath at 37°C with gentle shaking (60–70 rpm) for 60–90 min (*see* **Note 3**).

3. After 60 min, the mix should appear cloudy and muscle pieces should be digested. If small muscle fragments are still present, they will disappear after Step 7.
4. Vortex the tube for 5 s.
5. Clean the tube with alcohol and from now work under a sterile hood.
6. Add 15 mL of FAP proliferation medium.
7. Using a 30 mL syringe with a 20-gauge needle, go back and forth ten times to triturate the muscle suspension (*see* **Note 4**). Filter the cell suspension through a cell strainer (70 μm).
8. Rinse the cell strainer with another 10 mL of FAP proliferation medium.
9. Centrifuge at 450 g for 5 min at room temperature (RT).
10. Remove the supernatant and resuspend the pellet in 12 mL of FAP proliferation medium.

3.2.3 MPs

1. Add 10 mL of MP digestion medium (*see* **Note 2**).
2. Seal the tube with Parafilm and incubate horizontally in the water bath at 37°C with gentle shaking (60–70 rpm) for 60–90 min (*see* **Note 3**).
3. Add 2.5 mL FBS, homogenize by pipetting and place the tube on ice. From now on, keep the samples at 4°C.
4. Filter the cell suspension through a 100 μm cell strainer placed on top of a 30 μm cell strainer.
5. Rinse the cell strainers twice with another 5 mL of MP washing medium.
6. Centrifuge at 400 *g* for 10 min at 4°C.
7. Discard the supernatant and lyse red blood cells with 0.5 mL of ACK Lysing Buffer. Vortex until the solution gets a pink/red color (3–5 s) and stop the lysis with 5 mL of ACK STOP solution.
8. Centrifuge at 400 g for 10 min at 4°C.
9. Remove the supernatant and resuspend the pellet into 5 mL MACS buffer.
10. Count the cells and evaluate their viability (e.g., with Trypan blue dye).

3.3 Cell Isolation

3.3.1 MuSCs

1. From Step 15 of Sect. 3.2.1, centrifuge the cells at 400 g for 10 min at 4°C (*see* **Note 5**).

2. Remove the supernatant and resuspend cell pellet in 80 µL of MACS buffer.
3. Add 20 µL of satellite cell isolation kit.
4. Mix with a 100–200 µL tip and incubate for 15–30 min at 4°C.
5. Adjust the volume to 500 µL with MACS buffer.
6. Install an MS Miltenyi column on the magnetic stand with a 30 µm cell strainer on top and rinse the column with 1 mL of MACS buffer (*see* **Note 6**). Let the MACS buffer go through the column into a bin tube.
7. Pour the cell suspension onto the cell strainer on top of the column. Rinse the tube containing the cells with 0.5 mL of MACS buffer three times and transfer onto the column. For each wash, let the MACS buffer go completely through the column.
8. Collect the flow-through cell suspension that contains MuSCs and centrifuge at 400*g* for 10 min at 4°C.
9. Remove supernatant and resuspend the pellet in 10 mL of MuSC proliferation medium, and seed the cells for culture experiments (see Sect. 3.4 co-culture section below) (*see* **Note 7**).
10. To also recover FAPS, take off the column from the magnet and put it on top of a new collection tube. Add 1 mL of MACS buffer and gently flush with the plunger to recover the non-MuSCs and proceed to Sect. 3.3.2 below.

3.3.2 FAPs

1. From Step 10 of Sect. 3.2.2, transfer the cell suspension into a 75 cm² flask.
2. Incubate in the incubator at 37°C with 5% CO_2 for 3 h, which is the time necessary for FAPs to adhere.
3. Wash once with PBS and add 12 mL of FAP proliferation medium (*see* **Note 8**).

3.3.3 MPs

1. From Step 9 of Sect. 3.2.2, centrifuge the cells at 400*g* for 10 min at 4°C (*see* **Note 5**).
2. Resuspend the pellet with a 100–200 µL tip and add 2 µL of FcR blocking reagent.
3. Mix with a 100–200 µL tip and incubate for 15 min at 4°C.
4. Add 2 µl of anti-CD64-AF647 antibody (*see* **Note 9**).
5. Mix with a 100–200 µL tip and incubate for 20 min at 4°C.
6. Add 5 mL of MACS buffer, centrifuge for 10 min at 400*g*, and homogenize the pellet in 90 µl of MACS buffer.
7. Add 10 µL of AntiCy5/anti-AF647 microbeads.

8. Mix with a 100–200 μL tip and incubate for 15 min at 4°C.
9. Add 5 mL of MACS buffer, centrifuge for 10 min at 400*g*, and homogenize the pellet in 1 mL of MACS buffer.
10. Install an MS column on the magnetic stand with a 30 μm cell strainer on top and rinse the column with 1 mL of MACS buffer (*see* **Note 6**). Let the MACS buffer go through the column into a bin tube.
11. Pour the cell suspension onto the cell strainer on top of the column. $CD64^{pos}$ cells are retained in the column while $CD64^{neg}$ cells pass through into a "$CD64^{neg}$ collection tube."
12. Rinse the tube containing the cells with 0.5 mL of MACS buffer three times and transfer onto the column. For each wash, let the MACS buffer go completely through the column.
13. Take off the column from the magnet and put it on top of a new collection tube. Add 1 mL of MACS buffer and gently flush with the piston to recover the $CD64^{pos}$ cells.
14. In order to also recover FAPs, $CD64^{neg}$ cells can be centrifuged for 10 min at 400*g*. Resuspend them in 12 mL of FAP proliferation medium and proceed to Sect. 3.3.2.
15. $CD64^{pos}$ cells are MPs and can be directly seeded for experiments (see Sect. 3.4 below) (*see* **Note 10**).

3.4 Co-cultures

3.4.1 Direct co-culture of MPs with MuSCs

1. Coat the cell culture support with Matrigel coating solution (*see* **Note 11**) and incubate for 30 min at 37°C, 5% CO_2. Aspirate excess liquid and rinse carefully with PBS.
2. Seed MuSCs in MuSC proliferation medium at a density of 3000 cells/cm^2 for proliferation assay in a 96-well plate; or 30,000 cells/cm^2 for myogenesis assay in an eight-chamber slide.
3. Let MuSCs adhere to the culture support for at least 6 h at 37°C, 5% CO_2.
4. MPs may be isolated during the adhesion time.
5. For MuSC proliferation assay, resuspend MPs in MuSC proliferation medium. For MuSC myogenesis assay, resuspend MPs in MuSC differentiation medium. Calculate the cell concentration to obtain a ratio of 3:1 MPs:MuSCs in the adequate volume of medium (100 μL in a 96-well or 250 μL in an eight-chamber slide).
6. Carefully remove the medium in the MuSC cultures, gently rinse once with PBS and add MPs onto MuSCs.
7. Incubate the cells at 37°C for 24 h for proliferation assay and 48 h for differentiation assays, and proceed to immunostaining.

3.4.2 Direct co-culture of MPs with FAPs

1. Seed FAPs in FAP proliferation medium, at 10,000 cells/cm^2 (for instance in a 48-well plate or an eight-chamber slide).
2. Let FAPs adhere to the culture support for about 6 h at 37°C, 5% CO_2.
3. MPs may be isolated during the adhesion time.
4. For proliferation or fibroblastic differentiation assay, resuspend MPs in DMEM low serum medium. For FAP adipogenic assay, resuspend MPs in FAP adipogenic medium. Calculate the cell concentration to obtain a ratio of 3:1 MPs:FAPs in the adequate volume of medium (200 µL in a 48-well or 250 µL in an eight-chamber slide).
5. Carefully remove the medium in the FAP cultures, gently rinse once with PBS and add MPs onto FAPs.
6. Incubate cells at 37°C, 5% CO_2 for 24 h (for proliferation and fibroblastic differentiation assay) or 5–8 days (for the adipogenic assay, change medium every 2 days) and proceed for immunostaining.

3.4.3 Indirect co-culture of MPs with MuSCs

1. Coat the cell culture support with Matrigel coating solution (*see* **Note 11**) and incubate for 30 min at 37°C, 5% CO_2. Aspirate excess liquid and rinse carefully with PBS.
2. MuSCs are seeded in a 24-well plate in MuSC proliferation medium at a density of 3000 cells/cm^2 for proliferation assay; or 30,000 cells/cm^2 for myogenesis assay.
3. Let MuSCs adhere to the culture support for at least 6 h at 37°C, 5% CO_2.
4. MPs are seeded in 300 µL of DMEM low serum medium in inserts of 0.4 µm for 24-well plates to obtain a ratio of 3:1 MPs:MuSCs.
5. Wash MuSCs once with PBS and add 500 µL of DMEM low serum medium.
6. Place the inserts containing the MPs into the wells with the MuSCs with clean forceps.
7. Incubate the cells at 37°C for 24 h for proliferation assay and 48 h for differentiation assays, and proceed to immunostaining.

3.4.4 Indirect co-culture of MPs with FAPs

1. Seed FAPs in FAP proliferation medium, at 10,000 cells/cm^2 in a 24-well plate.
2. Let FAPs adhere to the culture support for about 6 h at 37°C, 5% CO_2.
3. MPs are seeded in 300 µL of DMEM low serum medium in inserts of 0.4 µm for 24-well plates to obtain a ratio of 3:1 MPs:FAPs.

4. Wash FAPs once with PBS and add 500 µL of DMEM low serum medium (for proliferation and fibroblastic assay) or FAP adipogenic medium (for adipogenic assay).
5. Place the inserts containing the MPs into the wells with the FAPs with clean forceps.
6. Incubate cells at 37°C, 5% CO_2 for 24 h (for proliferation and fibroblastic differentiation assay) or 5–8 days (for the adipogenic assay, change medium every 2 days) and proceed for immunostaining.

3.5 Immunostaining

1. If EdU labeling is not required, proceed directly to Step 3.
2. For EdU proliferation assay, add directly 10 µM of EdU into the culture medium. Directly mix well by gently swirling the plate and incubate 1 h at 37°C, 5% CO_2.
3. Wash once with PBS. Washes should be performed very carefully to avoid detaching the cells, even after fixation. The cell preparation must never dry out.
4. Fix the cells with Fixation solution for 10 min at RT.
5. Wash with PBS (3 × 5 min).
6. Permeabilize the cells with 0.5% Permeabilization solution for 10 min at RT.
7. Wash with PBS (3 × 5 min).
8. Block nonspecific binding sites with Blocking solution for 1 h at RT.
9. Wash with PBS (3 × 5 min).
10. Incubate with primary antibodies diluted in PBS (Fig. 2) overnight at 4°C in a humid chamber.
11. Wash with PBS (3 × 5 min).
12. Incubate with secondary antibodies diluted in PBS (Fig. 2) for 1 h at 37°C in a humid chamber.
13. Wash with PBS (3 × 5 min). If EdU labeling is not required, proceed directly to Step 17.
14. Add 100 µL of EdU revealing solution.
15. Incubate for 30 min at RT in dark.
16. Wash with PBS (3 × 5 min).
17. Incubate for 10–20 s with Nuclei staining solution.
18. Wash once with PBS (1 × 5 min).
19. Add Fluoromount medium.
20. Store at +4°C until imaging.

Examples of immunostaining for MuSCs are provided in Fig. 3: proliferation assay (Fig. 3A), myogenesis assay (Fig. 3B).

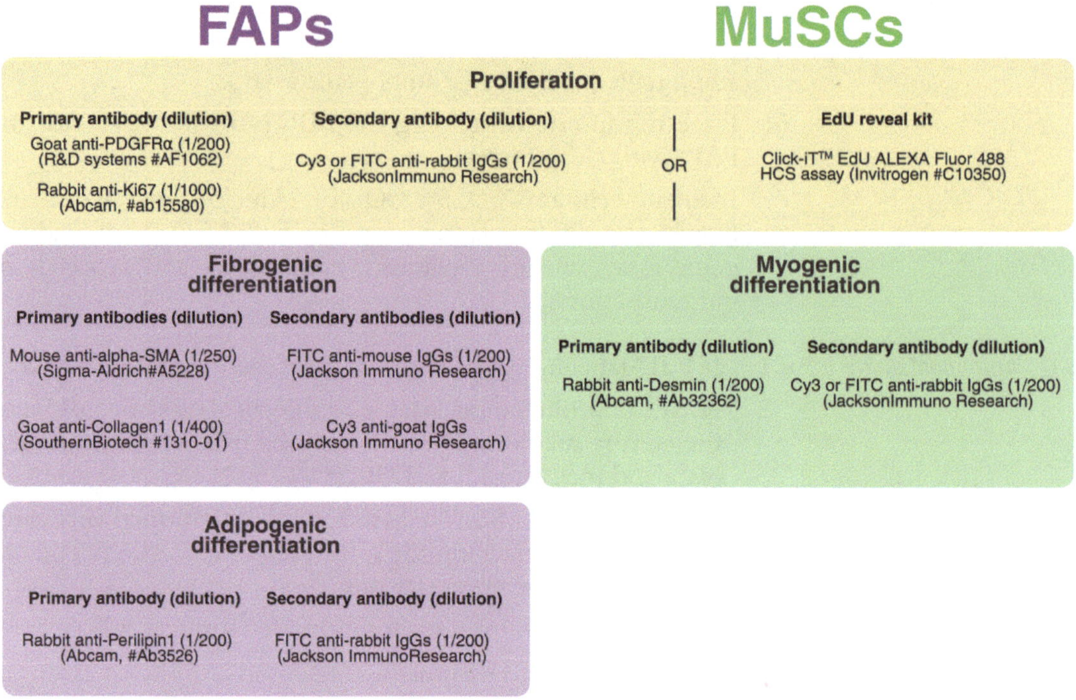

Fig. 2 Overview of proliferation and differentiation readouts for FAPs and MuSCs in macrophage co-culture assays. Schematic representation of the experimental readouts used to assess FAP and MuSC function. Proliferation is evaluated in both cell types using shared readouts (Ki67 or EdU incorporation). Lineage-specific differentiation is assessed by analyzing fibrogenic and adipogenic differentiation in FAPs, and myogenic differentiation in MuSCs, as described in the protocols presented in this chapter

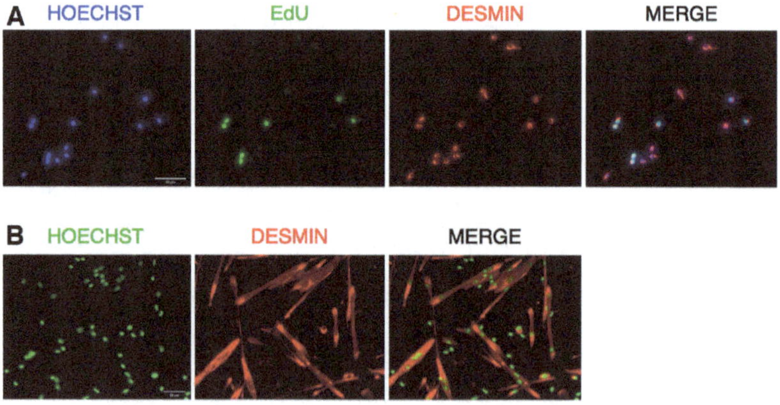

Fig. 3 Examples of immunostainings for MuSCs. **A,** Proliferation analysis of MuSC culture after EdU (in green) incorporation followed by Desmin (in red) and Hoechst (in blue) labeling. MuSCs (Desmin^{pos} cells) harboring a nucleus positive for EdU are considered in proliferation. **B,** Myogenesis assay performed by Desmin (in red) and Hoechst (in green) labeling allowing to determine the fusion index, which represents the percentage of nuclei in myogenic cells (Desmin^{pos} cells) harboring more than two nuclei. Scale bar: 50 μm

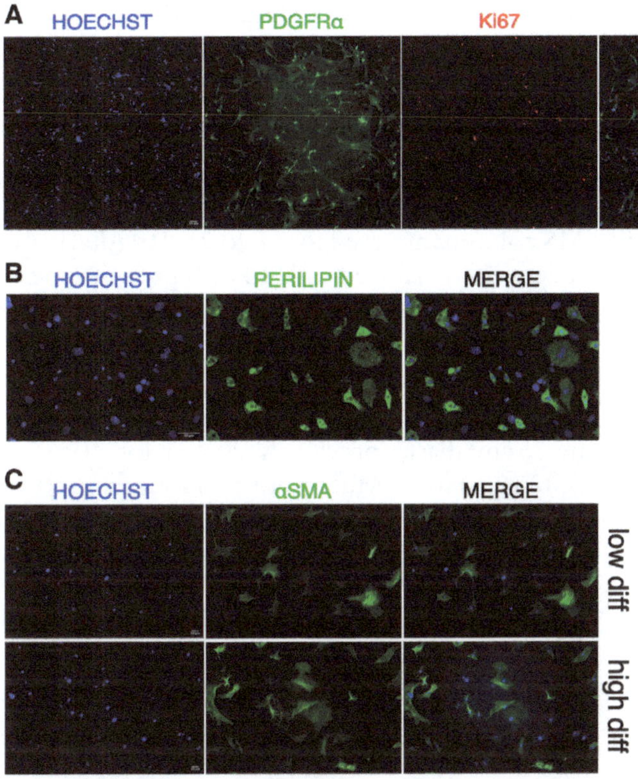

Fig. 4 Examples of immunostainings for FAPs. **A,** Proliferation analysis of FAP culture after PDGFRα (in green), Ki67 (in red), and Hoechst (in blue) labeling. FAPs (PDGFRα^{pos} cells) harboring a nucleus positive for Ki67 are considered in proliferation. **B,** FAP adipogenic differentiation determined by Perilipin (in green) and Hoechst (in blue) labeling. **C,** Fibroblastic differentiation assayed by αSMA (in green) and Hoechst (in blue) labeling in a condition of low (top panel) or high (bottom panel) fibroblastic differentiation efficiency. Scale bar: 50 μm

Examples of immunostaining for FAPs are provided in Fig. 4: proliferation assay (Fig. 4A), adipogenic assay (Fig. 4B), and fibroblastic assay (Fig. 4C).

4 Notes

1. Be careful not to pick up hair during dissection, to avoid contamination and use different scissors for cutting the skin, and for cutting tendons, and forceps to collect all muscles.
2. Digestion media are prepared extemporaneously and filtered under sterile conditions using a 0.22 μm filter.
3. Digestion duration has to be adjusted depending on the type of muscle. Regenerating muscle is easier to digest than dystrophic or fibrotic muscle. Monitor the digestion process:

the muscle fragments should appear as loose, filamentous structures.

4. Flush against the wall of the tube to avoid foam formation.
5. In cases of dystrophic or fibrotic muscle, an additional washing step with 5 mL of MACS buffer is required to remove matrix components that may saturate the column.
6. MS columns are used for up to 2×10^8 total cells (up to 1×10^7 magnetically labeled cells). For higher numbers of cells, samples can be separated using LS columns. Volumes are given for an MS column.
7. If MuSCs need to be expanded, they should be seeded as soon as possible after sorting in MuSC proliferation medium in 75 cm^2 flask (previously coated for 20 min at 37°C with 0.02% Gelatin). MuSCs form clusters due to their high proliferation rate. Therefore, cells must be trypsinized regularly (every 2–3 days) during the expansion phase to avoid the formation of myotubes in the clusters, and seeded at a density of 2500 cells/cm^2. After two passages, the cell number expected can be up to 1,000,000 cells.
8. FAPs can be expanded in FAP proliferation medium and passaged up to three times if seeded at a density of 5000 cells/cm^2. Usually, cells are expanded for 4–7 days before co-culture, which allows to obtain up to 2,000,000 cells for one mouse.
9. If necessary, specific pro- and anti-inflammatory subsets can be FACS-sorted based on the expression of the Ly6C marker [9].
10. MPs are used directly after isolation. The number of MPs recovered from one non-injured mouse is around 10,000–20,000 cells and can reach up to 1,000,000 cells per tibialis anterior muscle after an injury induced by the snake toxin cardiotoxin.
11. Preparation of Matrigel coating solution has to be performed strictly at 4°C to prevent Matrigel polymerization. To do so, thaw Matrigel aliquot on ice and use cold DMEM-F12 with pipet tips precooled at −20°C to dilute it.

Competing Interests

The authors declare no competing interests.

References

1. Wosczyna MN, Rando TA (2018) A muscle stem cell support group: coordinated cellular responses in muscle regeneration. Dev Cell 46(2):135–143
2. Arnold L, Henry A, Poron F et al (2007) Inflammatory monocytes recruited after skeletal muscle injury switch into anti-inflammatory macrophages to support myogenesis. J Exp Med 204:1071–1081.
3. Saclier M, Yacoub-Youssef H, Mackey AL et al (2013) Differentially activated macrophages orchestrate myogenic precursor cell fate during human skeletal muscle regeneration. Stem Cells 31:384–396
4. Lemos DR, Babaeijandaghi F, Low M et al (2015) Nilotinib reduces muscle fibrosis in chronic muscle injury by promoting TNF-mediated apoptosis of fibro/adipogenic progenitors. Nat Med 21:786–794
5. Juban G, Saclier M, Yacoub-Youssef H et al (2013) AMPK activation regulates LTBP4-dependent TGF-beta1 secretion by pro-inflammatory macrophages and controls fibrosis in duchenne muscular dystrophy. Cell Rep 25:2163–2176
6. Mounier R, Theret M, Arnold L et al (2013) AMPKalpha1 regulates macrophage skewing at the time of resolution of inflammation during skeletal muscle regeneration. Cell Metab 18:251–264
7. Perdiguero E, Sousa-Victor P, Ruiz-Bonilla V et al (2011) p38/MKP-1-regulated AKT coordinates macrophage transitions and resolution of inflammation during tissue repair. J Cell Biol 195:307–322
8. Tonkin J, Temmerman L, Sampson RD et al (2015) Monocyte/macrophage-derived IGF-1 orchestrates murine skeletal muscle regeneration and modulates autocrine polarization. Mol Ther 23:1189–1200
9. Panci G, Kneppers AM, Mounier R et al (2023) Co-cultures of macrophages with muscle stem cells with fibroadipogenic precursor cells from regenerating skeletal muscle. Methods Mol Biol 2640:57–71

Part II

In Situ and Ex Vivo Analysis of Muscle Stem Cell Quiescence and Self-Renewal

Chapter 4

Tissue Clearing and Immunostaining of Mouse and Human Skeletal Muscles for In Situ Analysis of Muscle Stem Cell Spatial Organization

Marie Quétin, Julien Mignot, Juliette Berthier, Christelle Dubois, Nathalie Didier and Marianne Gervais

Abstract

Skeletal muscle is essential for voluntary movement, postural maintenance, respiration, and thermoregulation. It consists of large, multinucleated myofibers and has strong regenerative capacity driven by resident muscle stem cells (MuSCs) located beneath the myofiber basal lamina. Traditional histological methods using longitudinal or cross sections have improved our understanding of muscle architecture, but their two-dimensional nature limits insight into the tissue's complex three-dimensional (3D) organization and cellular interactions. To better understand MuSC dynamics, 3D imaging of entire tissue volumes is necessary. However, deep tissue imaging remains challenging due to light scattering and autofluorescence within muscle tissue. While advances in microscopy and computational imaging have improved spatial resolution, optical tissue clearing has become essential to enhance tissue transparency. Yet, existing clearing methods, though effective, often cause tissue deformation or fluorescence loss, compromising data reliability. Moreover, in situ visualization of MuSCs frequently relies on fluorescent reporter models, which is incompatible with human tissues. This chapter introduces an optimized tissue-clearing protocol for skeletal muscle and MuSC visualization that preserves both morphology and fluorescence. Compatible with mouse and human samples, it enables high-resolution 3D imaging of MuSCs within their native microenvironment, providing a powerful tool to study muscle architecture, regeneration, and disease mechanisms.

Keywords Skeletal muscle, Muscle stem cells, Tissue clearing, 3D reconstruction

1 Introduction

Muscle stem cells (MuSCs), commonly referred as "satellite cells," are the primary drivers of skeletal muscle regeneration. Under physiological conditions, these cells remain quiescent. However, upon injury, they rapidly activate, proliferate, and differentiate to repair damaged myofibers while simultaneously replenishing the stem cell pool. To fully understand how MuSCs behave, interact with their

The authors "Marie Quétin, Julien Mignot, Juliette Berthier" are co-first authors.

Nathalie Didier and Caroline E. Brun (eds.), *Muscle Stem Cells*, Methods in Molecular Biology, vol. 3093,
https://doi.org/10.1007/978-1-0716-5643-3_4,

niche, and orchestrate regeneration, researchers require not only advanced molecular and genetic tools but also precise spatial visualization within their native three-dimensional (3D) environment [1, 2].

Traditionally, studies of MuSCs have relied on two-dimensional histological sections of muscle tissue, obtained through thin longitudinal or cross-sectional cuts of frozen or paraffin-embedded samples. While these methods have provided valuable insights into MuSC identity and localization, they offer only a fragmented view of their spatial organization within the entire muscle and fail to capture their dynamic interactions with myofibers, vasculature, and other niche components, which are critical for understanding their function in situ. Fluorescence microscopy, particularly confocal and multiphoton imaging, has opened new possibilities for visualizing MuSCs within intact tissue. Yet, the dense structure and high autofluorescence of skeletal muscle, arising from collagen, elastin, lipids, and NADH, pose significant challenges for deep fluorescence imaging [3]. Recent advances in optical clearing have mitigated some of these limitations. Techniques such as CLARITY [4], 3DISCO [5], and CUBIC [6] enhance tissue transparency and light penetration, but they often require long incubation periods and can induce irreversible morphological distortions, or fluorescence quenching. This limits their suitability for quantitative volumetric analysis of MuSCs [7]. Moreover, in situ visualization of MuSCs frequently relies on the use of fluorescent reporter models, which is incompatible with human tissues [8–11].

In this chapter, we present a novel optical clearing method based on RapiClear® technology (Sunjin Lab, 2014), designed for 3D imaging of skeletal muscle and MuSC visualization. Unlike existing clearing methods, this approach is non-toxic, rapid, and user-friendly. It preserves tissue morphology, maintains reporter fluorescent intensity, and is fully compatible with immunolabelling of both mouse and human muscles. This protocol enables high-resolution mapping of MuSC spatial organization and interactions across large muscle volumes, providing a more comprehensive and integrated view of the stem cell niche. We provide a step-by-step protocol ranging from sample preparation to 3D image processing using two different softwares (i.e., Fiji and Imaris®).

2 Materials

2.1 Specific Equipment

1. Tissue chopper or scalpel.
2. Glass microscope slides (25 × 75 × 1 mm).
3. Glass cover slips.

4. Adhesive imaging spacers (0.25 mm deep).
5. Dissection tools: curved micro scissors, micro dissection forceps.
6. Spinning disk confocal microscope, with long distance immersion objective, compatible with the use of immersion oil.

2.2 Solutions

1. Fixation solution: 4% Paraformaldehyde (PFA) in PBS 1X. Store at 4°C.
2. Permeabilization solution: 2% Triton-100X in PBS 1X.
3. Washing solution: 0.2% Triton-100X, 3% NaCl in PBS 1X.
4. Dilution buffer: 0.2% Triton-100X, 3% goat serum, 0.2% sodium azide in PBS 1X.
5. Sample conservation solution: 0.2% sodium azide in PBS 1X.
6. RapiClear® 1.52 (Sunjin Lab).
7. 1 µg/mL DAPI solution in PBS 1X.
8. Immersion oil.
9. Dental paste (Eco-sil speed, #1300 7100, Picodent) (optional).

2.3 Antibodies

1. Primary antibodies: Chicken anti-GFP 1:100 (#ab13970, Abcam) and Sheep anti-M-CADHERIN 1:50 (#AF4096, R&D Systems).
2. Alexa Fluor™ conjugated secondary antibodies: Alexa Fluor 488-coupled goat anti-chicken IgY (H+L) 1:500 and Alexa Fluor 647-coupled donkey anti-sheep IgG (H+L) 1:500.

2.4 Softwares

1. Fiji
2. Imaris®

3 Methods

Mice used in this study were 2-month-old *Tg:Pax7-nGFP* males [12]. The methods described below are also suitable for human skeletal muscles.

3.1 Tibialis Anterior Harvest from Adult Mouse

1. Excise the skin and expose the Tibialis Anterior (TA) by delicately removing the connective tissue with a thin forceps along the tibia, from the ankle to the knee.
2. Carefully take the distal TA tendon with forceps and cut it as close as possible to the foot.
3. Lift the TA cautiously and cut the proximal insertion as close as possible to the knee using curved micro scissors (*see* **Note 1**).

3.2 Sample Fixation

1. Fix the TA mouse muscle (or the human muscle biopsy) in 4% PFA at 4°C, for 2 h, under agitation (*see* **Note 2** and Fig. 1a).
2. Cut a muscle slice (<1 mm thick) parallel to the orientation of the myofibers using a tissue chopper or a scalpel (*see* **Notes 3 and 4**).
3. Postfix the muscle slice in 4% PFA at room temperature, for 2 h, under agitation.
4. At this step, the muscle slice can be conserved at 4°C in PBS 1X supplemented with 0.2% sodium azide.

3.3 Immunostaining of the Muscle Stem Cells

All following steps are performed in dark with gentle agitation, placing the muscle slice in 0.5 mL tube (Fig. 1a).

1. Wash three times in the washing solution (10 min/wash), at room temperature.
2. Incubate the sample in the permeabilization solution, overnight, at room temperature.

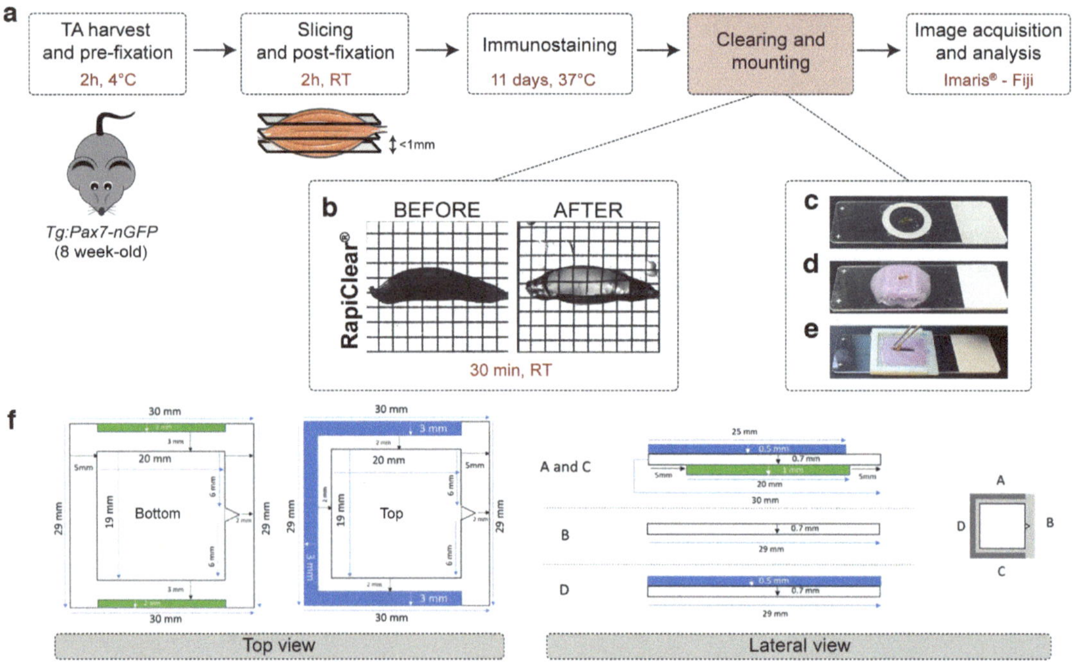

Fig. 1 Sample clearing and mounting protocols for 3D imaging of mouse Tibialis Anterior (TA) muscle. **a,** Experimental procedure. **b,** Picture of mouse TA slice before and after clearing in RapiClear®. **c–e,** Photos of sample mounting on glass microscope slides using adhesive spacers (**c**), homemade dental paste mold (**d**), or dental paste cast in a 3D-printed cassette (**e**). **f,** Top or lateral view of 3D printed cassette fabrication plans

3. Incubate with primary antibodies diluted in the dilution buffer, for 7 days at 37°C.
4. Wash three times in the washing solution (1 h/wash), at room temperature.
5. Incubate with secondary conjugated antibody diluted in the dilution buffer, for 4 days at 37°C.
6. Wash three times in the washing solution (1 h/wash), at room temperature.
7. Incubate with DAPI for 2 h at 37°C.
8. Wash three times in the washing solution (1 h/wash), at room temperature.
9. The muscle can be conserved at 4°C in PBS 1X supplemented with 0.2% sodium azide until image acquisition.

3.4 Muscle Clearing

1. Prewarm the RapidClear® solution at 37°C.
2. Place the immunostained muscle slice in 200–400 μL of pre-warmed RapiClear® for 30 min at 37°C before acquisition. At the end of the incubation, the sample should be entirely transparent (Fig. 1b), although some yellow reflections may be visible.

3.5 Sample Mounting

3.5.1 Using Adhesive Imaging Spacers (e.g., Sunjin Lab iSpacer 1-Circ Well, 0.3 mm deep, #IS317)

1. Stick the adhesive spacer on a glass microscope slide (*see* **Note 5** and Fig. 1c).
2. Place the cleared muscle slice in the central hole of the spacer, in pre-warmed RapiClear®.
3. Cover with a glass coverslip, taking care not to create bubbles.

3.5.2 Using Homemade Mold

1. Prepare the dental paste mixture (mixing ratio 1:1) following supplier's recommendations in a Petri dish.
2. Place the mixture on a glass microscope slide, let it solidify and then shape a cavity adapted to the dimensions of the sample (Fig. 1d).
3. Another option is to use a reusable 3D-printed plastic mounting cassette to create a dental paste mold for the muscle slices (Fig. 1e, f).
4. Place the cleared muscle slice into the cavity in pre-warmed RapiClear®.
5. Cover with a glass coverslip, taking care not to create bubbles.

3.6 Image Acquisition

The acquisition procedure described here applies to a Spinning-Disk CSU-X1 (Zeiss) under ZEN blue 2.6 software. However, any spinning-disk confocal with equivalent performance can be used.

1. Perform image acquisition with a spinning disk confocal microscope, using preferably a long-distance 25×/0.8 mm immersion objective. Use immersion oil for imaging to match with the refractive index of the RapiClear® (1.52).
2. Set the focus on the DAPI channel.
3. For multicolor acquisitions, make the entire z-stack of one channel (e.g., DAPI), then switch to another one using *"Full z-Stack per Channel"* mode.
4. Using the z-stack *"First/Last"* window, define the depth of the acquisition by selecting upper (*"Set first"*) and lower (*"Set last"*) thresholds.
5. Return to the center of the z-stack, adjust intensity and exposure time for each channel, and proceed to the acquisition.
6. Save the image in .CZI or .TIFF format for post-treatment and/or quantification using an image analysis software. By default, images are generated as hyperstacked images.

3.7 3D Image Processing (see Note 6)

3.7.1 3D Visualization and Volume Reconstruction Using Imaris® 10.0 Software (Fig. 2)

1. Drag the .TIFF file in the folder headed *"Arena"* (Fig. 2a and *see* **Note 7**).
2. Double-click on the generated image icon to convert the image into Imaris file and then double-click again to open it.
3. Set the voxel size by selecting *"Edit"* tab, and then *"Image properties"* (Fig. 2b).
4. Define the color and the contrast for each channel using "*Edit*" tab and then *"Show Display Adjustment"* (Fig. 2c).
5. Default settings open a 3D representation of the image. However, if it is not the case, select "*3D View*" icon.
6. For 3D volume reconstruction: in the "*Scene-Properties*" window (top left), select the "*Volume*" folder, which will change the window heading to *"Volume-Properties"*. Then, select *"Blend"* (Fig. 2d).

3.7.2 3D Visualization and Volume Reconstruction Using Fiji Software (Fig. 3)

1. Download Fiji from https://imagej.net/Fiji/Downloads.
2. Open image in.TIFF format.

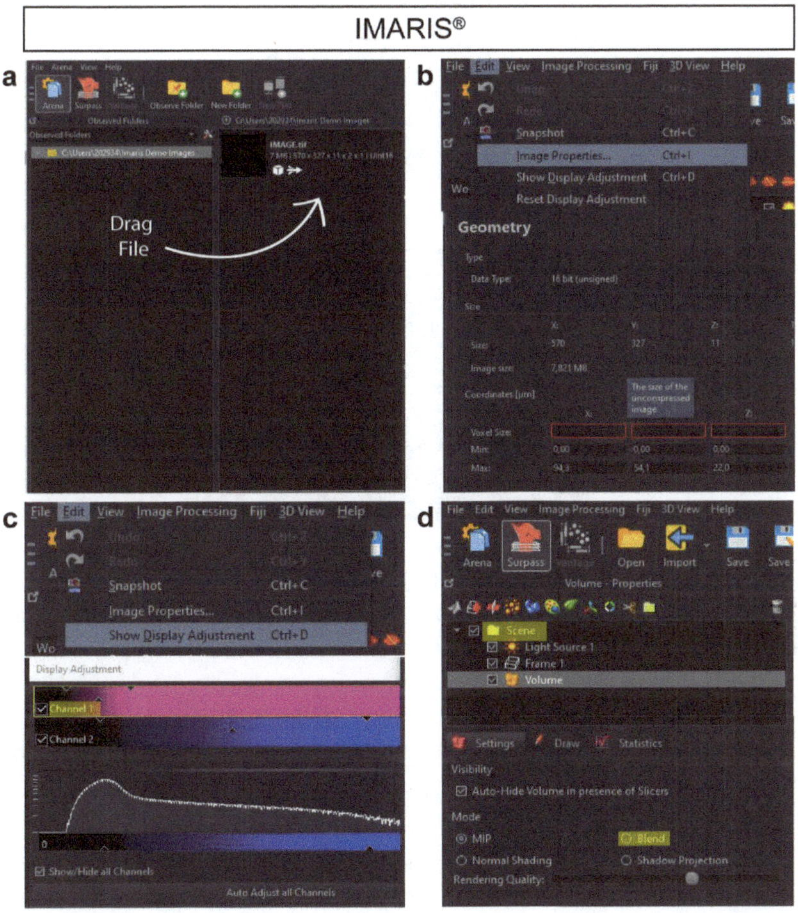

Fig. 2 Procedure for 3D image processing using Imaris®. **a–d,** Screenshots showing the software interface and the sequential steps to follow to obtain a 3D visualization of the image (**a–c**) or a volume reconstruction (**d**)

3. Set voxel size in *"Image"* tab, then select *"Properties"*. A window opens. Set Pixel width, Pixel height and Voxel depth, based on the calibration of the microscope used (Fig. 3a).
4. Adjust brightness and contrast for each channel, if necessary, in *"Image"* tab, then select *"Adjust"* and *"Brightness/Contrast"*. On the top of the image window are indicated: the channel (c), the stack position (z), and the pixel size. The channel and the z position can be chosen using the sliding bars at the bottom of the image (Fig. 3a).
5. For 3D visualization (Fig. 3b): Merge the channels by selecting *"Image"*, *"Color"*, and then *"Stack to RGB"*. Then, in *"Plugins"* tab, select *"3D Viewer"*. A window headed *"Add"* opens. If

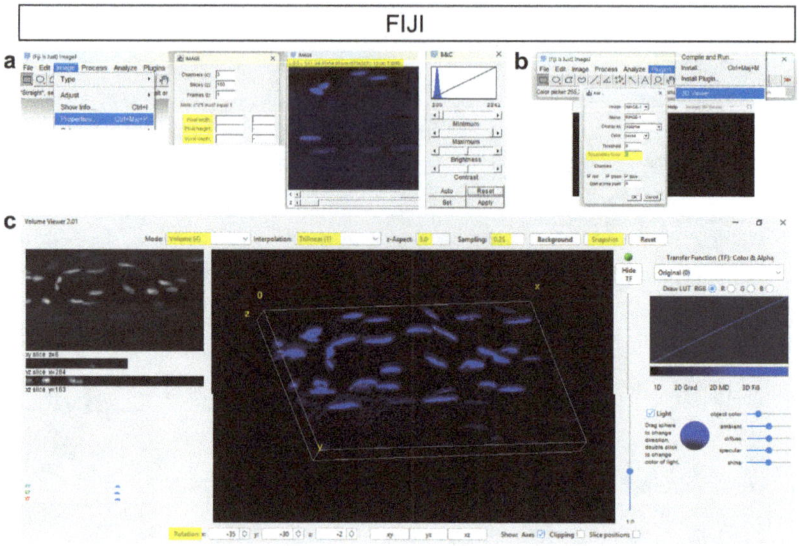

Fig. 3 Procedure for 3D image processing using Fiji. **a–c,** Screenshots showing the software interface and the sequential steps to follow to obtain a 3D visualization of the image (**a and b**) or a volume reconstruction (**c**)

needed, set the size of the 3D image in *"Resampling factor"* (original size = 1, desired size = 1/×, *see* **Note 8**).

6. For volume reconstruction (Fig. 3c): Do not merge channels (*see* **Note 9**). In *"Image"* tab, select *"Color"* and then *"Split Channels"* Images of each channel open separately. For each image, in *"Plugins"* tab, select *"Volume Viewer"*. A new window headed *"Volume Viewer"* opens (Fig. 3c). Next, set the following parameters using the tabs on the top: in *"Mode"* tab, select *"Volume"*; in *"Interpolation"* tab, select *"Trilinear"*; in *"z-Aspect"* tab choose the desired z depth visualization; in *"Sampling"* tab, set the desired resolution. On the right of the image, click on *"Light"* and set the light orientation using the drag sphere. At the bottom of the image, choose the desired orientation of the 3D image by setting rotation x, y, and z parameters. Click on *"Snapshot"* and save the generated image as .TIFF. Repeat the same procedure for each channel. Once done, in *"Image"* tab, select *"Color"*, and then *"Merge channels"*. Save the merge as .TIFF.

Representative images of M-CADHERIN immunostaining on mouse and human muscles cleared with RapiClear®, processed using Imaris® and Fiji are provided in Figs. 4 and 5, respectively.

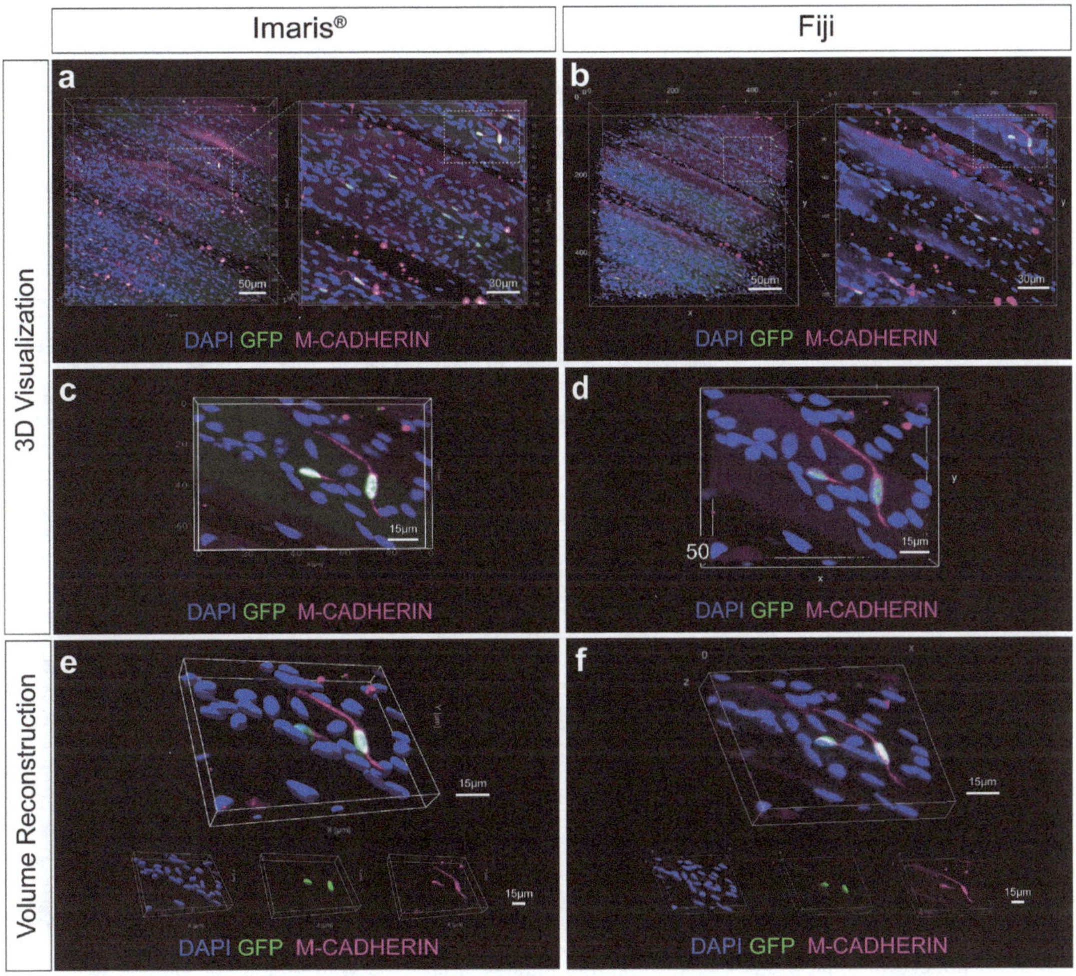

Fig. 4 3D images of muscle stem cells in TA muscle from a Tg:Pax7-nGFP adult mouse, processed with Imaris® or Fiji software. **a and b,** Representative images of a TA slice immunostained for GFP (PAX7) and M-CADHERIN, cleared with RapiClear® and processed using Imaris® (**a**) or Fiji (**b**). Nuclei were stained with DAPI. Dashed squares show selected Region Of Interest (ROI). Images were acquired with a Spinning Disk CSU-X1 (Zeiss), using LD LCI Plan-Apochromat 25×/0.8 Imm Korr DIC M27 objective (#420852-9871, Zeiss). **c–f**, 3D visualization (**c and d**) and volume reconstruction (**e and f**) of a ROI, performed with Imaris® or Fiji

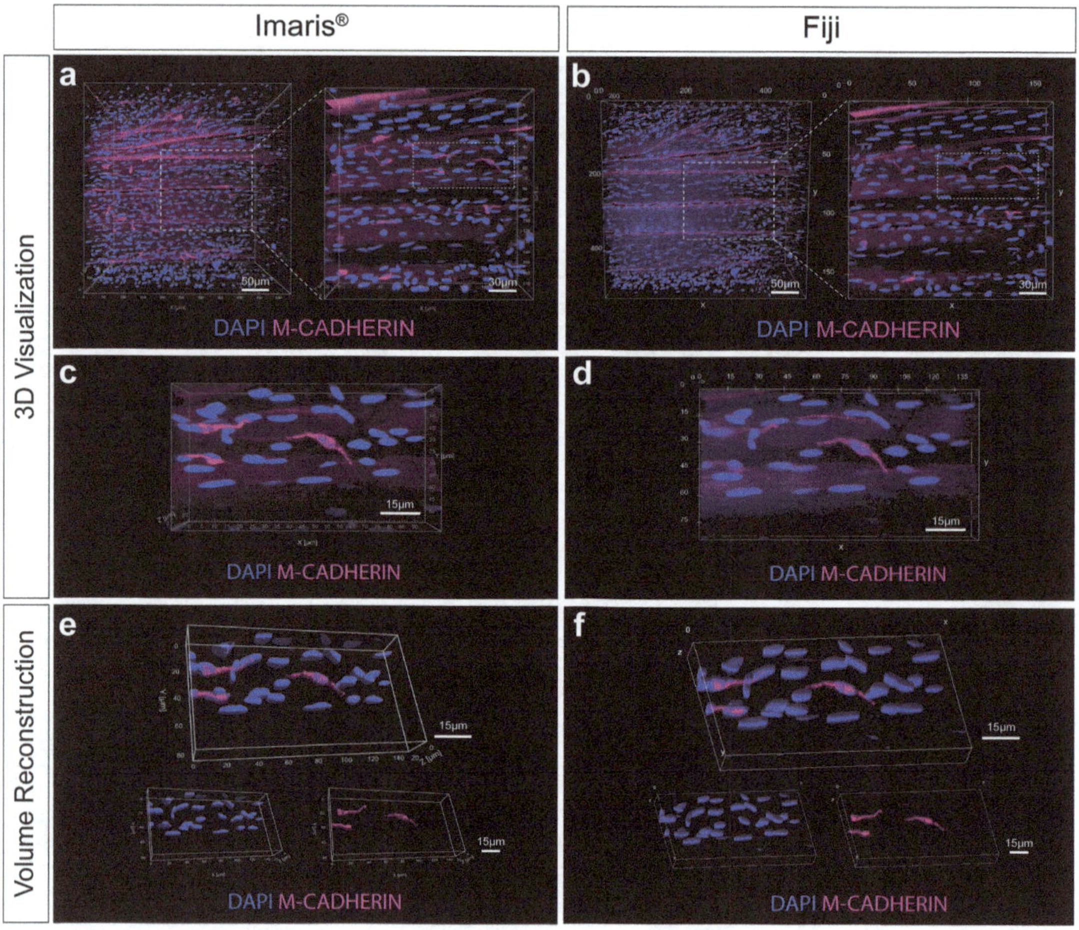

Fig. 5 3D images of muscle stem cells in human muscle, processed with Imaris® or Fiji software. **a and b,** Representative images of human Gastrocnemius muscle immunostained for M-CADHERIN, cleared with RapidClear® and processed using Imaris® (**a**) or Fiji (**b**). Nuclei were stained with DAPI. Dashed squares show selected Region Of Interest (ROI). Images were acquired with a Spinning Disk CSU-X1 (Zeiss), using LD LCI Plan-Apochromat 25×/0.8 Imm Korr DIC M27 objective (#420852-9871, Zeiss). **c–f,** 3D visualization (**c and d**) and volume reconstruction (**e and f**) of a ROI, performed with Imaris® or Fiji. Human muscle biopsy (*see* **Note 10**) was obtained from brain-dead organ donors with informed consent of the relatives and collected in accordance with the French Personal Protection Committee (CPP: 2018-A02491-54)

4 Notes

1. It is important to harvest the tendon of the TA as distally as possible (at the level of the mouse's foot) to avoid taking the Extensor Digitorum Longus (EDL) muscle at the same time.
2. For initial fixation, preferably use a flat-bottom tissue culture plate (12, 24, or 48 wells, depending on the sample size).

3. Place the muscle on parafilm to make cutting easier and to ensure that the thin longitudinal cuts align parallel to the orientation of the myofibers.
4. For small mouse muscles, such as EDL, Biceps, or Soleus, this step is not required.
5. If needed, stack two or three spacers together, depending on the sample thickness. In this case, make sure that the mounting depth is compatible with the depth of field of the objective lens used for image acquisition.
6. Fiji has the advantage of being a free and open-source software. However, mastering Fiji software may be more time-consuming and less intuitive than Imaris®.
7. Note that it is possible to select a Region Of Interest (ROI) in Fiji, save it in .TIFF format and then perform a 3D reconstruction with Imaris®.
8. Ensure that the performances of the computer used are compatible with the processing of large 3D images for the plugin to run effectively.
9. Unlike Imaris®, Fiji do not support hyper stacks for image analysis.
10. The use of an electric scalpel is not recommended as it can damage the tissues.

Acknowledgments

We thank Institute Jacques Monod's ImagoSeine platform for technical support in image acquisition and analysis. We also thank Thomas Guilbert from Cochin's Institute for cassette's 3D printing.

Competing Interests The authors declare no competing interests.

References

1. Dinulovic I, Furrer R, Handschin C (2017) Plasticity of the muscle stem cell microenvironment. Adv Exp Med Biol 1041:141–169
2. Evano B, Tajbakhsh S (2018) Skeletal muscle stem cells in comfort and stress. NPJ Regen Med 3:24
3. Croce AC, Bottiroli G (2014) Autofluorescence spectroscopy and imaging: a tool for biomedical research and diagnosis. Eur J Histochem EJH 58:2461
4. Chung K, Deisseroth K (2013) CLARITY for mapping the nervous system. Nat Methods 10:508–513
5. Renier N, Wu Z, Simon DJ et al (2014) iDISCO: a simple, rapid method to immunolabel large tissue samples for volume imaging. Cell 159:896–910
6. Susaki EA, Tainaka K, Perrin D et al (2015) Advanced CUBIC protocols for whole-brain and whole-body clearing and imaging. Nat Protoc 10:1709–1727
7. Padgett CA (2023) Tissue clearing: an overview of methods, considerations, and applications. Microsc Today 31:31–35
8. Verma M, Murkonda BS, Asakura Y, Asakura A (2016) Skeletal muscle tissue clearing for LacZ and fluorescent reporters, and

immunofluorescence staining. Methods Mol Biol Clifton NJ 1460:129–140

9. Verma M, Asakura Y, Murakonda BSR et al (2018) Muscle satellite cell cross-talk with a vascular niche maintains quiescence via VEGF and notch signaling. Cell Stem Cell 23:530–543
10. Collins BC, Shapiro JB, Scheib MM et al (2024) Three-dimensional imaging studies in mice identify cellular dynamics of skeletal muscle regeneration. Dev Cell 59:1457–1474
11. Karthikeyan S, Asakura Y, Verma M, Asakura A (2025) Protocol for the three-dimensional analysis of rodent skeletal muscle. STAR Protoc 6:103549
12. Sambasivan R, Gayraud-Morel B, Dumas G et al (2009) Distinct regulatory cascades govern extraocular and pharyngeal arch muscle progenitor cell fates. Dev Cell 16:810–821

Chapter 5

Automated Bioinformatic Analysis of Skeletal Muscle Immunofluorescence for the Quantification of Muscle Stem Cells

Rémi Kurdian, Martina Canatelli-Mallat, Daniel Stockholm, Colin Crist and Alicia Mayeuf-Louchart

Abstract

Muscle stem cells (MuSCs) are essential for skeletal muscle regeneration. They express the transcription factor PAX7 and reside beneath the basal lamina of muscle fibers in a quiescent state. Disruption in MuSCs quiescence or activation impairs regenerative capacity, as observed in myopathies and during ageing. Consequently, the accurate quantification of PAX7-expressing MuSCs constitutes a key readout in myology research. Here, we present a complete and standardized workflow enabling the automated quantification of MuSCs immunolabeled with anti-PAX7 antibodies by immunofluorescence on skeletal muscle sections. This methodology comprises three main steps: cardiotoxin-induced muscle injury followed by muscle dissection, muscle freezing, cryosectioning, and immunofluorescence staining with subsequent MuSCs quantification using MuscleJ2, a freely available high-content analysis Fiji plugin that we previously developed. This automated pipeline ensures reproducible, unbiased, and high-throughput assessment of MuSCs abundance, providing a robust tool for studies investigating muscle homeostasis, pathology, and regeneration.

Keywords Muscle stem cell (MuSC), Satellite cells, PAX7, Skeletal muscle, Regeneration, MuscleJ2, Histology, Bioinformatic tool

1 Introduction

Adult muscle stem cells (MuSCs), or satellite cells, are resident stem cells responsible for skeletal muscle regeneration. MuSCs reside in a quiescent state beneath the basal lamina and sarcolemma of muscle fibers and are characterized by the expression of the transcription factor PAX7 [1]. Upon injury or under pathological conditions, MuSCs

The authors "Rémi Kurdian and Martina Canatelli-Mallat" are coauthors.

Supplementary Information The online version contains supplementary material available at https://doi.org/10.1007/978-1-0716-5643-3_5

Nathalie Didier and Caroline E. Brun (eds.), *Muscle Stem Cells*, Methods in Molecular Biology, vol. 3093,
https://doi.org/10.1007/978-1-0716-5643-3_5,

activate, proliferate, differentiate, and fuse with damaged fibers to restore tissue integrity and maintain homeostasis [2]. Histological examination of skeletal muscle is a powerful approach to evaluate its pathophysiological state in diverse contexts such as myopathies, aging, exercise, and regeneration after injury. These analyses are particularly valuable for diagnosis, for monitoring disease progression, and for assessing therapeutic strategies. These experiments and analyses are inherently time-consuming, which has prompted the development of a range of semi-automatic and automatic tools over the years, based on diverse computational platforms and analytical strategies [3–7]. Among these methods, we developed MuscleJ and subsequently MuscleJ2 [8, 9], a high-content skeletal muscle analysis bioinformatic tool implemented as a Fiji (Fiji is Just ImageJ) plugin. They include an integrated module for quantifying fiber-associated MuSCs, providing a robust and automated approach for assessing MuSCs abundance [8, 9]. MuscleJ/MuscleJ2 segments muscle fibers based on immunofluorescent laminin staining and defines multiple regions of interest (ROIs), including the MuSC/satellite cell region (ROI^{SC}) located beneath the basal lamina. Nuclei double-labeled with PAX7 and DAPI are then automatically identified and quantified, with the results presented in both a detailed report and an illustrative cartography.

High-quality images are essential for accurate automated quantification. Image quality depends on multiple critical steps, including tissue freezing, cryosectioning, and immunostaining, with specific technical details at each stage significantly affecting the final outcome. Here, we present a step-by-step protocol with practical recommendations to optimize image quality, enabling reliable and reproducible quantification of MuSCs.

2 Materials

2.1 Skeletal Muscle Injury and Dissection

1. Isoflurane and anesthesia induction chamber.
2. Cardiotoxin 10 μM (Latoxan, L8102): Dissolve 1 mg Cardiotoxin into 14.7 mL of sterile 1× PBS. Filter with a 0.2 μm filter. Store aliquots at −20°C.
3. 70% ethanol.
4. Heating pad or a heat lamp.
5. 1 mL syringe.
6. Dissection board.
7. T-pins.
8. Dissection tools (fine forceps, scissors, micro-scissors).

2.2 Skeletal Muscle Freezing

1. Isopentane (2-methylbutane).
2. A glass beaker positioned on an elevated support within a container of liquid nitrogen.

3. Liquid nitrogen.
4. Pieces of cork (approximately 5×5 mm).
5. Tragacanth gum (10% (w/v) tragacanth in distilled water: Add 1 g of tragacanth powder to 10 mL of distilled water to obtain a thick, sticky paste. Store at 4°C for up to 2 weeks).
6. Cryotubes (2 mL).

2.3 Cryosections

1. Tissue-Tek optimum cutting temperature (O.C.T.) compound.
2. Forceps and brushes.
3. Superfrost Plus microscope slides.
4. Box of histological slides.
5. Cryostat.

2.4 Immunolabeling

1. Hydrophobic PAP pen.
2. Mouse on Mouse (M.O.M.) blocking reagent (Vector laboratories, MKB-2213).
3. Fixation solution: 4% paraformaldehyde (PFA) in 1× PBS.
4. Methanol.
5. Antigen retrieval buffer (10 mM sodium citrate, pH6): Add 0.65 g of sodium citrate tribasic dihydrate to 250 mL of distilled water. Adjust pH to 6.0 by adding HCl. Store at 4°C for up to 1 month.
6. Blocking buffer (5% bovine serum albumin (BSA) in 1× PBS): Add 5 g of BSA to 100 mL of 1× PBS. Store at 4°C for up to 1 week.
7. Primary antibodies: Mouse monoclonal IgG1 anti-PAX7 (DHSB, supernatant), Rabbit polyclonal anti-Laminin (Sigma, L9393).
8. Secondary antibodies: Goat anti-mouse IgG1 Alexa Fluor 488 (ThermoFisher scientific), Donkey anti-rabbit IgG Alexa Fluor 647 (ThermoFisher scientific).
9. DAPI (1 mg/mL stock solution).
10. Fluorescent mounting medium (Dako Agilent).
11. Coverslips.
12. Nail polish.
13. Equipment: 37°C incubator, −80°C freezer.

2.5 Acquisition and Bioinformatic Analysis

1. Microscope (Slide scanner, apotome, or confocal with mosaic mode option)
2. Software: ImageJ or Fiji with MuscleJ2 plugin (available at: https://github.com/ADanckaert/MuscleJ2/tree/Plugin)

3 Methods

3.1 Cardiotoxin-Induced Muscle Injury, Muscle Dissection, and Freezing (Figs. 1 and 2, ESM Videos 1–2)

1. Anaesthetize 8- to 12-week-old mouse according to ethical guidelines and place it supine on a heating pad or under a heat lamp (*see* **Note 1**).
2. Shave the mouse leg with an electric shaver and clean skin with 70% ethanol.
3. Fill a 0.5 mL syringe with 10 μM cardiotoxin and inject 50 μL into TA muscle from the proximal tendon towards the knee, by following the position of the tibia bone as guidance (*see* **Note 2**) (Electronic Supplementary Video 1).
4. After the procedure, place the mouse in a heated chamber (at 37°C) until it is awake, as the anesthesia and ethanol reduce body temperature. Monitor the mouse for 2–3 days. The mouse shouldn't show any sign of distress and must be able to walk in a normal way after a couple of hours.
5. Euthanize the mouse according to ethical guidelines.
6. To dissect the TA muscle, position the mouse supine on a dissection board and secure the leg with a T-pin. Spray the leg with 70% ethanol (Electronic Supplementary Video 2, Fig. 1a).
7. Using forceps and scissors, cut and peel off the leg skin to expose the hind limb muscles. Expose tendons of the TA muscle (Fig. 1b).
8. Gently remove the fascia covering the TA muscle (Fig. 1c).
9. Cut the distal TA tendon and use it to peel off the TA muscle. Carefully remove the TA muscle by cutting the TA proximal tendon (*see* **Note 3**) (Fig. 1d).
10. To dissect the *Soleus* (SOL) muscle, position the mouse prone on a dissection board and secure the leg with a T-pin. (Fig. 1e).
11. Cut the tendon of the *Gastrocnemius* (GA) muscle and use it to peel off the GA muscle until the exposure of the proximal SOL tendon (Fig. 1f–g).
12. Peel off the SOL muscle by its proximal tendon. Carefully remove the SOL muscle by cutting the SOL distal tendon (Fig. 1h).
13. Fill a beaker with isopentane and immerse its lower half into liquid nitrogen 15 min before experiment. When isopentane becomes slightly viscous and forms solid white laminate lining the inside of the beaker (temperature: −160°C) then it is ready for use (*see* **Note 4**).

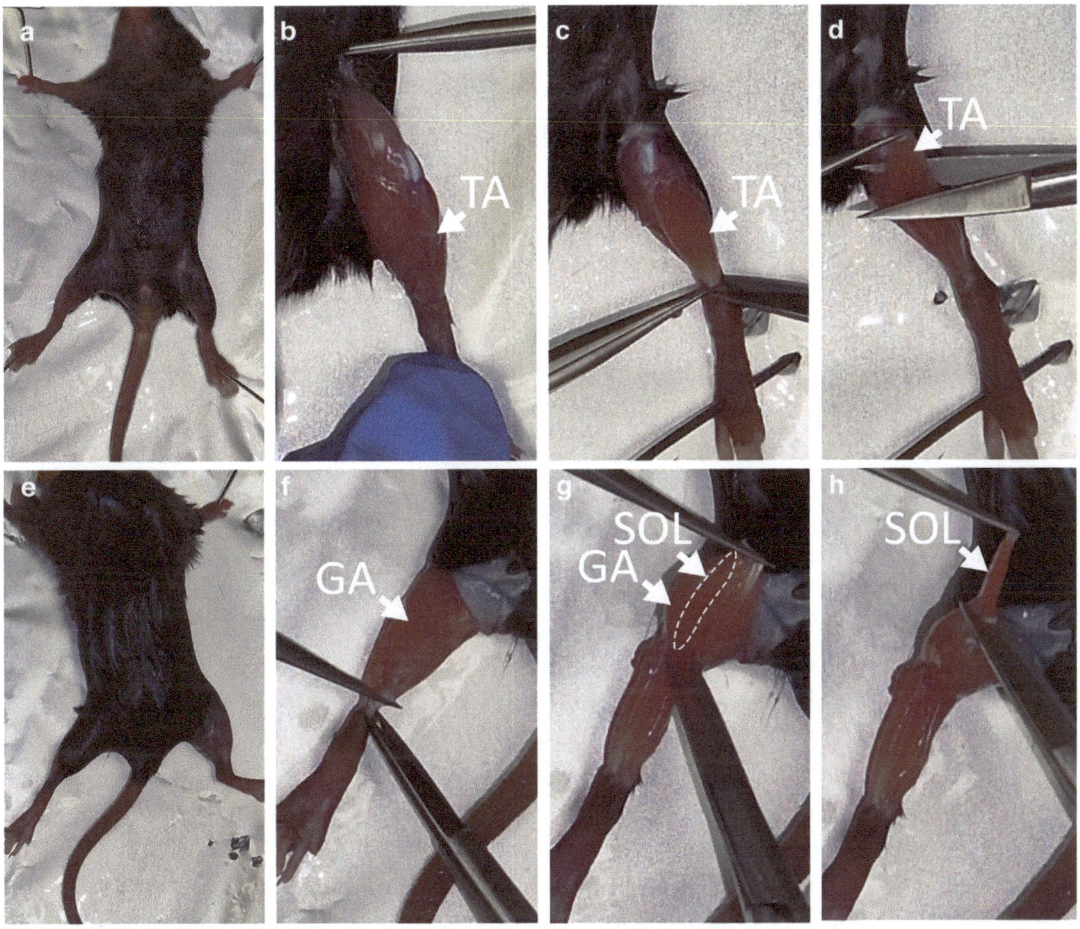

Fig. 1 Muscle dissection. **a and b,** Exposure of the *Tibialis Anterior* (TA) muscle after careful removal of overlying skin and fascia. **c and d,** TA muscle isolated by the tendon, immediately prior to transection. **e and f,** Exposure of the *Gastrocnemius* (GA) muscle following TA removal. **g and h,** *Soleus* (SOL) muscle exposed and ready for separation from the GA

14. Place a small quantity of tragacanth gum on an approximately 5 × 5 mm piece of cork (Fig. 2a).
15. Pick up the muscle by the tendon and carefully position the end of the fresh muscle on top of the tragacanth gum, holding it vertically relative to the cork (Fig. 2b).
16. Immerse the cork and attached muscle into the isopentane bath and leave it to float for at least 5 min until the muscle is completely frozen (*see* **Note 5**) (Fig. 2c, d).
17. Freeze a cryotube by immersing it in liquid nitrogen. Place the muscle on its cork support upside down (cork facing up) in the cold cryotube and immediately place the closed tube in liquid nitrogen (*see* **Note 6**).
18. Store the tubes at −80°C before cryosectioning.

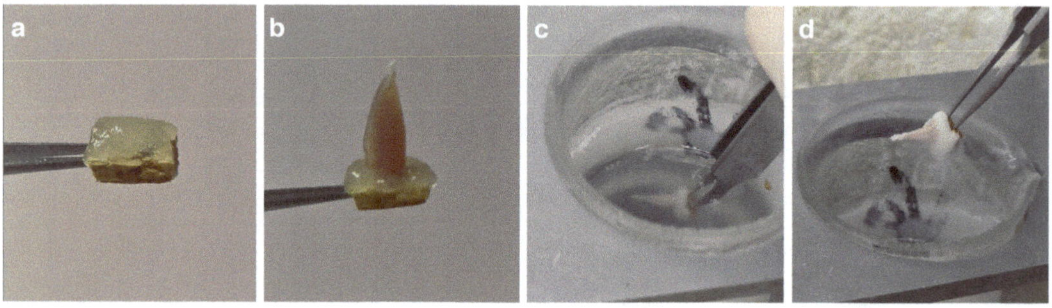

Fig. 2 Snap-freezing procedure of *Tibialis Anterior* (TA). **a,** Preparation of the embedding medium: cork base coated with tragacanth gum to secure the TA muscle. **b,** TA muscle embedded in the gum, positioned for optimal orientation during freezing. **c,** Immersion of the embedded TA muscle in pre-cooled isopentane to ensure rapid and uniform freezing. **d,** Frozen TA muscle immediately after isopentane immersion, ready for long-term storage in liquid nitrogen

3.2 Muscle Cryosections and Immunofluorescence Labeling of MuSCs (Figs. 3 and 4)

1. Set the working temperature of the cryostat inner chamber to −20°C (+/− 1°C) and the object temperature to −18°C (+/−1°C).
2. Remove the muscle from the −80°C freezer and transport it to the cryostat on dry ice. It is crucial that the muscle does not return to room temperature.
3. Place the tube in the cryostat at −20°C and open it to remove the muscle using cold forceps previously placed in the cryostat, touching only the cork and not the muscle.
4. Drop Tissue-Tek O.C.T. compound to a cold sample holder. As soon as it turns white, position the cork with the muscle in the center, perpendicular to the mounting axis (*see* **Note 7**) (Fig. 3a–c).
5. Fill with Tissue-Tek O.C.T. compound if the cork is not securely attached, applying it only to the cork and not to the muscle. Make 10 μm thick muscle sections, placing them on Superfrost Plus glass slides at room temperature as they are cut (Fig. 3d, e).
6. Store slides at −80°C in a dedicated box until further analysis.
7. If muscle remains, Tissue-Tek O.C.T can be gently cut, and the muscle returned to the cryotube to be stored at −80°C.
8. Allow the slides to dry at room temperature for 5 min (Fig. 4).
9. Surround slides with a hydrophobic pen.
10. Rehydrate sections in 1× PBS for 5 min at room temperature.
11. Fix sections with 4% PFA for 10 min at room temperature.
12. Wash three times with 1× PBS, 5 min each time.
13. Permeabilize sections with cold (−20°C) methanol for 6 min at −20°C.

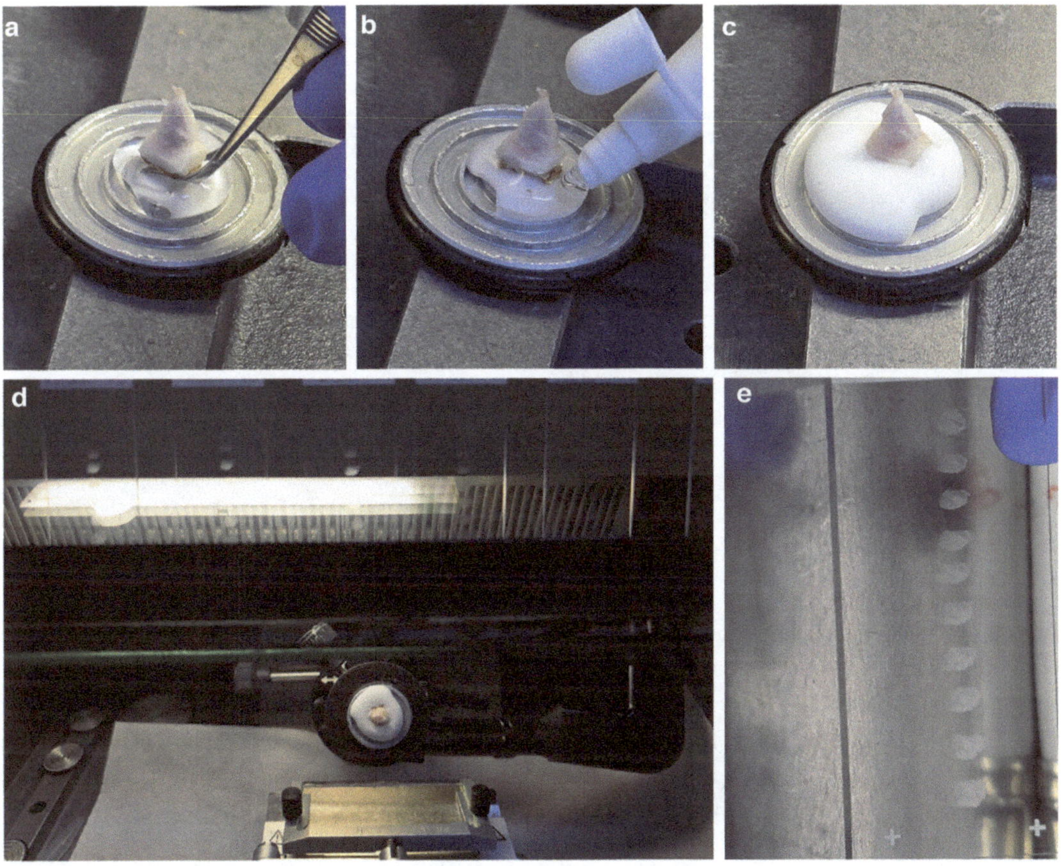

Fig. 3 Cryosectioning of the frozen *Tibialis Anterior* (TA) muscle. **a–c,** Positioning and attachment of the frozen TA muscle to the cryostat platform using O.C.T. compound, ensuring minimal manipulation of the muscle. **d and e,** Cryostat in operation, generating serial cross-sections of the TA muscle, which are collected onto glass slides for subsequent analyses

14. Wash three times with 1× PBS, 5 min each time.
15. Perform antigen retrieval by boiling the sections in 10 mM sodium citrate buffer pH6 for 6 min and cooling them slowly while uncovered for 20 min at room temperature.
16. Wash three times with 1× PBS, 5 min each time.
17. Incubate sections with 1× PBS-5% BSA for 1 h at room temperature in a humid chamber, to block nonspecific sites.
18. Incubate sections with M.O.M blocking reagent diluted in 1× PBS (1:40) for 30 min at room temperature, to block endogenous mouse immunoglobulins and decrease background.
19. Incubate sections overnight with primary antibodies diluted in 1× PBS-5% BSA overnight at 4°C in a humid chamber: mouse anti-PAX7 (1:5), rabbit anti-laminin (1:200).
20. Wash three times with 1× PBS, 5 min each time.

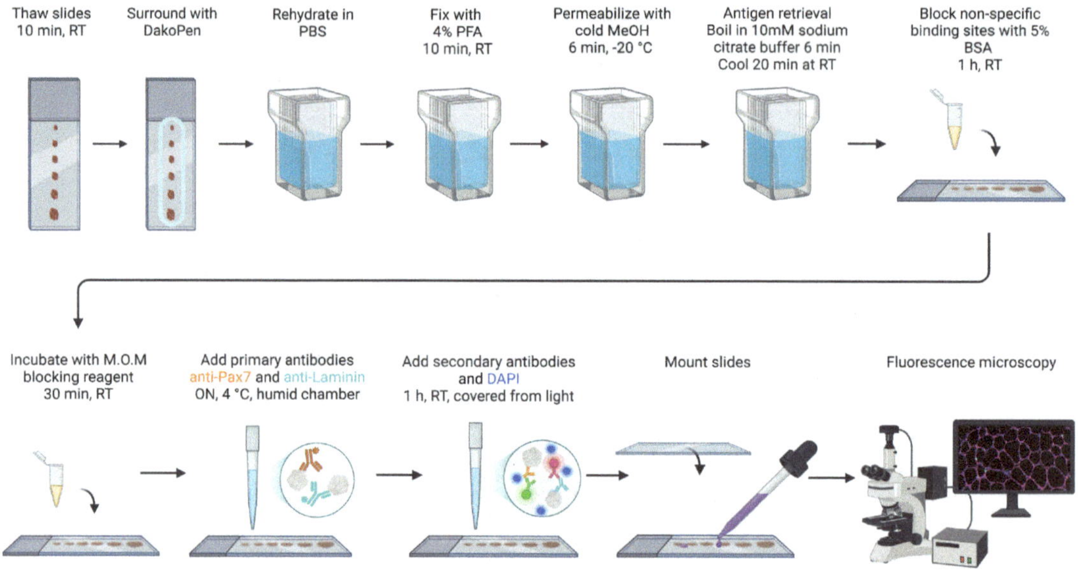

Fig. 4 Schematic representation of the immunofluorescence staining protocol on muscle slides. Step-by-step illustration of the immunofluorescence workflow applied to muscle cryosections, in order to analyze fiber morphology and MuSCs (Figure created with BioRender.com, #YY295AMM6R)

21. Incubate slides with fluorescent secondary antibodies and DAPI diluted in 1× PBS-5% BSA for 1 h at room temperature in a humid chamber and protected from light: goat anti-mouse IgG1 Alexa Fluor 488 (1:500), donkey anti-rabbit IgG Alexa Fluor 647 (1:500), DAPI (1:1000).
22. Wash three times with 1× PBS, 5 min each time.
23. Add two drops of fluorescent mounting medium and carefully put the coverslip to avoid air bubbles. Let dry for a few hours and seal the border of the coverslip with nail polish. For long-term storage, store slides at 4°C protected from light.
24. Images can be acquired using any microscope as MuscleJ2 supports most image formats (.TIFF, .CZI, etc.). Here, we used a ZEISS Slide Scanner Axioscan 7 (Zeiss, Germany) for whole section imaging, but a confocal or apotome microscope can be used to image a specific region of interest or to create a mosaic (*see* **Note 8**).

3.3 MuSCs Quantification on Images with MuscleJ2 (Figs. 5 and 6)

1. Download Fiji from https://imagej.net/Fiji/Downloads and follow the instructions or update your Fiji version to the latest version.
2. Download from MuscleJ2 github space (https://github.com/ADanckaert/MuscleJ2/tree/Plugin) the file named "*MuscleJ_.jar*," corresponding to the latest version of MuscleJ2. A user guide is also available to help you run MuscleJ2.

3. In Fiji, click on *Plugins* menu and select *Install...* command, then open the "*MuscleJ_.jar*" file and save it in "*Plugins*" folder. Restart Fiji, MuscleJ should now appear in the *Plugins* menu (*see* **Note 9**).
4. Before starting MuscleJ2, ensure that the input folder and the images do not have any spaces or symbols in their name. The hyphens must be replaced by underscores. It is possible to use renaming programs such as *Antrenamer* or *NameChanger* to do this on all images at once.
5. All images contained in the input folder must have the same acquisition parameters (number of channels, channel order, staining and acquisition settings) and the same sample properties (same muscle type and same pathophysiological state). Images containing multiple scenes (e.g., CZI images acquired with a ZEISS Slide Scanner Axioscan 7) must be split into individual images before starting the analysis.
6. To run MuscleJ, click on *MuscleJ2* in the *Plugins* menu. A dialog box appears (Fig. 5a) (*see* **Note 10**).
7. In the *Sample Data* panel, enter the pathophysiology state ("*Healthy*" or "*Damaged*") of the muscles being analyzed in the *Pathophysiology* menu and select "*Limb*" in the *Muscle Types* menu.
8. In the *Data Acquisition* panel, under *Volume*, check "*Single Z*" or "*Z stack*" depending on whether you have acquired one or more plans of your images, respectively. Under *Scanned Area*, select "*Entire section*" if the whole section has been scanned, otherwise select "*Crop*." Under *Artefact Detection (% area min)*, set the minimum threshold for the fiber-covered area detection. This is used for the initial quality check of the fiber shape signal (*see* **Note 11**) (Fig. 5a).
9. In the *Data Analysis by Section* panel, leave the checkboxes unchecked as it is not required for the analysis of MuSCs.
10. In the *Data Analysis by Fiber* panel, select "*Fiber Morphology*" and "*Satellite Cells*" to quantify morphological criteria (cross-sectional area, Ferret diameters) and the number of PAX7+ MuSCs, respectively, for each muscle fiber (*see* **Note 12**) (Fig. 5a).
11. In the *Data Cartographies* panel, select "*Fiber Area*" and "*Satellite Cells*" to generate cartographies of fiber size and the number of MuSCs per fiber, respectively (Fig. 5b). Check "*Multi Cartography Montage*" to generate a single montage with the different cartographies. Under *Legend*, select "*Yes*" if you want the legend to appear in the upper left corner of the cartographies. You can choose to display the scale bar and its location under *Scale Bar* (Fig. 5a).

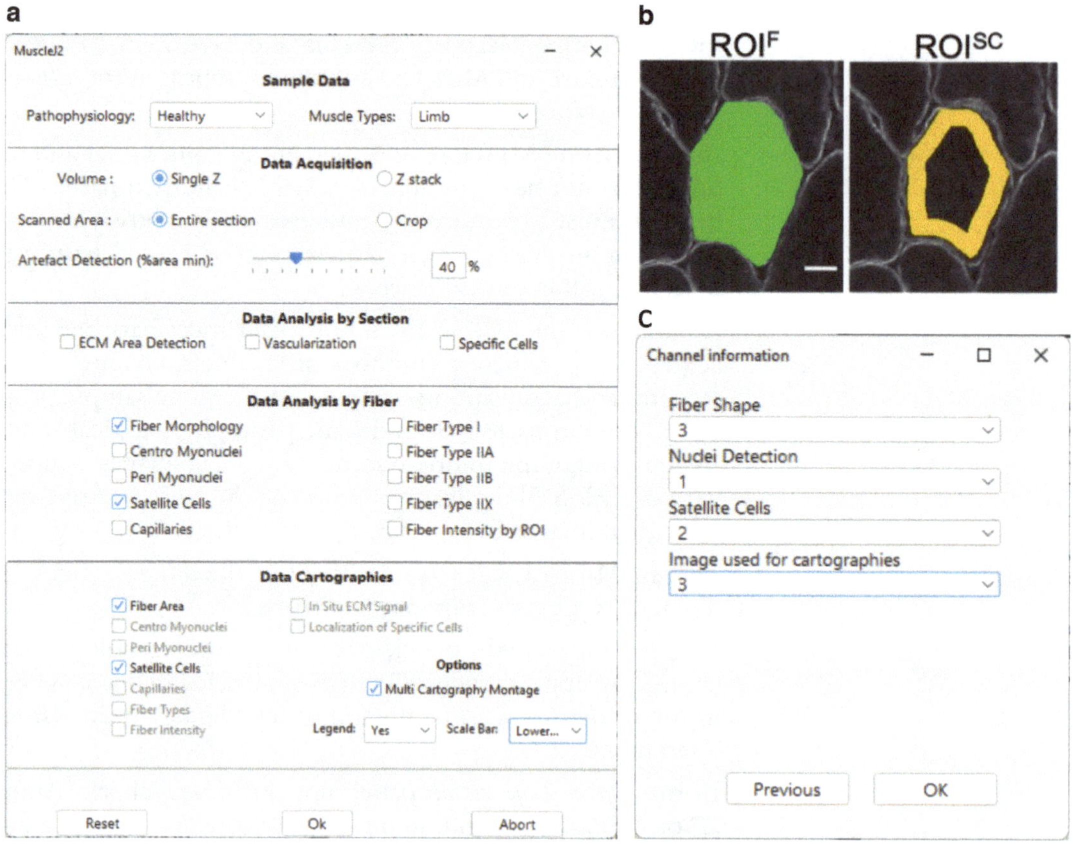

Fig. 5 MuscleJ2 plugin interface. **a,** The main MuscleJ2 dialog box is divided into five sections in which the user can make a selection from a drop-down menu or check boxes. Here are the choices to be made when analyzing MuSCs. **b,** Representation of the regions of interest corresponding to the fiber (ROI^F) in green and satellite cell (ROI^{SC}) in yellow. **c,** The channel information dialog box is used to specify the channel number for each analysis requested.

12. Click *Ok*, the *Channel information* dialog box appears (Fig. 5c).
13. For each parameter, select the channel number using the drop-down menus. The channel numbers correspond to the order of the channels in the original image. For *Fiber Shape* and *Image used for cartographies*, select the channel number corresponding to the Laminin staining. For *Nuclei Detection*, select the channel number corresponding to the DAPI staining. For *Satellite Cells*, select the channel number corresponding to the PAX7 staining (Fig. 5c).
14. Click *OK*, a dialog box appears and asks you to select a repository for the batch. Click *OK* and select the input folder containing the images to be analyzed. Then, select the output folder to save cartographies and results. MuscleJ2 starts the automatic analysis image by image.

15. Once the analyses are complete, the selected result folder contains four different folders named "*Artefacts—Cartography—Results_byfile—ROI.*" The "*Artefacts*" folder contains all images that have not been analyzed and have been rejected by the initial quality check (*Artefact Detection (%area min)*). The "*Cartography*" folder contains all generated cartographies. The "*Results_byfile*" folder contains.txt files with all results for each image analyzed. The "*ROI*" folder contains all regions of interest created and analyzed by MuscleJ2. This file is useful for reviewing the analyses. Finally, a.txt file of global results per batch is created with the following nomenclature "*ImageFolderName_GlobalResults_FM_SC*.txt". The results indicate the number of fibers with satellite cells and the total number of satellite cells detected on muscle section. Examples of results are provided in Fig. 6.

a

b

Output for all analyses

Fiber Morphology output

FileName	Whole Section Area	Nb Segmented Fiber	Fiber Area Mean	Fiber Feret Mean	<250 µm2	250-500 µm2	500-750 µm2	750-1000 µm2
XXXX	2692351.2	1056	2264.219	67.4325	11	53	43	28

Fiber Morphology output / *Satellite Cells* output

1000-2000 µm2	2000-3000 µm2	3000-4000 µm2	4000-5000 µm2	5000-6000 µm2	>6000 µm2	Fibers with SatCells	Nb Sat Cells (Pax7+)
225	487	183	21	5	0	82	82

C

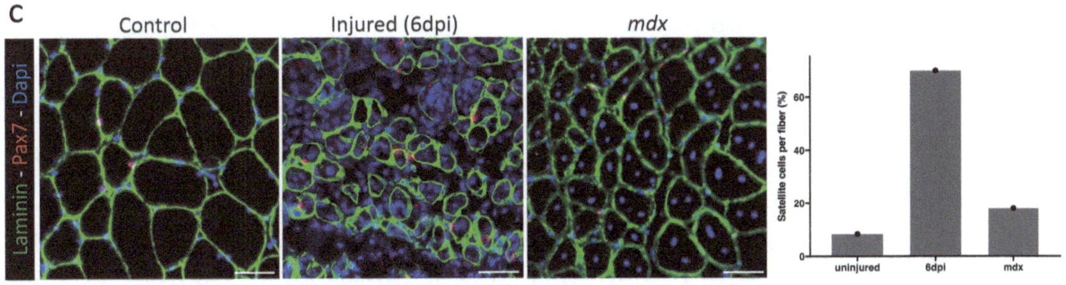

Fig. 6 Quantification of MuSCs by MuscleJ2. **a,** Representative image of a cross-section of the S*oleus* muscle immunolabeled with antibodies against laminin (green), DAPI (blue), and PAX7 (red) (left image) and respective cartographies for fiber morphology (middle image) and MuSCs quantification (right image) generated by MuscleJ2. **b,** Raw data obtained from MuscleJ2. **c,** Representative images of a cross-section of the *Tibialis Anterior (TA)* muscle immunolabeled with antibodies against laminin (green), DAPI (blue), and PAX7 (red) in control, injured mice [6 days post-injury (dpi)], and *mdx* mice (dystrophic mice), and their respective quantification of MuSCs performed by MuscleJ2 (scale bars, 50 µm).

4 Notes

1. This part describes the induction of an acute injury of the *Tibialis Anterior (TA)* muscle by the intramuscular injection of cardiotoxin, a snake venom toxin that induces a rapid necrosis of myofibers and the activation of an inflammatory response, leading to the loss of muscle architecture and the activation of the regeneration process [10]. Here, mice were anaesthetized and maintained unconscious by isoflurane inhalation, but intraperitoneal injection of ketamine/xylazine can be performed instead, according to ethical guidelines.
2. When filling the syringe with cardiotoxin, make sure there are no air bubbles. Insert the needle into the center of the TA muscle, 2–3 mm deep and at an approximately 20° angle, avoiding going too deep or beyond the muscle. After injection, wait 2–3 s before pulling out the needle to prevent leakage of cardiotoxin drops from the muscle and gently massage the limb to diffuse the cardiotoxin (Electronic Supplementary Video 1).
3. Always grasp the muscle by the tendon to avoid damaging the myofibers.
4. Isopentane vapors are toxic and should be handled in a fume hood.
5. Hold the assembly in cold isopentane with forceps for a few seconds until the muscle turns white. Then detach the assembly from the forceps and leave in isopentane for at least 5 min.
6. Using long forceps, dip the open cryotube into liquid nitrogen until the cooling noise stops. Empty the cryotube of all liquid nitrogen and immediately place the frozen muscle inside. Close the tube and place immediately in liquid nitrogen.
7. The Tissue-Tek O.C.T. must cover the cork to ensure stability, but must not touch the muscle, as this would heat it up. Keeping the muscle at the lowest possible temperature and avoiding reheating it throughout the experiment is essential to ensuring the quality of the sections.
8. File formats that are not supported are .JPEG, .PNG, .TIFF by channel and time series. Images obtained from any microscope magnification (10×, 20×, 25×, 40×, 63×) can be used. The limitations are relative to the quality of image staining. For MuSCs assessment, a minimum magnification of *20×* is recommended to ensure sufficient resolution.
9. The amount of computer memory devoted to Fiji can be increased by selecting *Edit* menu, then *Memory & Threads*

without allocating more than 75% of the computer's total memory, as Fiji may become slow and unstable.

10. MuscleJ2 allows the analysis of numerous parameters at the same time. In this section, we detail the steps for MuSCs analysis on muscle sections. To perform other analyses or for more information, please refer to the MuscleJ2 user guide and the two original publications [8, 9].
11. *Artefact Detection (% area min)* is used for the initial quality check of the fiber shape signal (relying on Laminin staining). Example: *%area min=X*, if less than X% of the total surface of the section does not contain segmentable fibers (artefact, hole, weak staining intensity, etc.), the analysis will not be performed for this section. We recommend leaving the default threshold at 40%, but it can be adapted based on the quality of the section.
12. Regardless of the analyses performed, the "*Fiber Morphology*" analysis is mandatory, as this is the process during which the regions of interest (ROIs) are created and saved.

Acknowledgments

We would like to express our sincere thanks to Didier Montarras, who taught us these techniques with patience and passion. We would also like to thank Virginia Zoglio for her advice on immunostaining. The authors acknowledge funding support from French National Institute of Health and Medical Research (INSERM), Agence Nationale pour la Recherche (ANR-23-CE13-0044), French Society of Myology (SFM), Association Française contre les Myopathies (AFM-Téléthon #22281), and Canadian Institutes for Health Research.

Competing Interests The authors have no competing interests to declare.

References

1. Montarras D, L'Honoré A, Buckingham M (2013) Lying low but ready for action: the quiescent muscle satellite cell. FEBS J 280:4036–4050
2. Dumont NA, Bentzinger CF, Sincennes MC, Rudnicki MA (2015) Satellite cells and skeletal muscle regeneration. Compr Physiol 5:1027–1059
3. Smith LR, Barton ER (2014) SMASH—semi-automatic muscle analysis using segmentation of histology: a MATLAB application. Skelet Muscle 4:1–15
4. Kostrominova TY, Reiner DS, Haas RH, Ingermanson R, McDonough PM (2013) Automated methods for the analysis of skeletal muscle fiber size and metabolic type. In: International review of cell and molecular biology. Elsevier Inc, pp 275–332
5. Sertel O, Dogdas B, Chiu CS, Gurcan MN (2011) Microscopic image analysis for quantitative characterization of muscle fiber type composition. Comput Med Imaging Graph 35:616–628
6. Yuan Wen X, Murach KA, Vechetti Jr IJ, Fry CS, Vickery C, Peterson CA, McCarthy

JJ, Campbell KS (2018) INNOVATIVE METHODOLOGY MyoVision: software for automated high-content analysis of skeletal muscle immunohistochemistry. J Appl Physiol 124:40–51

7. Bergmeister KD, Gröger M, Aman M, Willensdorfer A, Manzano-Szalai K, Salminger S, Aszmann OC (2017) A rapid automated protocol for muscle fiber population analysis in rat muscle cross sections using myosin heavy chain immunohistochemistry. J Visualized Exp 121:1–15
8. Mayeuf-Louchart A, Hardy D, Thorel Q, Roux P, Gueniot L, Briand D, Mazeraud A, Bouglé A, Shorte SL, Staels B, Chrétien F, Duez H, Danckaert A (2018) MuscleJ2: a high-content analysis method to study skeletal muscle with a new Fiji tool. Skelet Muscle 8:1–11
9. Danckaert A, Trignol A, Le Loher G, Loubens S, Staels B, Duez H, Shorte SL, Mayeuf-Louchart A (2023) MuscleJ22: a rebuilding of MuscleJ2 with new features for high-content analysis of skeletal muscle immunofluorescence slides. Skelet Muscle 13:1–14
10. Chargé SBP, Rudnicki MA (2004) Cellular and molecular regulation of muscle regeneration. Physiol Rev 84:209–238

Chapter 6

Spatial and Quantitative Immunofluorescence Analysis of Human Myogenic Cells in Mouse Skeletal Muscle Xenografts

Mona Bensalah, Laura Muraine, Pierre Klein, Ingo Riederer, Anne Bigot and Elisa Negroni

Abstract

Immunofluorescence analysis remains a cornerstone technique for evaluating the fate of transplanted human cells in xenograft models, as it provides essential spatial and phenotypic information unavailable through molecular quantification alone. This chapter outlines a standardized procedure for the preparation of human myogenic cells prior to intramuscular injection and then describes a comprehensive immunofluorescence-based method to identify, localize, and quantify human myogenic cells engrafted into regenerating skeletal muscle of immunodeficient mice. Using human-specific antibodies combined with extracellular matrix markers, this approach enables discrimination and classification of engrafted human cells as myonuclei, interstitial cells, or incorporated into the satellite cell niche. Quantitative strategies are presented to estimate total human cell numbers and assess longitudinal and transverse dispersion within host muscle tissue.

Keywords Xenotransplantation, Immunofluorescence, Human myogenic cells, Skeletal muscle regeneration, Human-specific antibodies, Satellite cells, Cell therapy assessment

1 Introduction

Xenotransplantation of human myogenic progenitors into immunodeficient mice provides a preclinical platform to study human muscle biology, pathology, and regeneration in vivo, enabling investigation of disease mechanisms and therapeutic interventions. Immunofluorescence (IF) analysis is indispensable for assessing cell localization, tissue integration, and contribution to regeneration in xenografted tissues. It enables detailed evaluation of the spatial integration, differentiation status, and niche occupancy of transplanted cells. In skeletal muscle xenotransplantation models, IF allows discrimination between human cells that have fused with host myofibers, those that remain within the interstitial space, and those that have colonized the satellite cell (SC) niche beneath the basal lamina. These distinct

Nathalie Didier and Caroline E. Brun (eds.), *Muscle Stem Cells*, Methods in Molecular Biology, vol. 3093,
https://doi.org/10.1007/978-1-0716-5643-3_6,

cellular fates have important functional and therapeutic implications, particularly in the context of muscular dystrophies [1] where repeated cycles of muscle regeneration may occur.

In addition to post-transplantation analysis, accurate interpretation of engraftment outcomes requires careful control of pretransplantation parameters. This chapter therefore presents a robust immunofluorescence-based methodology for analyzing human myogenic cells transplanted into immunodeficient mouse skeletal muscle, including a standardized procedure for cell preparation prior to intramuscular injection, to ensure reproducible transplantation conditions. The approach combines systematic serial sectioning, human-specific nuclear and cytoplasmic markers, and quantitative image analysis to provide reproducible and biologically meaningful data [2].

Injected muscles are cryosectioned along their entire length, enabling both transversal and longitudinal analysis of human cell distribution. The analysis method relies on detection of human cells within mouse skeletal muscle using antibodies that specifically recognize human proteins absent from murine tissues. Human lamin A/C, a nuclear envelope protein, is used as a highly sensitive marker to identify human nuclei regardless of differentiation state. Additional markers, including human spectrin and Pax7, are employed to assess myogenic differentiation and SC niche occupation, respectively [2, 3]. Quantification strategies account for section thickness, nuclear size, and inter-section spacing, permitting estimation of total human cell numbers and dispersion within the muscle. This approach provides qualitative and quantitative information essential for interpreting engraftment efficiency beyond simple cell counts.

2 Materials

2.1 Preparation of Human Myogenic Cells for Transplantation

1. Human myogenic cells: expanded under standard culture conditions and suitable for intramuscular transplantation.
2. Cell culture medium: appropriate growth medium for maintenance and expansion of human myogenic cells.
3. Phosphate-buffered saline (PBS): sterile phosphate-buffered saline used for washing and resuspension of cells prior to transplantation.
4. 15 mL conical tubes: sterile 15 mL polypropylene conical tubes used for all steps of cell preparation to minimize handling and cell loss.

2.2 Intramuscular Grafting of Human Cells

1. Immunodeficient mice: adult immunodeficient mice permissive for human cell engraftment (i.e., Rag2−/− Il2rb−/−), 8–12 weeks of age, male or female, housed under specific

pathogen-free conditions and handled in accordance with institutional and national ethical guidelines for animal experimentation (*see* **Note 1**).

2. Anesthetic/analgesic agents and materials required for surgical procedures and induction of muscle regeneration are selected and used in accordance with institutional and national ethical guidelines for animal experimentation (*see* **Note 2**).
3. Electric shaver suitable for removal of hair from the skin overlying the TA muscle prior to intramuscular injection. Use of a shaver is preferred over a scalpel to avoid accidental skin injury during hair removal.
4. 25-μL Hamilton syringe fitted with a removable 30-gauge needle (RN type, 2.5 cm length; Hamilton Company, catalogue no. 7803-07).

2.3 Tissue Processing

1. Tragacanth gum: tragacanth gum (G1128, Sigma-Aldrich) is prepared as a 6% (w/v) aqueous solution in distilled water. Thymol [5-méthyl-2-(1-méthyléthyl)phénol, 5-méthyl-2-isopropylphénol, T0501, Sigma-Aldrich] is added to the solution as a few crystals to prevent mould and microbial growth. Tragacanth gum is used for embedding skeletal muscle tissue to preserve orientation during freezing and cryosectioning.
2. Cork support for mounting and orienting the muscle prior to freezing.
3. Isopentane and liquid nitrogen: analytical-grade isopentane used as a freezing medium, precooled in liquid nitrogen to ensure rapid snap-freezing of muscle samples, minimizing ice crystal formation, and preserving tissue morphology.

2.4 Primary Antibodies

1. Mouse anti-human lamin A/C: mouse monoclonal antibody recognizing human lamin A/C and not cross-reactive with murine lamin A/C, is used for identification of human nuclei in mouse skeletal muscle xenografts (mouse monoclonal IgG2b, 1:400, NCL-Lam-A/C, Novocastra, Leica Biosystems, or mouse monoclonal IgG1, 1:400, clone JOL2, AbCam) (*see* **Note 3**).
2. Mouse anti-human spectrin: mouse monoclonal antibody directed against human spectrin, used to detect human protein expression within regenerated skeletal muscle fibers following xenotransplantation (mouse monoclonal IgG2b, 1/50, clone RBC2/3D5 Novocastra, Leica Biosystems) (*see* **Note 3**).
3. Mouse anti-Pax7: mouse monoclonal antibody recognizing Pax7, used for identification of satellite cells in skeletal muscle sections (mouse monoclonal IgG1,1:50, Pax7 Developmental Studies Hybridoma Bank) (*see* **Note 3**).

4. Rabbit anti-laminin: anti pan-laminin rabbit polyclonal antibody, used to delineate the basal lamina and muscle fiber architecture in skeletal muscle cryosections (1:400, rabbit polyclonal, Z0097, Dako) (*see* **Note 3**).

2.5 Secondary Antibodies

1. Alexa Fluor–conjugated goat anti-mouse IgG: fluorophore-conjugated goat secondary antibodies specific for mouse IgG subclasses (e.g., IgG1 and IgG2b), used for selective detection of mouse monoclonal primary antibodies in multiplex immunofluorescence experiments. Antibodies are chosen to allow simultaneous visualization of multiple antigens without cross-reactivity between detection channels.
2. Alexa Fluor–conjugated goat anti-rabbit IgG: fluorophore-conjugated goat secondary antibody recognizing rabbit IgG, used for immunofluorescence detection of rabbit polyclonal primary antibodies and compatible with multicolor fluorescence imaging.

2.6 Staining and Mounting Reagents

1. Hoechst 33258: DNA-binding fluorescent dye used for nuclear counterstaining, allowing visualization of total nuclei in skeletal muscle cryosections (Hoechst staining 0.5 μg/ml).
2. Bovine serum albumin (BSA) (Fraction V) used as a blocking agent to reduce nonspecific antibody binding and background fluorescence during immunofluorescence staining (*see* **Note 4**).
3. Phosphate-buffered saline (PBS) is used for tissue washing, antibody dilution, and maintenance of physiological pH during immunofluorescence procedures.
4. Fluorescent aqueous mounting medium optimized for fluorescence microscopy, is used to preserve fluorescence signal intensity and tissue morphology after staining and coverslip mounting.

2.7 Equipment

1. Bench-top centrifuge compatible with 15 mL conical tubes for pelleting cells.
2. Hemocytometer or automated cell counter device for accurate determination of cell number during preparation.
3. Cryostat equipped with temperature control suitable for sectioning frozen skeletal muscle tissue into reproducible thin transverse sections (5 μm thickness).
4. Epifluorescence microscope equipped with appropriate excitation and emission filter sets for Hoechst and Alexa Fluor dyes, allowing multichannel image acquisition.
5. High-resolution digital camera coupled to the fluorescence microscope for acquisition of immunofluorescence images with sufficient dynamic range for quantitative analysis.

6. Image analysis software capable of measuring nuclear dimensions, counting labeled cells, and calculating spatial dispersion parameters (i.e., ImageJ, https://imagej.net/ij/).

3 Methods

3.1 Preparation of Human Myogenic Cells for Transplantation

1. Human myogenic cells are expanded under standard culture conditions, passaged as required, and harvested for transplantation (*see* **Note 5**).
2. Cells are counted to determine the total cell number.
3. Cells are pelleted by centrifugation for 5 min at 1200 rpm (centrifugation parameters may be adjusted depending on the cell type), and the culture medium is carefully aspirated.
4. The cell pellet is resuspended in PBS (e.g., 10 mL) to wash the cells, and cells are counted again to obtain the most accurate final cell number possible.
5. Cells are pelleted a second time by centrifugation, and PBS is carefully removed, leaving a minimal residual volume at the bottom of the tube to avoid disturbing the nonvisible pellet (*see* **Note 6**).
6. The remaining PBS volume is measured using a pipette. Based on this volume, an appropriate amount of PBS is added to obtain a final concentration of 100k cells per 15 μL (*see* **Note 7**). When possible, the same pipette tip is retained to verify the residual volume. PBS is added using a new pipette tip, and the original tip is then used to gently resuspend the cell pellet by slow pipetting to minimize cell loss.
7. All steps of cell preparation are performed using a single 15 ml conical tube to reduce handling and limit cell loss.
8. The final cell suspension is kept on ice and used for intramuscular injection.

3.2 Injury and Cell Grafting into Mouse Muscle

1. Surgical procedures and induction of muscle regeneration should be performed using protocols approved by and conducted in accordance with institutional and national ethical guidelines for animal experimentation (*see* **Note 2**).
2. Under anesthesia, 15 μL of the cell suspension is aspirated into a Hamilton syringe and injected into the tibialis anterior (TA) muscle. The skin covering the TA is shaved and disinfected beforehand using Vetadine or an equivalent antiseptic, in accordance with institutional and national ethical guidelines for animal experimentation.

3. The needle is inserted at the proximal end of the TA muscle at an insertion angle of 30°–45° relative to the muscle surface, with the needle oriented parallel to the tibia, and advanced toward the mid-belly of the TA. The cell suspension is then slowly injected while keeping the needle firmly in position. Once the injection is complete and the suspension fully delivered, the needle is gently withdrawn (*see* **Note 8**).

3.3 Harvesting and Freezing

1. At defined time points after transplantation of human myogenic cells, mice are euthanized according to institutional and national guidelines for animal experimentation.
2. The skin covering the hind limb is incised, and the TA muscle is exposed for dissection (*see* **Note 9**).
3. The TA muscle is carefully dissected using fine scissors, taking care to avoid compression, stretching, or tearing of the tissue.
4. Muscles are immediately oriented longitudinally and mounted on a small amount of 6% (w/v) tragacanth gum on a cork support, with the distal tendon of the TA oriented upward to preserve the proximal–distal axis for longitudinal analysis, and are then rapidly frozen by immersion in isopentane precooled in liquid nitrogen, ensuring uniform freezing and minimizing ice crystal formation.
5. Frozen muscles are stored at −80°C until cryosectioning.

3.4 Cryosectioning Strategy

1. Frozen TA muscles are equilibrated to the cryostat temperature prior to sectioning.
2. Muscles are oriented in the cryostat to obtain transverse sections perpendicular to the longitudinal axis of the muscle fibers.
3. Serial transverse cryosections are cut at a thickness of 5 μm along the entire length of the muscle using a cryostat.
4. Sections are collected sequentially on glass microscope slides, maintaining the original distal-to-proximal order of the TA muscle.
5. A systematic sampling strategy is applied in which analyzed sections are collected at regular intervals along the muscle length, typically every 500 μm (Fig. 1A).
6. The position of each analyzed section along the muscle is recorded to allow reconstruction of longitudinal cell distribution.
7. Sections are air-dried at room temperature (RT) prior to immunofluorescence staining or stored at −80°C until further processing.

3.5 Immunofluorescence Staining

1. Cryosections are removed from −80°C storage and allowed to equilibrate to RT.

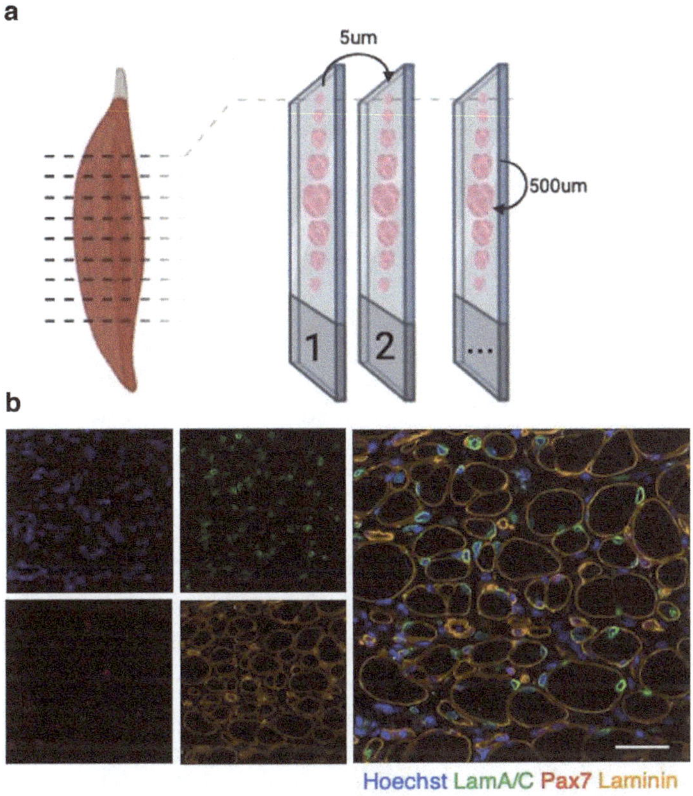

Fig. 1 Identification of human myoblasts in TA muscle by immunofluorescence. **A,** Schematic representation of the processing of a TA muscle. Each section (5 μ m) on the same slide is separated from the previous and from the following one by 500 μm. **B,** Typical immunofluorescence analysis of human myogenic cells 21 days after transplantation into immunodeficient mice. Cells were injected 24 h after notexin-induced muscle injury. Transverse sections are stained using antibodies against human lamin A/C (green), Pax7 (red), and laminin (orange). Nuclei were counterstained with Hoechst (blue). Scale bar = 50 μm

2. Sections are air-dried for 10–15 min and rehydrated in PBS.
3. Nonspecific binding sites are blocked by incubating sections in PBS containing 2% (w/v) bovine serum albumin (BSA) for 30 min at RT (*see* **Note 4**).
4. Sections are incubated with primary antibodies diluted in blocking solution for 1 h at RT in a humidified chamber.
5. Following primary antibody incubation, sections are washed three times for 5 min each in PBS.
6. Sections are incubated with appropriate Alexa Fluor–conjugated secondary antibodies diluted in blocking solution for 45 min at RT in a dark humidified chamber.

7. Sections are washed three times for 5 min each in PBS.
8. Nuclei are counterstained with Hoechst 33258 for 5 min.
9. Sections are washed three times for 5 min each in PBS and mounted using an aqueous fluorescent mounting medium for microscope observation (Fig. 1B).

3.6 Identification of Human Cells and Tissue Context

1. Human cells are identified based on nuclear immunoreactivity for human lamin A/C (*see* **Note 3**).
2. Muscle architecture and basal lamina of the fibers are visualized by laminin immunostaining (*see* **Note 3**).
3. Human nuclei located within laminin-positive muscle fibers are classified as human myonuclei.
4. Human spectrin identify host myofibers that have incorporated human myoblasts.
5. Human lamin A/C–positive nuclei located outside the basal lamina are classified as interstitial human cells.
6. Human cells occupying the satellite cell position are identified as lamin A/C–positive nuclei located beneath the laminin-positive basal lamina.
7. Co-immunostaining for Pax7 is used to identify satellite cells and to confirm occupation of the satellite cell niche by transplanted human cells (*see* **Note 3**).

3.7 Quantification of Human Cells

3.7.1 Measurement of Human Nuclear Diameter

1. The diameter of human nuclei is measured on lamin A/C–positive nuclei using image analysis software.
2. Nuclear diameter is measured on representative sections distributed along the muscle length.
3. The average nuclear diameter (D) is determined by measuring lamin A/C-positive nuclei using an image analysis software from a minimum of 500 human nuclei per experimental condition. Measurements have to be performed on a large number of nuclei to obtain a representative average nuclear size.
4. This mean value is used for correction of nuclear overcounting across serial cryosections.

3.7.2 Calculation of Corrected Linear Nuclear Density

1. The number of lamin A/C–positive human nuclei (N) is counted on each analyzed transverse section.
2. Raw nuclear counts are corrected for section thickness ($s = 5\ \mu$ m) and average nuclear diameter (D) to obtain the linear density (n) of human nuclei along the muscle length (Fig. 2).
3. The linear density of human nuclei (n) is obtained by dividing the number of lamin A/C-positive nuclei (N) counted in a section over the section size s (5 μm). But given its finite size D, each cell could be visible in a number of sections. The linear

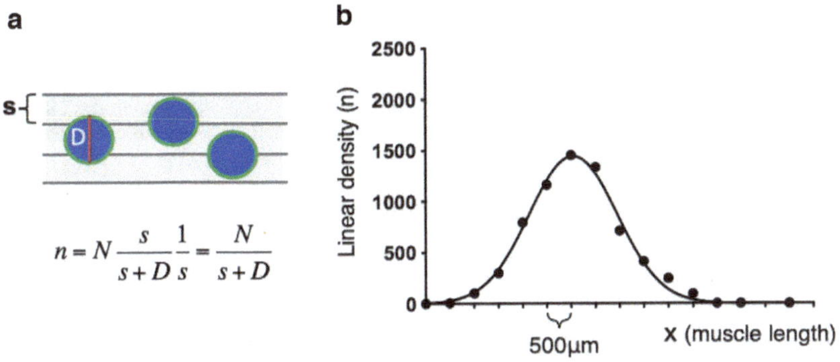

Fig. 2 Quantification of human myoblasts in TA muscle by immunofluorescence method. **A,** Schematic representation illustrating the distribution of human nuclei in the muscle sections and calculation of linear density (n) of the nuclei in the muscle length. s indicates the size of muscle section, D the nucleus diameter, n the linear density, N the number of cells counted in a muscle section. **B,** Representative curve obtained by quantification of human myoblasts 21 days after a single intramuscular injection performed at the mid-belly of the TA

density (n) should, therefore, be weighted by the average number of sections over which each nucleus can be detected [(D+ s)/s], as follows

$$n = N\frac{s}{s+D}\frac{1}{s} = \frac{N}{s+D}$$

3.7.3 Estimation of Total Human Cell Number Per Muscle

1. The total number of human cells per muscle is estimated by integrating the linear density n of the cells along the entire length (x) of the muscle (Fig. 3).
2. Corrected linear nuclear density values can be plotted as a function of the longitudinal position along the muscle to visualize the distribution within the host muscle.
3. The number of human cells between consecutive analyzed sections is estimated by linear interpolation.
4. The total number of human cells per muscle is calculated by numerical integration of corrected linear density values over the entire muscle length. Human nuclei are counted only on 5 µm sections collected at 500 µm intervals (L). The number of cells in the intermediate regions is estimated by linear interpolation, and numerical integration is performed using the formula given below:

$$\sum_{\text{length of muscle}} \text{Min}(n_x; n_{x+1})L + \text{abs}(n_x - n_{x+1})L/2$$

where Min represents the minimum value between n_x and n_{x+1} and abs the absolute value.

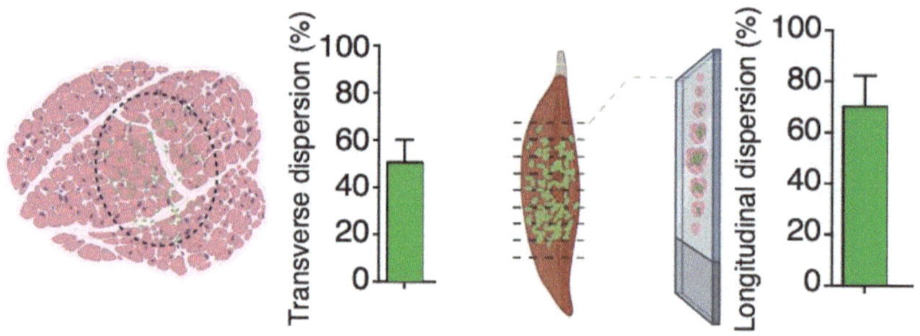

Fig. 3 Dispersion of injected human cells within the host muscle. Representative transverse mouse muscle section and quantification of the transverse dispersion of injected human cells 21 days after injury and transplantation. Schematic longitudinal representation of the TA muscle and quantification of the longitudinal dispersion of injected human cells 21 days after injury and transplantation

3.8 Analysis of Transverse and Longitudinal Dispersion

1. The transverse section containing the highest number of lamin A/C–positive human nuclei along the entire muscle is identified.
2. The smallest ellipse area encompassing all human nuclei on this section is determined using image analysis software.
3. Transverse dispersion is expressed as the ratio of this area to the total TA muscle cross-sectional area (Fig. 3).
4. Longitudinal dispersion is assessed by calculating the proportion of the total muscle length (i.e., number of sections) containing at least one lamin A/C–positive human nucleus, and is expressed as a percentage of the total muscle length (Fig. 3).

4 Notes

1. Adult immunodeficient mice are permissive for human cell engraftment (i.e., SCID, Rag2−/− Il2rb−/−, Rag2−/− Il2rb−/− C5−/−, *NOD, NSG*), 8–12 weeks of age, male or female. The choice of immunodeficient mouse strain should be guided by the experimental objective, the origin and immunogenicity of the transplanted cells (murine vs human), and the planned duration of the study. Residual innate immune activity, differences in genetic background, and animal availability may all influence engraftment efficiency and experimental variability. For comparative studies, the same mouse strain and age, should be used consistently within an experimental series. Notably, a previous study comparing male and female recipient mice reported no significant difference in engraftment efficiency following intramuscular transplantation of murine satellite cells [4].

2. We routinely induce muscle regeneration by notexin injection performed 24 h prior to cell transplantation. However, alternative injury models may be used depending on the experimental objective, with appropriate adjustment of the timing and nature of the injury. For a detailed comparison of commonly used muscle injury models and their impact on regeneration, readers are referred to [5].
3. The antibodies described in this chapter were selected based on several experimental considerations, including robustness, specificity, and compatibility with multiplex immunofluorescence. Human lamin A/C provides a sensitive and differentiation-independent nuclear marker for identifying human cells in mouse skeletal muscle during muscle regeneration, while laminin costaining is essential for accurate anatomical discrimination between intrafiber myonuclei, interstitial cells, and satellite cell positions beneath the basal lamina. Identification of satellite cell localization relies primarily on nuclear position with Pax7 expression. Antibody specificity should always be verified on non-transplanted control muscles to exclude cross-reactivity with murine tissues, and careful selection of secondary antibodies and fluorophores is required to avoid spectral overlap and cross-detection of mouse IgG subclasses. Importantly, antibody choice should be adapted to the specific experimental context, as optimal marker combinations may vary depending on the biological question, tissue condition, and imaging setup.
4. A standard concentration of BSA is used in the blocking solution, as highly immunodeficient mice do not produce immunoglobulins, thereby limiting nonspecific tissue background. This blocking step may need to be optimized when different mouse genetic backgrounds are used or when murine cells are transplanted.
5. This model enables assessment of the myogenic potential of human myogenic cells, including satellite cells, satellite cell–derived myoblasts, and other myogenic progenitors [6]. The same approach can be extended to investigate the behavior of additional human cell populations, such as fibro-adipogenic progenitors [7]. Furthermore, freshly isolated murine satellite cells or other murine cell types can be analyzed using the same experimental framework, thereby avoiding potential immune rejection associated with the expression of highly immunogenic proteins, such as GFP.
6. Ethical committees often require that cells be injected using a physiological solution, such as 0.9% NaCl. When necessary, phosphate-buffered saline (PBS) can be replaced at this step with 0.9% NaCl to meet the requirements of the relevant ethical committees.

7. This cell number corresponds to the standard dose used in our laboratory when transplanting precious and limited human cell populations, such as myoblasts or fibro-adipogenic progenitors. A previous study from our group compared injected cell numbers ranging from 10^5 to 10^6 cells per muscle and showed that donor cell recovery did not increase when doses above 10^6 cells were applied [8]. In contrast, lower cell numbers are generally used for murine cells, typically in the range of $1–2 \times 10^4$ cells per muscle [9–11].
8. Single injection site is preferable when dispersion parameters (either transverse or longitudinal) are among the outcomes to be measured, as it facilitates accurate interpretation of cell distribution within the muscle. To further improve precision, fluorescent microspheres may be added to the cell suspension at the time of injection to mark the exact injection site and allow accurate assessment of cell dispersion or migration relative to this reference point. Fluorescent polystyrene microspheres (1.0 μm; FluoSpheres™, Invitrogen™, F13080) can be used for this purpose [12].
9. The tibialis anterior (TA) muscle is one of the most used recipient muscle for cell transplantation due to its superficial location, well-defined anatomy, which facilitate surgical access, standardized injury, and intramuscular injection. Nevertheless, other skeletal muscles may be used depending on the experimental objective.

Acknowledgments

The authors gratefully thank Capucine Trollet for fruitful discussions and critical reading of the manuscript. They also thank G. Butler-Browne and Vincent Mouly for their support and for many valuable discussions over the years. The group is supported by funding from the Institut National de la Santé et de la Recherche Médicale (INSERM), Sorbonne Université, Association Institut de Myologie, Fondation pour la Recherche Médicale FRM (EQUIPE FRM EQU201903007784), Agence Nationale pour la Recherche (ANR, «EpiMUSE» grant # ANR-22-CE14-0068; "FibroDys" grant #ANR-24-CE14-5209; "CPJ" grant #ANR-24-CPJ1-0122-01).

Competing Interests The authors declare no competing or financial interests.

References

1. Negroni E, Bigot A, Butler-Browne GS et al. (2016) Cellular therapies for muscular dystrophies: frustrations and clinical successes. Hum Gene Ther 27:117–126
2. Bensalah M, Klein P, Riederer I et al. (2019) Combined methods to evaluate human cells in muscle xenografts. PLoS One 14:e0211522
3. Negroni E, Riederer I, Chaouch S et al. (2009) In vivo myogenic potential of human CD133+ muscle-derived stem cells: a quantitative study. Mol Ther 17:1771–1778
4. Neal A, Boldrin L, Morgan JE (2012) The satellite cell in male and female, developing and adult mouse muscle: distinct stem cells for growth and regeneration. PLoS One 7:e37950
5. Hardy D, Besnard A, Latil M et al. (2016) Comparative study of injury models for studying muscle regeneration in mice. PLoS One 11:e0147198
6. Mamchaoui K, Trollet C, Bigot A et al. (2011) Immortalized pathological human myoblasts: towards a universal tool for the study of neuromuscular disorders. Skelet Muscle 1: 34
7. Bensalah M, Muraine L, Boulinguiez A, et al. (2022) A negative feedback loop between fibroadipogenic progenitors and muscle fibres involving endothelin promotes human muscle fibrosis. J Cachexia Sarcopenia Muscle 13(3):1771–1784. https://doi.org/10.1002/jcsm.12974 Epub 2022 Mar 22. PMID: 35319169; PMCID: PMC9178170
8. Silva-Barbosa SD, Butler-Browne GS, Di Santo JP, Mouly V (2005) Comparative analysis of genetically engineered immunodeficient mouse strains as recipients for human myoblast transplantation. Cell Transplant 14:457–467
9. Montarras D, Morgan J, Collins C et al. (2005) Direct isolation of satellite cells for skeletal muscle regeneration. Science 309:2064–2067
10. Boldrin L, Zammit PS, Muntoni F, Morgan JE (2009) Mature adult dystrophic mouse muscle environment does not impede efficient engrafted satellite cell regeneration and self-renewal. Stem Cells 27:2478–2487
11. Sacco A, Doyonnas R, Kraft P et al. (2008) Self-renewal and expansion of single transplanted muscle stem cells. Nature 456:502–506
12. Bentzinger CF, von Maltzahn J, Dumont NA et al. (2014) Wnt7a stimulates myogenic stem cell motility and engraftment resulting in improved muscle strength. J Cell Biol 205:97–111

Chapter 7

Ex Vivo Analysis of Muscle Stem Cells on Single EDL Myofibers, from Quiescence to Self-Renewal

Ludivine Rotard, Hsiao-Fan Lo, Anna Rausch De Traubenberg, Robert S. Krauss and Caroline E. Brun

Abstract

Skeletal muscle regeneration relies on normally quiescent muscle stem cells that, upon tissue injury, activate, proliferate, and either differentiate to repair the myofibers or self-renew to reconstitute the stem cell pool. Quiescence regulation, myogenic lineage progression or self-renewal are tightly controlled by intrinsic factors and extrinsic signals released from the muscle stem cell microenvironment. Here, we provide a method for isolating individual *extensor digitorum longus* myofibers and visualizing muscle stem cells in their quiescent state and during their progression into the myogenic lineage.

Keywords Skeletal muscle, Muscle stem cell, Myofiber isolation, Myofiber culture, Immunostaining, Quiescence, Stem cell fate, Cytoskeleton

1 Introduction

Adult muscle stem cells (MuSCs) serve as a reservoir of cells that can generate progenitors dedicated to skeletal muscle regeneration and retain their stem cell identity by self-renewal throughout life [1]. In homeostatic muscle, MuSCs all express the transcription factor Pax7, and they reside quiescent in a niche between the myofiber and the surrounding basal lamina. Quiescence is a state of reversible cell cycle arrest in which MuSCs retain the ability to re-enter the cell cycle, i.e., to activate, under appropriate stimuli such as muscle injury [2]. MuSC quiescence is actively maintained by intrinsic factors and extrinsic cues sourcing from their microenvironment, the so-called niche [3, 4]. The niche thus confers cellular polarity to the quiescent MuSCs: their apical side accumulates Cadherins (e.g., M-Cadherin) to promote MuSC adhesion to their adjacent myofiber [5, 6], while the basal side is enriched for Integrins (e.g., Integrin alpha-7/beta-1) that bind the Laminins, anchoring the MuSCs underneath the basal

Nathalie Didier and Caroline E. Brun (eds.), *Muscle Stem Cells*, Methods in Molecular Biology, vol. 3093,
https://doi.org/10.1007/978-1-0716-5643-3_7,

lamina [7]. These cell adhesion molecules also contribute to shaping the cytoskeleton and stabilizing the cytoplasmic projections specifically observed in quiescent MuSCs [5, 8–10]. Therefore, the niche highly contributes in maintaining MuSC quiescence, which prevents stem cell exhaustion and preserves tissue replenishment capacity. However, upon a muscle injury that will inevitably disrupt the niche, MuSCs exit quiescence and become activated. This transition from quiescence to activation triggers multiple mechano-property and molecular changes, starting with the retraction of cytoplasmic projections and rearrangement of the cytoskeleton necessary for migration and cell cycle entry. Then, activated MuSCs proliferate as myogenic progenitors that will eventually differentiate into myofibers or self-renew and return to quiescence to replenish the stem cell pool.

Many methods (in vivo, ex vivo, and in vitro systems) are available to explore MuSC behavior at the molecular and cellular levels, from activation to myogenic lineage commitment or self-renewal [11]. Yet, most, if not all, methods require mechanical dissection and/or enzymatic digestion of the muscle tissue, which disturbs MuSC quiescence due to their sensitivity to the niche perturbations [8, 12]. Consequently, the distinctive parameters of quiescent MuSCs, such as cellular polarity or cytoskeleton-associated morphology, cannot be properly assessed with such protocols that trigger precocious MuSC activation. Although muscle sections fairly preserve the tissue's integrity, the muscle section thickness and the MuSC confinement within its niche barely allow visualizing the entire cytoskeleton of quiescent MuSCs [13]. Alternative protocols, such as tissue clearing, two-photon microscopy, or mechanical teasing of fiber bundles from fixed muscles, are time-consuming and usually require extensive troubleshooting of the conditions for MuSC immunolabelling when possible, or the use of mouse models having a MuSC-specific fluorescent reporter [9, 14, 15].

To overcome these obstacles, we and others have developed an ex vivo approach based on the isolation of single myofibers and their associated MuSCs [8, 16–19]. This approach commonly uses the *extensor digitorum longus* (EDL) muscle of mice but can be easily transposed to any muscle that can be isolated from tendon to tendon (for instance, the *flexor digitorum brevis* (FDB) or the *soleus* muscles [18, 19]). It starts with a gentle enzymatic digestion of the muscles, thus minimizing the disruption of the MuSC immediate niche, followed by rapid but gentle mechanical dissociation to release individual, straight myofibers. Single myofibers can be immediately fixed for visualization of the anatomical and molecular hallmarks of quiescent MuSCs, or they can be cultured up to 3 days to track the quiescence-to-activation, proliferation, differentiation,or self-renewal of the associated MuSCs.

2 Materials

2.1 Mouse Model

For the purpose of this study, we used wild-type male C57B/6J mice from 8- to 16-week-old.

2.2 Dissection Tools

1. Iris scissors
2. Straight Cohan-Vannas Spring scissors
3. Half-curved micro-forceps
4. Half-curved Iris forceps
5. Straight Dumont forceps
6. 70% Ethanol

2.3 EDL Muscle Digestion and Myofiber Isolation

1. Stereo dissecting microscope
2. 37°C shaking water bath
3. Cell culture incubator with a humidified atmosphere containing 5% CO_2 at 37°C
4. Bunsen burner
5. Glass Pasteur pipettes and 2 mL pipette bulbs
6. Diamond pen
7. 50 mL centrifuge tube
8. 15 mL centrifuge tube
9. Sterile 100 mm culture dish for one mouse
10. 4× sterile 60 mm culture dishes for one mouse
11. Sterile 6-well culture plates
12. Sterile 12-well culture plate
13. Heat-inactivated Horse Serum (HS)
14. Fiber plating medium: DMEM (Dulbecco's modified Eagle's medium; 1 g/L D-glucose, L-glutamine, sodium pyruvate) supplemented with 1% 1 M HEPES, 10% HS and 1% Penicillin/Streptomycin, and filtered through a 0.22-μm filter (*see* **Note 1**)
15. Collagenase solution: 2.5–3 mg/mL Type II Collagenase diluted in fiber plating medium and filtered through a 0.22-μm filter (*see* **Note 2**)
16. Sterile Phosphate-Buffered Saline (PBS), pH 7.4
17. Paraformaldehyde 8% in PBS (PFA 8%/PBS), pH 7.4

2.4 Materials and Reagents for EDL Myofiber Immunostaining

1. Sterile Phosphate-Buffered Saline (PBS), pH 7.4.
2. Permeabilization buffer: 0.1 M glycine, 0.1% Triton X-100 in PBS (*see* **Note 3**).

3. Blocking buffer: 10% goat serum, 0.1% Triton X-100 in PBS (*see* **Note 4**).
4. Primary antibodies used: mouse IgG1 anti-Pax7 (DSHB, Hybridoma supernatant, 1:10); mouse IgG1 anti-Myogenin (5FD, Santa Cruz, 1:500); rabbit anti-Myogenin (EPR4789, Abcam, 1:500); rabbit anti-M-Cadherin (D4B9L, Cell Signaling, 1:250); mouse IgG1 anti-alpha-Tubulin (DM1A, Santa Cruz, 1:500); mouse IgG2b anti-acetylated alpha-Tubulin (6-11B-1, Santa Cruz, 1:500).
5. Alexa-fluor conjugated secondary antibodies: Donkey anti-Mouse IgG Secondary Antibody, Alexa Fluor 488, Donkey anti-Mouse IgG Secondary Antibody, Alexa Fluor 555, Donkey anti-Rabbit IgG Secondary Antibody, Alexa Fluor 488, Donkey anti-Rabbit IgG Secondary Antibody, Alexa Fluor 555.
6. 25 × 75 mm microscope slides.
7. 24 × 60 mm coverslips.
8. ReadyProbes hydrophobic barrier PAP pen.
9. Fluoromount-G™ Mounting Medium (Thermo Fisher).

3 Methods

3.1 Before Starting the EDL Muscle Dissection

1. Fill the 50 mL centrifuge tube with 5 mL sterile and heat-inactivated Horse Serum (HS). It will be used for coating the Glass Pasteur pipets.
2. Set shaking water bath at 37°C with low agitation (80 rpm).
3. For EDL myofiber isolation from one mouse, coat one sterile 100 mm culture dish and four sterile 60 mm culture dishes with 4 and 2 mL of HS, respectively. Ensure that the entire surface of the dishes is covered and incubate at room temperature for 10 min. Remove the serum and let the dishes dry for approximately 15 min. Finally, add 10 and 4 mL of plating medium in the 100 and 60 mm dishes, respectively, and keep the dishes in a cell culture incubator (humidified, 37°C, 5% CO_2) (*see* **Note 5**).
4. For EDL myofiber culture, coat with HS one sterile 6-well culture plate for each culture time point that will be analyzed. Add 1 mL of HS per well to ensure proper covering of the entire well surface and incubate at room temperature for 10 min. Remove the serum and let the 6-well culture plate dry for approximately 15 min. Finally, add 1–2 mL of plating medium per well and keep the 6-well plate in a cell culture incubator.

5. For EDL myofiber isolation from one mouse, prepare a wide-bore glass Pasteur pipette by cutting the tip of the pipette using a diamond pen to obtain an opening size of 4–5 mm diameter. Using a Bunsen burner, flame-polish the opening to smooth the pipette edges. Add a pipette bulb and coat the pipette with sterile HS in the 50 mL centrifuge tube by pipetting up and down. The wide-bore glass pipette will be used for triturating the EDL muscle (*see* **Note 6**).
6. Additionally, prepare a curved glass Pasteur pipette. Flame-polish the opening of a glass Pasteur pipette and curve the tip (around 2 cm from the edge). Add a pipette bulb and coat the pipette with sterile HS in the 50 mL centrifuge tube by pipetting up and down. The curved glass pipette will be used for transferring single-isolated EDL myofibers.
7. For EDL myofiber isolation from one mouse, prewarm at 37°C 2 mL of collagenase solution in a 15 mL centrifuge tube (*see* **Note 7**).
8. Clean the dissection tools with 70% Ethanol.

3.2 EDL Muscle Dissection and Digestion

1. Anesthetize the mouse with isoflurane in a specific induction chamber prior to cervical dislocation euthanasia.
2. Place the mouse in supine position on a clean polystyrene board and spray hindlimbs with 70% Ethanol.
3. Using the Iris scissors, incise the skin all around the ankle, then cut the skin from the ankle to the knee to facilitate the removal of the skin without damaging the underlying muscles and, finally, using the half-curved micro-forceps, pull out the skin along the entire length of the hindlimb (Fig. 1a).
4. Carefully remove the fascia surrounding the *tibialis anterior* (TA) muscle with the half-curved Iris forceps, from the ankle joint to the knee joint. The fascia appears as a thin and transparent membrane surrounding the muscle (Fig. 1b).
5. At the ankle joint, visually locate the two distal tendons of the TA and *extensor digitorum longus* (EDL) muscles (Fig. 1c). The medial arises from the TA muscle and the lateral one connects to the EDL muscle. Grasp the distal tendon of the TA muscle with the straight Dumont forceps and cut it using the Cohan-Vannas Spring scissors (Fig. 1d).
6. Carefully lift the TA muscle towards the knee and cut the proximal insertion of the TA muscle using the Iris scissors. The EDL muscle should be now entirely exposed (Fig. 1e).
7. To isolate the EDL muscle without stretching or damaging it so that myofibers retain their integrity and muscle stem cells preserve their quiescence features, cut the proximal tendon first. As the proximal tendon is embedded in connective

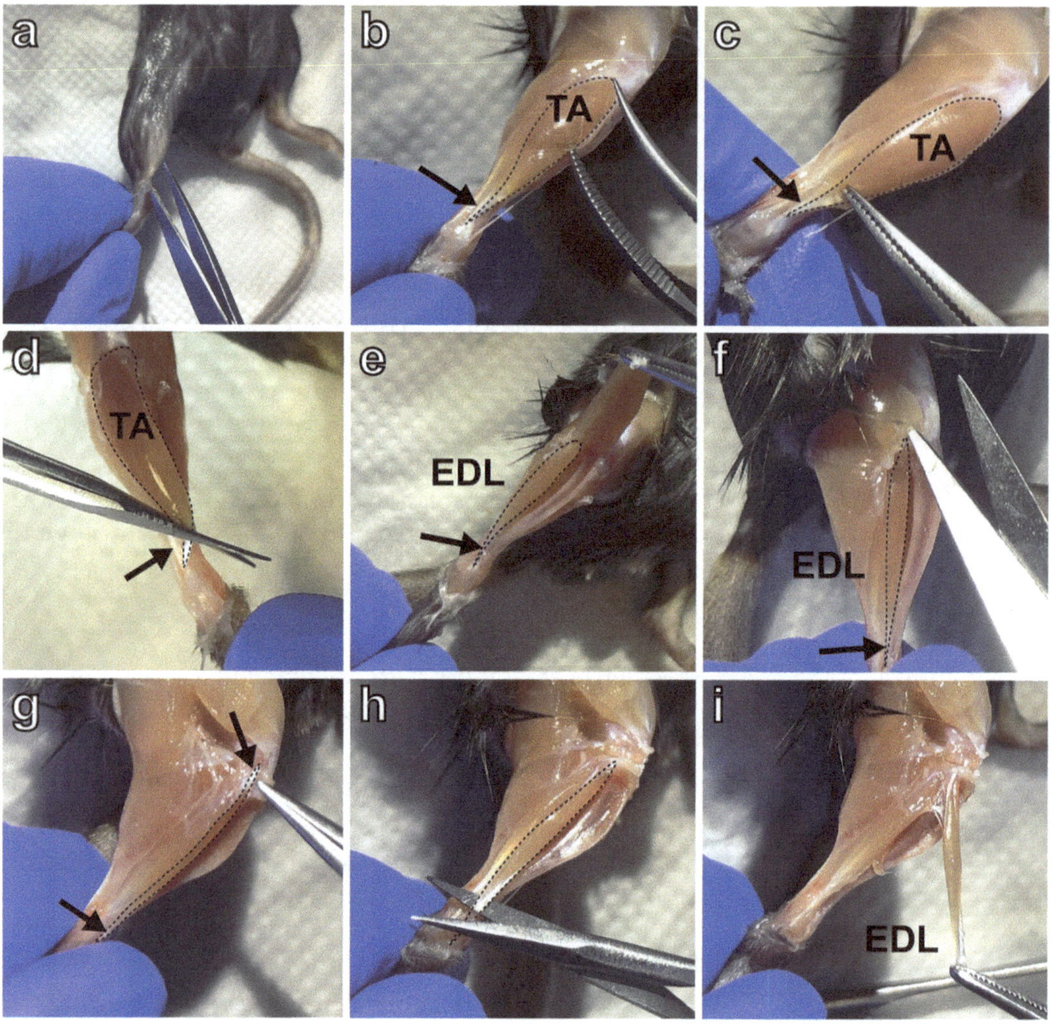

Fig. 1 Key steps of the EDL dissection and isolation. **a,** The skin is incised around the ankle and up to the knee to facilitate the removal of the skin and expose the muscles. **b,** The fascia surrounding the *tibialis anterior* (TA delineated by the dot line) muscle is removed. The black arrow shows the distal tendon of the *extensor digitorum longus* (EDL) muscle. **c,** Once the fascia is removed, the tendons of the EDL and TA muscles are visible. **d,** The tendon of the TA muscle can be cut. **e,** The TA muscle is grabbed by the distal tendon and pulled out to expose the EDL muscle (delineated by the dotted line). **f,** An incision lateral from the knee is performed to expose the EDL proximal tendon. **g,** Both distal and proximal tendons of the EDL are visible. **h,** The distal tendon of the EDL is cut and (**i**) grabbed to lift out the EDL muscle as parallel to the bone as possible. The distal tendon can also be grabbed with the straight Dumont forceps prior to cutting it

tissues around the patella and the knee joint, perform a profound incision, lateral from the knee, using the Iris scissors and cut around it to expose the EDL proximal tendon (Fig. 1f, g). Cut it using the Cohan-Vannas Spring scissors.

8. Cut the EDL distal tendon with the Cohan-Vannas Spring scissors, grab it with the straight Dumont forceps and lift it

off delicately towards the knee-end (Fig. 1h, i). The EDL should appear elongated, S-shaped, with the two tendons visible (*see* **Note 8**).

9. Transfer the EDL muscle in a 15 mL centrifuge tube containing the pre-warmed collagenase solution and place the tube in the 37°C shaking water bath with low agitation.
10. Repeat Steps 3–9 to isolate the second EDL muscle.
11. Incubate the 15 mL centrifuge tube in the 37°C shaking water bath with low agitation for 53–55 min (*see* **Note 9**) (Fig. 2a).

3.3 EDL Myofiber Isolation

1. At the end of the digestion time, remove the 100 mm culture dish containing plating medium from the incubator and use the HS-coated wide-bore pipette to transfer digested EDL muscles. All the following steps are performed under the stereo dissecting microscope using bright field illumination.
2. Pipet up and down against the wall of the dish using the HS-coated wide-bore pipette to triturate the EDL muscles and continue until reaching an appropriate number of myofibers. Do not exceed 5 min trituration (*see* **Note 10**). Individual myofibers should fall off the muscle (Fig. 2b).
3. After trituration, place the 100 mm culture dish in the incubator for 10 min. This step, named the recovery time, will allow single EDL myofibers to settle and straighten out (Fig. 2c).

3.4 EDL Myofiber Fixation for Analyzing Muscle Stem Cell Quiescence

1. After the myofiber recovery time, remove the 100 mm culture dish containing the single EDL myofibers from the incubator, as well as one 60 mm dish filled with plating medium, and label it "quiescent MuSCs." We note that MuSCs initiate the activation process upon dissection of the EDL. By following the myofiber isolation protocol outlined here, many of the properties of quiescent MuSCs in vivo can be at least partially

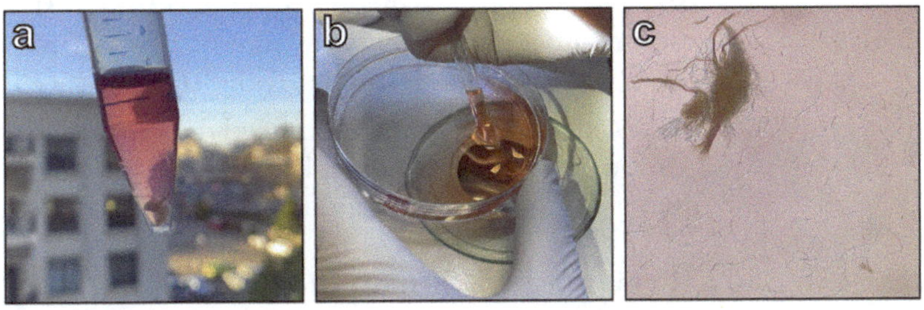

Fig. 2 Digestion and trituration of EDL muscles. **a,** Following isolation, the two EDL muscles are transferred into a 15 mL conical tube containing 2 mL of collagenase solution. **b,** Following digestion, the two EDL muscles are mechanically triturated in a 100 mm dish containing plating medium. **c,** The trituration allows single EDL myofibers to fall off

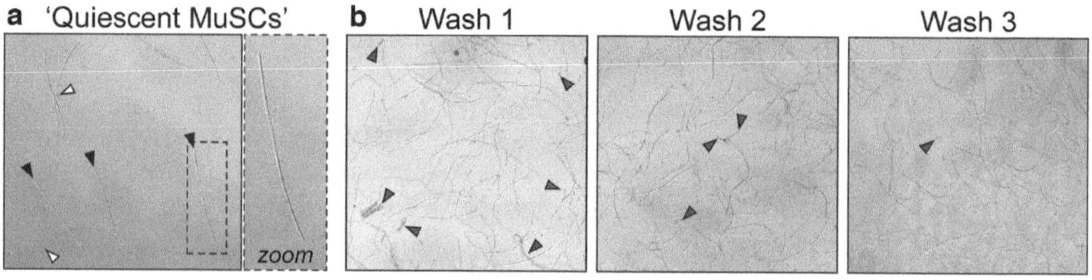

Fig. 3 Selection of single EDL myofibers. **a,** Picture showing the myofibers selected directly after the EDL trituration for the analysis of quiescent muscle stem cells (MuSCs). The black arrowheads indicate the straightest myofibers that should be selected. The white arrowheads show the irregularly shaped myofibers on which the MuSCs will inevitably lose their quiescent features. **b,** Pictures showing the result of successive washes to clear out the debris and dead, hypercontracted, or short myofibers (red arrowheads)

preserved ex vivo. Therefore, we refer to these MuSCs as "quiescent MuSCs."

2. Using the HS-coated curved glass pipette under the stereo dissecting microscope, delicately aspirate the single EDL myofibers and transfer them slowly into the 60 mm dish (*see* **Note 11**). Select the straightest myofibers that do not show any wavy-shape, tears, bends, or angles along their length (Fig. 3a, **"Quiescent MuSCs"**). If you desire to keep the remaining EDL myofibers of the 100 mm dish for further culture, bring back the 100 mm dish in the incubator.
3. Use the HS-coated curved pipette to remove approximately 2–3 mL of plating medium from the 60 mm dish containing the selected individual EDL myofibers.
4. Add 5 mL of pre-warmed PBS in the 60 mm dish slowly to dilute the plating medium and let the fibers settle down for a few minutes (*see* **Note 12**).
5. Remove the mix of plating medium/PBS from the 60 mm dish. Leave a small amount (around 500 μL) in the dish to prevent EDL myofibers from drying out.
6. Add 3 mL of pre-warmed PBS in the 60 mm dish slowly and fix the myofibers in 4% PFA by adding 3 mL of prewarmed PFA 8%/PBS slowly for 10 min at room temperature.
7. Remove PFA and wash the fixed EDL myofibers by adding 5 mL of PBS. At this point, EDL myofibers can be used immediately for immunostaining or stored at 4°C for up to 2 months.

3.5 EDL Myofiber Culture in Floating Condition and Fixation

1. Remove the 100 mm culture dish containing the EDL myofibers and one 60 mm dish filled with plating medium from the incubator.
2. Using the HS-coated curved glass pipette, transfer live single EDL myofibers from the 100 mm to the 60 mm dish. Live myofibers look straight and translucent (Fig. 3b, **wash 1**).

3. Once all the live EDL myofibers have been selected, take a new 60 mm dish from the incubator and transfer selected EDL myofibers from the first 60 mm dish to the new one (Fig. 3b, **wash 2**). Repeat this step once (Fig. 3b, **wash 3**). Transferring the myofibers from one dish to the next one will help neutralize the collagenase and remove debris and dead, hypercontracted, or short myofibers.
4. Incubate the last 60 mm dish in the incubator for 30 min to 1 h. This incubation time will allow myofibers to recover from the procedure and help select the live myofibers for the next step.
5. For myofiber culture, transfer the live myofibers into the wells of a 6-well plate containing 2 mL of plating medium. We find that a maximum of 50–100 EDL myofibers per well is ideal to prevent myofiber clumps during culture. One 6-well plate should be prepared for each time point to analyze.
6. To study muscle stem cell transition from quiescence to activation, culture EDL myofibers from 1 to 30 h in the incubator. To observe muscle stem cell proliferation, culture myofibers for 48 h. To monitor muscle stem cell differentiation, culture myofibers for 72 h. Although myofibers can be cultured for longer periods, we find that after 96 h of culture, most of the muscle stem cells fall off myofibers, which makes their analysis not accurate.
7. To fix EDL myofibers at each desired time point. Use the HS-coated curved pipette to remove plating medium from each well of the 6-well plate, ensuring that a small amount of liquid (around 200 μL) remains in the bottom of the well to prevent myofibers from drying out.
8. Add 2 mL of prewarmed PBS in each well of the 6-well plate and let the fibers settle down for 1–2 min.
9. Remove PBS from the wells.
10. Add 1 mL of prewarmed PBS in each well and fix the myofibers by adding 1 mL of prewarmed PFA 8%/PBS slowly. Incubate 10 min at room temperature.
11. Remove PFA and wash the fixed EDL myofibers two times with 2 mL of PBS. At this point, EDL myofibers can be used immediately for immunostaining or stored at 4°C for up to 2 months.

3.6 EDL Myofiber Immunostaining

1. Using a HS-coated curved glass pipette under a stereo dissecting microscope, transfer 20–50 fixed individual EDL myofibers into a well of a 12-well culture plate pre-filled with 1 mL of PBS.

2. Remove PBS. For all the following steps, always leave a small amount of liquid (around 200 μL) in the bottom of the well to prevent myofibers from drying out.
3. Add 1 mL of PBS per well and incubate 5 min to wash the EDL myofibers.
4. Remove PBS and add 1 mL of permeabilization buffer per well. Incubate 10–15 min at room temperature.
5. Remove permeabilization buffer and add 1 mL of blocking buffer per well. Incubate for at least 1 h at room temperature.
6. Remove blocking buffer and add the primary antibodies diluted in blocking buffer. Incubate overnight at 4°C. For a 12-well culture plate, we find that 250 μL of primary antibody solution per well is enough to incubate properly the myofibers without drying them out. For the purpose of this experiment, we used antibodies directed against M-Cadherin, Pax7, and Myogenin. M-Cadherin labels all muscle stem cells (MuSCs), independently of their progression in the myogenic lineage. Pax7 stains the quiescent, proliferating, and self-renewing MuSCs, while Myogenin labels the differentiating cells (Fig. 4). Alpha-Tubulin and acetylated alpha-Tubulin stain the microtubule cytoskeleton of the myofiber and MuSCs and the quiescent projections, respectively.
7. Remove the primary antibody solution and wash the myofibers by adding 1 mL of PBS in each well for 5 min.
8. Remove PBS and wash two additional times with 1 mL of PBS for 5 min.
9. Remove PBS and add the appropriate secondary antibodies diluted in 250 μL of blocking buffer per well of a 12-well culture plate. Incubate at least 1 h at room temperature. In this protocol, we used Alexa-fluor conjugated secondary antibodies, all diluted at 1:1000.
10. Remove the secondary antibody solution and wash the myofibers by adding 1 mL of PBS in each well for 5 min.
11. Remove PBS and wash two additional times with 1 mL of PBS for 5 min.
12. Remove PBS and add 500 μL of DAPI solution per well. Incubate 5 min.
13. Remove DAPI solution and add 1 mL of PBS.
14. On a microscope slide, delineate the edges of the slide using a hydrophobic PAP pen to create a water-repellent barrier. Let it dry for 10 min. Using a HS-coated curved glass pipette under a stereo dissecting microscope, transfer myofibers within the area delimited by the hydrophobic pen.

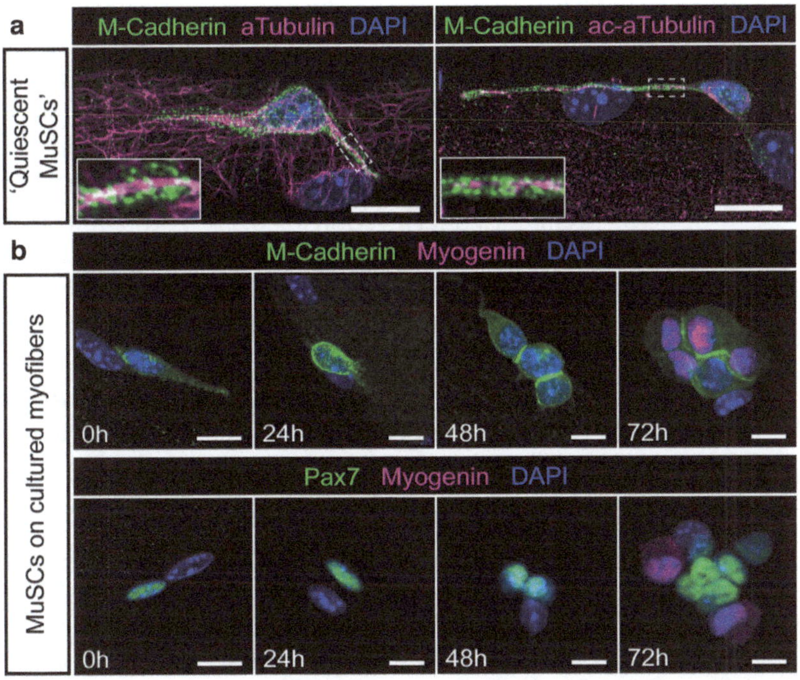

Fig. 4 Immunofluorescence pictures showing muscle stem cells on single EDL myofibers. **a,** Representative immunofluorescence images of muscle stem cells (MuSCs) stained for M-Cadherin (green) on an EDL myofiber fixed directly after trituration. Note that M-Cadherin allows visualizing typical features of the quiescent MuSCs, such as the cytoplasmic projections. The alpha-Tubulin (aTubulin) labels the microtubule network within the myofiber and the MuSC. The acetylated alpha-Tubulin (ac-aTubulin) marks the microtubule skeleton of the quiescent projections. **b,** Representative immunofluorescence of muscle stem cells stained on single EDL myofibers cultured for 24, 48, or 72 h. While M-Cadherin labels all MuSCs, Pax7 and Myogenin staining allow distinguishing the self-renewing MuSCs expressing Pax7 only from the differentiating cells expressing Myogenin. Scale bars: 10 μm

15. Carefully aspirate excess PBS from the microscope slide. Apply two small drops of Fluoromount-G™ Mounting Medium at two opposite corners of the slide. Add coverslip (*see* **Note 13**). Let dry the slides overnight at room temperature prior to visualizing, and seal with clear nail polish for long-term storage. Slides can now be imaged (Fig. 4).

4 Notes

1. Fiber medium can be stored at 4°C up to 6 months. DMEM that includes pyruvate is required for the viability of the myofiber. HEPES will help buffer the medium throughout the myofiber trituration and digestion.

2. The collagenase solution is prepared extemporaneously. Collagenase concentration is adapted for digestion and trituration of myofibers from young and adult wild-type mice (8-week to 1-year-old mice) but should be adjusted for younger or aged mice or for specific mouse strains (for instance, mouse models for muscular dystrophies).
3. While a concentration of 0.1% Triton X-100 is enough for most of the antibodies tested, some can require higher concentrations. Although not ideal for membrane-associated proteins, Triton X-100 can be increased up to 1% to visualize intracellular proteins of the myofiber.
4. Blocking buffer can be adapted according to your experiment. Adding 0.1% Triton X-100 will increase antibody efficiency.
5. We obtain better results when the dishes are coated the day before and maintain in the cell culture incubator (humidified, 37°C, 5% CO_2).
6. Flame polish is essential as the glass will be too harsh to keep myofiber integrity without damaging.
7. Do not pre-warm the collagenase solution for more than 30 min before starting the digestion. If the two EDL muscles from one mouse need to be treated separately, each EDL muscle can be digested in a 1.5 mL microcentrifuge tube containing 1 mL of collagenase solution.
8. Do not tug the muscle, as this will stretch it and affect muscle stem cell quiescence. For instance, it will cause retraction of the muscle stem cell projections and increase the number of rounded activated muscle stem cells.
9. Duration of the collagenase digestion should be adapted according to the age (old mice) or the strain (mouse models of muscular dystrophies) of the mice. We noticed that leaving the water bath lid open during collagenase incubation improves muscle digestion and subsequent myofiber isolation.
10. The trituration step should not exceed 5 min to allow the release of individual live myofibers with appropriate shape and maximal integrity. This step requires to be monitored under the dissecting microscope.
11. As you proceed to myofiber isolation and washes, careful not to leave the dishes outside the incubator for more than 10 min. It will help maintain the preparation at a proper temperature (close to 37°C). Every 10 min, place back the preparation in the incubator for 10 min. We recommend performing this protocol with no more than four mice on a same day, to ensure that the time outside the incubator is kept to a minimum in between each step.

12. PBS should be pre-warmed at 37°C so it does not affect myofiber integrity while adding it to the isolated myofibers. Too cold PBS will trigger myofiber hyper-contraction.
13. We recommend mounting each slide one by one to prevent stained myofibers from drying out. For long-term storage, slides can be stored at 4°C.

Acknowledgments

The authors thank Jade Ravent for her invaluable assistance with data collection. L.R. is supported by a doctoral scholarship from the ED340 BMIC, University of Lyon. R.S.K. and H.F.L. were supported by a grant from the National Institute of Arthritis, Musculoskeletal and Skin Diseases (R01AR070231). C.E.B. receives support of grants from the Agence Nationale de la Recherche (ANR-22-CE14-0070 and ANR-25-CE14-2467) and the Association Française contre les Myopathies (AFM)-Téléthon (MyoNeurALP 2). This study was carried out with funding to C.E.B. and R.S.K. from a Tremplin International Inserm "First Step" grant [Q-Stem].

Competing Interests The authors declare no competing or financial interests.

References

1. Esper ME, Saber J, Rudnicki MA (2026) Satellite cells in regeneration and disease. Cold Spring Harb Perspect Biol 18(2), a041474.
2. Ancel S, Stuelsatz P, Feige JN (2021) Muscle stem cell quiescence: controlling stemness by staying asleep. Trends Cell Biol 31:556–568
3. Brun CE, Chevalier FP, Dumont NA et al. (2017) Chapter 10—the satellite cell niche in skeletal muscle. In: Vishwakarma A, Karp JM (Eds.)Eds. Biology and engineering of stem cell niches. Academic Press, Boston
4. Hung M, Lo HF, Jones GEL et al (2023) The muscle stem cell niche at a glance. J Cell Sci 136(24):jcs261200
5. Hung M, Lo, HF, Beckmann AG et al (2024) Cadherin-dependent adhesion is required for muscle stem cell niche anchorage and maintenance. Development. 151(7):dev202387
6. Goel AJ, Rieder MK, Arnold HH et al. (2017) Niche cadherins control the quiescence-to-activation transition in muscle stem cells. Cell Rep 21:2236–2250
7. Rozo M, Li L, Fan CM (2016) Targeting beta1-integrin signaling enhances regeneration in aged and dystrophic muscle in mice. Nat Med 22:889–896
8. Kann AP, Hung M, Wang W et al. (2022) An injury-responsive Rac-to-Rho GTPase switch drives activation of muscle stem cells through rapid cytoskeletal remodeling. Cell Stem Cell 29:933–947 e936
9. Ma N, Chen D, Lee JH et al. (2022) Piezo1 regulates the regenerative capacity of skeletal muscles via orchestration of stem cell morphological states. Sci Adv 8:eabn0485
10. Lo H-F, Lu Y, Sachs D et al (2025) Regulation of muscle stem cell dynamics and quiescence by netrin-1 cytoskeletal signaling. BioRXiv. 2025.2011.2001.686042
11. Evano B, Sarde L, Tajbakhsh S (2023) Temporal static and dynamic imaging of skeletal muscle in vivo. Exp Cell Res 424:113484
12. Machado L, Esteves de Lima J, Fabre O et al. (2017) In situ fixation redefines quiescence and

early activation of skeletal muscle stem cells. Cell Rep 21:1982–1993

13. Dumont NA, Rudnicki MA (2017) Characterizing satellite cells and myogenic progenitors during skeletal muscle regeneration. Methods Mol Biol 1560:179–188
14. Collins BC, Shapiro JB, Scheib MM et al. (2024) Three-dimensional imaging studies in mice identify cellular dynamics of skeletal muscle regeneration. Dev Cell 59:1457–1474 e1455
15. Schuler SC, Dumontier S, Rigaux J et al. (2021) Visualization of the skeletal muscle stem cell niche in fiber bundles. Curr Protoc 1:e263
16. Goel AJ, Krauss RS (2019) Ex vivo visualization and analysis of the muscle stem cell niche. Methods Mol Biol 2002:39–50
17. Brun CE, Wang YX, Rudnicki MA (2018) Single EDL myofiber isolation for analyses of quiescent and activated muscle stem cells. Methods Mol Biol 1686:149–159
18. Keire P, Shearer A, Shefer G et al. (2013) Isolation and culture of skeletal muscle myofibers as a means to analyze satellite cells. Methods Mol Biol 946:431–468
19. Collins CA, Zammit PS (2009) Isolation and grafting of single muscle fibres. Methods Mol Biol 482:319–330

Check for updates

Chapter 8

An Ex Vivo Cross-Grafting and Live-Imaging Assay to Analyze Myofiber Niche Impact on Muscle Stem Cells

Liza Sarde , Shahragim Tajbakhsh and Brendan Evano

Abstract

The isolation and culture of primary muscle fibers and their associated muscle stem cells (MuSCs) allow investigation of MuSC function in their myofiber niche, notably in disease conditions. However, this assay does not discriminate between stem cell-intrinsic and fiber-derived cues driving MuSC (dys)function. Here, we describe the development of an ex vivo cross-transplantation assay between healthy and mutant MuSCs and myofibers ex vivo, coupled with live imaging for the dynamic characterization of MuSC behavior. The progeny of up to 60 MuSCs can be analyzed per experiment, in four grafting conditions and dynamically over a period of 3 days. This approach allows assessment of the relative contribution of the myofiber niche to impaired MuSC functions in disease conditions, providing a framework to experimentally dissect niche-dependent mechanisms underlying muscle stem cell dysfunction.

Keywords Skeletal muscle, Satellite cell, Muscle stem cell, Myofiber, Grafting, Live imaging, Muscular dystrophy, Proliferation, Differentiation, Migration

1 Introduction

Deregulated stem cell function is a hallmark of many pathologies, yet how the dynamic interactions of stem cells with their microenvironment contribute to disease progression remains poorly addressed. Notably, it remains unclear whether stem cell dysfunctions observed in several skeletal muscle wasting diseases arise from stem cell-intrinsic defects or from environmental signals, from the muscle fibers.

Ex vivo culture of primary myofibers and associated stem cells has been used broadly as it partially maintains the MuSC microenvironment (fiber + basement membrane), and it recapitulates the main steps of myogenic progression as MuSCs exit quiescence, proliferate, and differentiate. This model has been used extensively in reverse genetics studies to investigate gene function in MuSCs in their fiber niche. Studies relying on germline gene mutations or RNA interference then analyze the combined effects of a gene loss-of-function in the fiber and in the stem cells on MuSC functions.

Nathalie Didier and Caroline E. Brun (eds.), *Muscle Stem Cells*, Methods in Molecular Biology, vol. 3093,
https://doi.org/10.1007/978-1-0716-5643-3_8,

Using conditional Cre recombinase-dependent mutant alleles and fiber- or MuSC-specific Cre driver alleles represents an alternative. However, conditional mutant alleles are not systematically available, and mouse lines are time- and cost-intensive to import in animal facilities. As an alternate strategy, we describe here the cross-grafting of wild-type (WT) and mutant MuSCs between WT and mutant fibers to discriminate between MuSC-intrinsic and -extrinsic effects of a gene mutation on MuSC function.

Characterization of MuSC functions ex vivo on myofibers includes recording of activation, division, differentiation, and migration kinetics. Notably, symmetric and asymmetric cell divisions are classically assessed by examining fates within pairs of cells, assuming these pairs correspond to sister cells that remained in proximity after mitosis. However, several reports indicated a high motility of MuSCs on cultured myofibers [1, 2] leading to misidentification of sister cells. A proper analysis of MuSCs modes of cell division then requires (i) recording by live imaging of cell divisions and cell relatedness and (ii) analysis of fate marker expression using live reporters or endpoint immunostaining. This is technically challenging as myofibers are generally cultured either in suspension (live-imaging challenging due to fluid movements; cannot link immunostained fibers to filmed fibers retrospectively), attached to a substrate (e.g., Matrigel, where MuSCs eventually migrate away from the fiber, analyzing then mixed effects from the fiber and the substrate), or embedded in gels (e.g., collagen [1, 2] or agarose [3], compatible with live imaging but poorly with subsequent immunostaining). To overcome these issues, we have developed a confinement system with microwells allowing the culture in suspension and live-imaging of fibers and associated MuSCs from the *Flexor digitorum brevis* (FDB) muscle up to 3 days, possibly followed by endpoint immunostaining [4].

Using MuSCs and FDB fibers from WT and *mdx* mice (*Dmd*$^{mdx\text{-}\beta Geo}$ allele [5], mouse model of Duchenne Muscular Dystrophy), the cross-grafting procedure we developed and our microwell system, we could analyze the migration and modes of cell division of WT and *mdx* MuSCs in healthy or pathological myofiber niches. This protocol involves five main procedures: (i) Microwell plate construction, (ii) FDB dissection, (iii) FDB fibers dissociation and MuSC isolation, (iv) cross-transplantation, and (v) live-imaging.

2 Materials

This protocol is designed for a cross-transplantation assay containing four experimental conditions: (1) WT MuSCs to WT fibers (WT control); (2) *mdx* MuSCs to *mdx* fibers (*mdx* control); (3) WT MuSCs to *mdx* fibers; and (4) *mdx* MuSCs to WT fibers. In our study [4], WT or *mdx* MuSCs from *Pax7*^{CreERT2} [6]; *R26*^{YFP} [7];

Myog^{ntdTom} [8] mice were engrafted to *mdx* or WT receiving FDB myofibers. Tamoxifen-induced recombination of the *R26*^{YFP} allele through the *Pax7*^{CreERT2} allele allows labeling permanently MuSCs and their progeny with a YFP fluorescent reporter. In parallel, the *Myog*^{ntdTom} live reporter mice allows real-time monitoring of cell fate decisions through to the expression of a nuclear tdTOMATO reporter that marks *Myogenin* expression. The cross-transplantation protocol requires the use of four mice: two mice for the grafted MuSCs (WT and *mdx*) and two mice providing the recipient fibers (WT and *mdx*) (*see* **Note 1**).

2.1 Microwell Plate Assembly

1. Two Plexiglas plates (Fig. 1a): Bottom plate: 128 mm × 85 mm × 10 mm; Top plate: 123 mm × 81 mm × 10 mm. Each plate contains four laser-cut rectangular openings (55.5 mm × 34 mm × 10 mm).
2. Four glass coverslips cut to the desired dimension with diamond pen (58 mm × 38 mm; adapted from Claritx Coverglass #1).
3. Silicone (MoldStar 20 T, Smooth-on).
4. 10 mL syringes and 200 μL pipette tips.
5. Silicone sheets (50 mm × 30 mm) containing microwells designed with a 3D-printed mold (Fig. 1b) (3 mm thickness; microwell volume ≈ 6 μL; approximately 200 microwells per glass-bottom chamber).
6. Lid from a six-well culture dish (used to close the microwell plate).

2.2 Dissociation and Isolation of FDB Fibers

1. Dissection tools: one pair of micro scissors, one fine forceps, and one scalpel.
2. 50 mL Falcon tube cap and adhesive tape.
3. Stereomicroscope.
4. Four Sterilin 7 mL Polystyrene Bijou Containers filled with 5 mL of prefiltered and prewarmed DMEM/0.2% Collagenase (Type 1; filtered at 0.22 μm; 37°C).
5. Cell culture incubator set at 37°C, 5% CO_2 (3% O_2 when available).
6. Water bath set at 37°C.
7. Glass pipettes and rubber bulbs.
8. Serum (same type as used in the culture medium, see below).
9. Plastic Pasteur pipettes.
10. Thirty Sterilin Petri deep dishes coated with serum and filled with prewarmed DMEM. For ten of them, and prior to coating, caps of 15 mL Falcon tubes are fixed in the center with silicone.

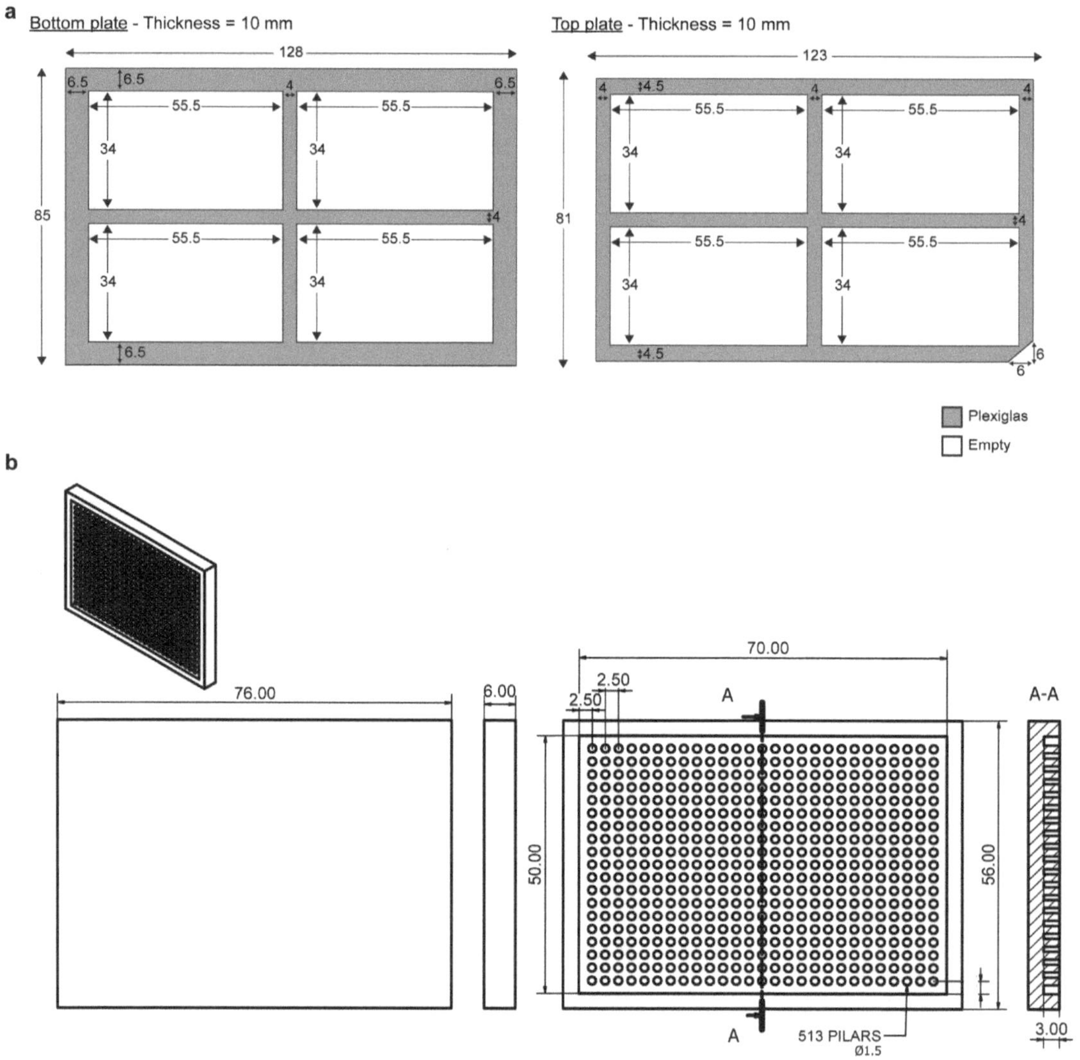

Fig. 1 Schematic showing exact dimensions of Plexiglas plates and microwell mold. **a,** Exact dimensions of top and bottom Plexiglas plates. **b,** Dimensions of 3D-printed mold used to fabricate the silicone sheets containing microwells. To produce the silicone sheet, a layer of Parafilm is tightly stretched and sealed over the mold. Two holes are pierced in the Parafilm on either side of the mold: one to inject silicone (Part A + Part B) slowly using a 10 mL syringe fitted with a P20 pipette tip, and one to allow air to escape as the silicone is introduced into the mold. After 40 min of curing, the parafilm can be removed, and the silicone sheet with microwells extracted. Carefully clean the mold with a toothbrush and soap to remove traces of silicone. Clean the mold with water and ethanol to sterilize

11. 200 mL of culture medium (88.5% DMEM, 10% Horse Serum, 1% Penicillin/Streptomycin, 0.5% Chick Embryo Extract).
12. Four 15 mL Falcon tubes with the base cut to form a cylinder and attached with silicone to the lid of a Sterilin dish.

2.3 MuSC Extraction from FDB Fibers

1. TrypLE Express.
2. 30 μm filter.

3 Methods

3.1 Microwell Plate Construction

Microwell plates must be assembled 4 days prior to use to allow complete silicone curing and ensure watertight sealing (as in Fig. S3a from [4],).

1. Wash glass coverslips and Plexiglas plates sequentially with soap, distilled water, and 70% ethanol, and allow to air-dry completely.
2. Prepare silicone by mixing equal volumes of Part A and Part B and load into a 10 mL syringe fitted with a pipette tip. Apply silicone at the contact zone between the two Plexiglas plates and assemble the plates. Let the silicone cure at least 20 min at RT.
3. Seal glass coverslips onto the outer surface of the bottom Plexiglas plate with silicone (applied with a 10 mL syringe), forming four independent glass-bottom chambers (*see* **Note 2**).
4. Place the silicone microwell sheets inside each chamber, ensuring tight contact between the sheets and the glass coverslips. Carefully seal the sheets with silicone along all borders to prevent leakage or flow underneath the sheet.
5. One day before muscle dissection, fill the chambers with PBS and remove trapped air bubbles using a P200 pipette. Store overnight at 4°C. This step also serves as a preliminary cleaning and leak-tightness verification.
6. On the day of muscle dissection, equilibrate the microwell plate to room temperature (*see* **Note 3**).

3.2 FDB Muscle Dissection (Fig. 2)

For each mouse, dissect both FDB muscles as follows:

1. Incise the skin at heel level with microscissors.
2. Remove carefully the skin from the site of incision to the toes.
3. Cut the main tendon at the heel level with a scalpel and use it to manipulate the FDB.
4. Gently lift the muscle by the main tendon using fine forceps while cutting adjacent connective tissues with the microscissors. As the muscle branches distally towards the toes, sever the main tendon to fully release the muscle without cutting into the muscle belly.

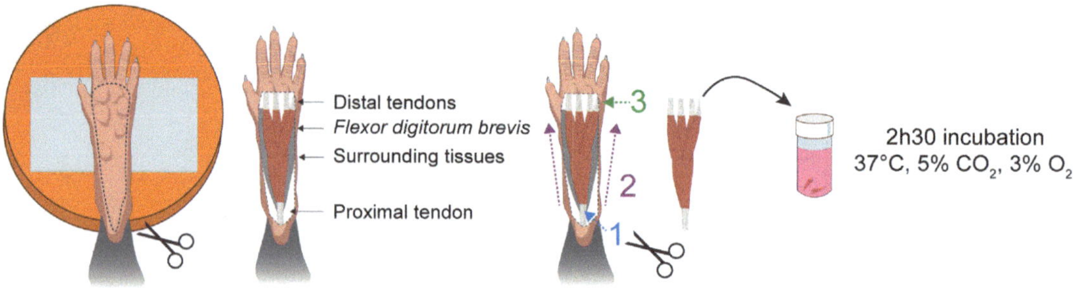

Fig. 2 FDB muscle dissection. The foot is cleaned with 70% ethanol and attached to a 50 mL Falcon tube cap using tape, allowing a proper foot elevation. The dissection is performed under a stereomicroscope. Following an initial skin incision at the heel level, the skin overlying the FDB is removed up to the toes. The proximal tendon is then cut using a scalpel (1), allowing the FDB to be gently lifted while the surrounding connective tissue on both sides is carefully cut to release the muscle (2). The muscle is fully detached after cutting the distal tendons at the level of the toes (3). One or two FDB muscles can be incubated in a Bijou containing 5 mL of prewarmed collagenase/DMEM for 2.5 h at 37°C

5. Completely detach the FDB muscle by cutting the distal tendons.
6. Incubate the two FDB muscles from a single mouse in one Bijou containing filtered collagenase/DMEM for 2.5 h at 37°C.

3.3 FDB Fiber Dissociation and MuSC Isolation (Fig. 3 and in Fig. S5 from [4])

1. During collagenase incubation, prepare 200 mL of culture medium, coat all Sterilin dishes with serum and fill with DMEM. Incubate at 37°C, 5% CO_2 (and 3% O_2 when available).
2. Wash the microwell plate three times with PBS (15 min per wash), fill with culture medium (10 mL per chamber) and incubate at 37°C, 5% CO_2, 3% O_2. After 30 min, remove residual air bubbles trapped in the microwells using a P200 pipette.
3. Transfer FDB muscles into Sterilin dishes using serum-coated glass pipettes.
4. Dissociate fibers by gently but firmly pipetting the muscle with ten back-and-forth flushes. Once the first fibers detach, transfer remaining tissue to a new dish and repeat this procedure sequentially until fibers are fully dissociated.

3.3.1 Recipient WT and mdx Fibers (Processed Independently)

1. Once four to five Sterilin dishes containing FDB fiber suspensions are obtained, allow fibers to settle and remove half of the DMEM using a plastic pipette. Pool dishes two by two.
2. Allow fibers to settle, carefully remove half of the DMEM with a plastic pipette, and replace with prewarmed DMEM. It is critical to obtain debris-free fiber suspensions; however, avoid excessive washing steps (maximum three washes).

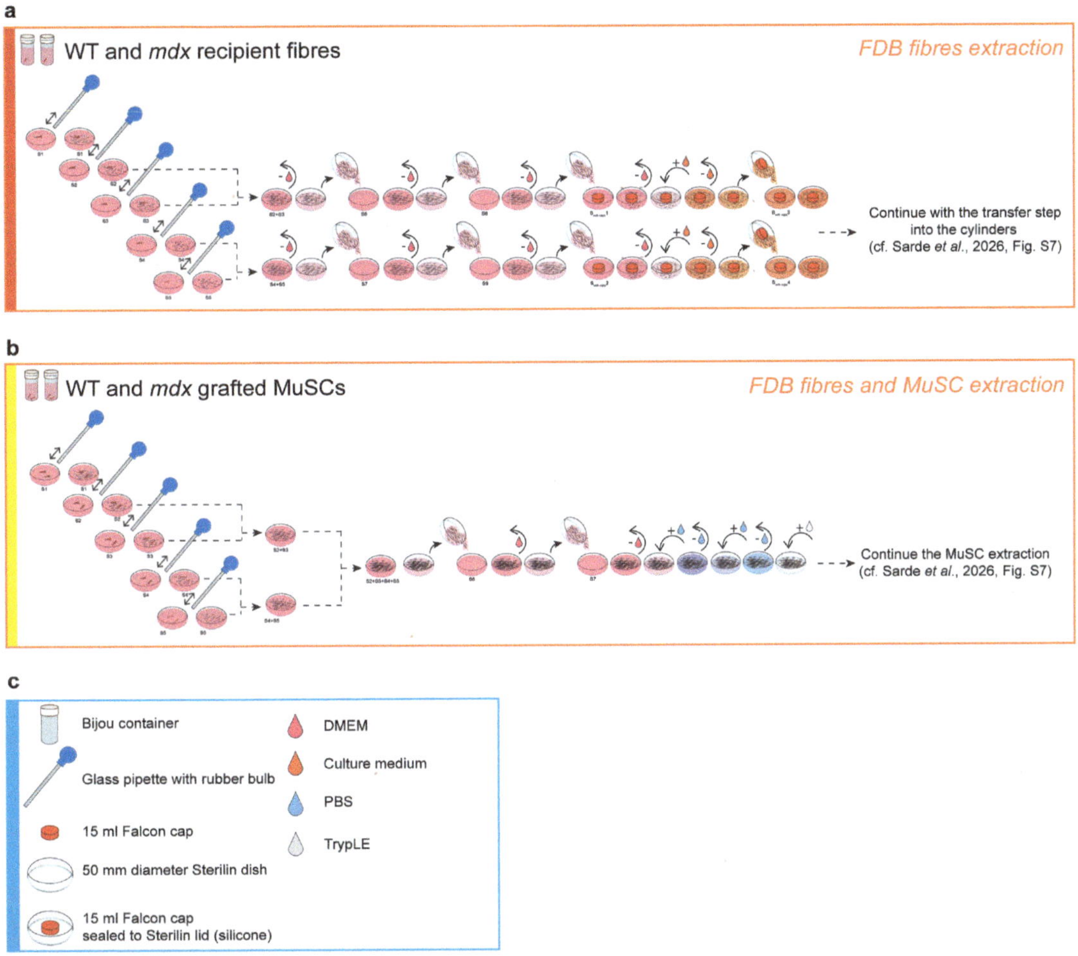

Fig. 3 FDB fiber extraction. **a,** For recipient FDB fibers, only one FDB muscle per mouse (WT and *mdx*) is dissected and incubated in collagenase in a Bijou container. After digestion, the muscle is transferred to a Sterilin dish (labeled S1, with subsequent dishes labeled S2–Sn). This is followed by a series of dissociation steps using a glass pipette fitted with a rubber bulb, consisting of ten aspiration/dispersion flushes per dish. Typically, the first Sterilin dish contains debris, blood vessels, and connective tissue, and can therefore be discarded. The remaining Sterilin dishes are pooled, and successive washes are performed until clean fiber suspensions are obtained (typically S8 and S9). The fibers are subsequently transferred into Sterilin dishes with caps. DMEM is then replaced with culture medium through two washing steps. Of note, for WT, approximately twice as many Sterilin dishes with caps may be required compared with *mdx*. The following steps of the cross-transplantation protocol are described in Fig. S5 from [4]. **b,** For grafted MuSCs, the objective is to recover as many MuSCs as possible; therefore, both FDB muscles from each mouse are digested. After digestion, the fibers are dissociated using the same aspiration/dispersion flushing technique with a glass pipette as for the recipient fibers. All fibers can be pooled to obtain a single Sterilin dish, which is then washed twice with DMEM and PBS before the addition of TrypLE. The following steps of the cross-transplantation protocol are described in Fig. S5 from [4]. **c,** Keys to symbols used in Fig. 3

3. Transfer the fibers into new Sterilin dishes with caps. As fiber yield is typically about twofold lower in *mdx* preparations, expect four dishes for WT and two for *mdx*.
4. Wash fibers twice with culture medium by allowing the fibers to settle, carefully removing half of the DMEM with a plastic pipette, and replacing with pre-warmed medium.
5. For each WT and *mdx* mouse, aliquot one quarter of the fiber suspension into each of two truncated Falcon cylinders. Incubate the cylinders at 37°C, 5% CO_2, 3% O_2 to allow fibers to settle. Carefully remove supernatant, leaving 1 mL of fiber suspension per cylinder (use graduations on Falcon tube).

3.3.2 Grafted WT and mdx MuSCs (Processed Independently)

1. Pool fibers from each mouse into a single Sterilin dish.
2. Wash fiber suspension twice with DMEM and twice with PBS (allow fibers to settle, carefully remove as much liquid as possible).
3. Remove as much as PBS as possible and add 3 ml of prewarmed TrypLE™. Incubate for 15 min at 37 °C, 5% CO_2.
4. Triturate fibers vigorously using a serum-coated P1000 pipette until fibers appear blue–translucent and begin to contract.
5. Filter suspension through a 30 μm cell filter into a serum-coated 15 mL Falcon.
6. Centrifuge WT and *mdx* samples at 550 g for 15 min at room temperature.
7. Aspirate supernatant and resuspend pellets in 500 μL of culture medium.

3.4 Cross-Transplantation

1. Split MuSC suspensions equally into cylinders containing WT or *mdx* recipient fibers to match experimental conditions.
2. Incubate fiber/MuSC cocultures for 5 h at 37°C, 5% CO_2, 3% O_2 (*see* **Note 4**).
3. Fill each cylinder with ~10 mL culture medium and transfer to Sterilin dishes with caps.
4. For each microwell plate chamber, pour half of the suspension directly into chamber.
5. Manually transfer remaining fibers into unoccupied microwells using a micropipette (*see* **Note 5**). With practice, ~50 fibers can be isolated within 30 min. Total time for filling the plate is ~ 3 h (*see* **Note 6**).
6. Incubate microwell plates overnight at 37°C, 5% CO_2, 3% O_2.

3.5 Live Imaging

1. Grafted fibers were imaged live using a Zeiss Observer Z1 equipped with a Plan-Apochromat 20×/0.8 M27 objective, Colibri 7 LEDs, a Hamamatsu Orca Flash four camera, and

piloted with Zen software (Carl Zeiss). Maintain cultures at 37°C, 5% CO_2, and 3% O_2 using a Pecon incubation chamber (*see* **Note 7**).

2. Acquire images every 10–14 min in two or three channels (Brightfield, YFP, DsRed), using a Z-stack of 50 μm (5 μm slices). Set motorised stage speed to 50% with 5% acceleration to minimize fiber displacement.

4 Notes

1. *mdx* FDB fibers are more difficult to isolate and more fragile than WT FDB fibers. Isolation of *mdx* fibers from mice older than 4 months is not recommended.
2. When sealing glass coverslips to the Plexiglas, care should be taken to avoid applying excessive silicone, as this may obstruct the objective during live imaging.
3. Pilot experiments can be performed using a simplified setup, using silicone microwells in Ibidi 2-well chambers.
4. Avoid grafting too many MuSCs onto fibers, as excessive cell numbers exert mechanical pressure and may induce fiber hyper-contraction.
5. When manually collecting FDB fibers using a P20 pipette, set the volume to 12 μl to aspirate sufficient medium for comfortable fiber uptake. Gently resuspend the fibers before aspiration to ensure that only fibers in suspension are collected, and not those settled at the bottom of the dish. Any material that encounters the fibers must be equilibrated with serum or medium to prevent the fibers from sticking. During loading of a chamber, cover the other chambers with Parafilm to ensure sterile conditions. Several factors can trigger fiber contraction during the isolation procedure. These include excessive fiber density within a dish, an insufficient number of fibers (leading to frequent resuspension of the dish), and contact with toxic substances such as ethanol or disinfectants on pipette tips. Temperature fluctuations in the room can also affect the pH of the culture medium in contact with ambient air. Incubation with 3% O_2 increases fiber survival and MuSC proliferation. When releasing a fiber into the microwell, avoid abrupt movements or forceful expulsion, which may damage the fiber. Instead, release the fiber gently from slightly above the microwell and allow it to settle. If necessary, aspirating a small volume of liquid adjacent to the microwell can help guide the fiber into position.

6. As the isolation of approximately 200 fibers in microwells can take more than 2 h, both the microwell plate and Sterilin dishes remain exposed to air for extended periods. It is recommended to prepare multiple Sterilin dishes containing the same batch of fibers and to switch to a fresh Sterilin every 30 min. If isolating more than 300 fibers (~3.5 h of manipulation), fiber viability is compromised, and hyper-contraction results likely from changes in medium pH. It is therefore preferable to end by isolating and plating the most sensitive fibers (*mdx*).
7. Each chamber (i.e., grafting condition) of the microwell plate contained ~200 recipient fibers isolated in microwells, ~25% of them grafted with one to two YFP-labeled MuSCs from donor mice. For each experiment, ~20 fibers per grafting condition were imaged live. When selecting fibers for imaging, preferentially choose those carrying one, or at most two, MuSCs. Avoid fibers that contain debris or show signs of bending or early contraction.

Acknowledgments

We acknowledge funding support from the Institut Pasteur, Agence Nationale de la Recherche (Laboratoire d'Excellence Revive, Investissement d'Avenir; ANR-10-LABX-73 to ST), Association Française contre les Myopathies (#23774 to BE), European Research Council (Advanced Research Grant #101055234 to ST), La Fondation ARC pour la Recherche sur le Cancer and the Centre National de la Recherche Scientifique. LS was supported by a PhD Fellowship from La Ligue Contre le Cancer.

Competing Interests The authors declare no competing interests.

References

1. Siegel AL, Atchison K, Fisher KE et al. (2009) 3D timelapse analysis of muscle satellite cell motility. Stem Cells 27:2527–2538
2. Siegel AL, Kuhlmann PK, Cornelison DDW (2011) Muscle satellite cell proliferation and association: new insights from myofiber time-lapse imaging. Skelet Muscle 1(1):7
3. Moyle LA, Cheng RY, Liu H et al. (2020) Three-dimensional niche stiffness synergizes with Wnt7a to modulate the extent of satellite cell symmetric self-renewal divisions. Mol Biol Cell 31:1703–1713
4. Sarde L, Letort G, Varet H et al. (2026) Impaired stem cell migration and divisions in Duchenne muscular dystrophy revealed by live imaging. Nat Commun 17: 1769
5. Wertz K, Füchtbauer EM (1998) Dmd(mdx-beta geo): a new allele for the mouse dystrophin gene. Dev Dyn Off Publ Am Assoc Anat 212:229–241
6. Murphy MM, Lawson JA, Mathew SJ et al. (2011) Satellite cells, connective tissue fibroblasts and their interactions are crucial for muscle regeneration. Development 138:3625–3637
7. Srinivas S, Watanabe T, Lin CS et al. (2001) Cre reporter strains produced by targeted insertion of EYFP and ECFP into the ROSA26 locus. BMC Dev Biol 1:4
8. Benavente-Diaz M, Comai G, Di Girolamo D et al. (2021) Dynamics of myogenic differentiation using a novel Myogenin knock-in reporter mouse. Skelet Muscle 11: 5

Part III

In Vitro Functional Assays and Fate Modeling

Check for updates

Chapter 9

Efficient In Vitro Generation and Isolation of Quiescent Pax7⁺ Human Muscle Reserve Cells

Diego Michel, Eléonore Parisel and Thomas Laumonier

Abstract

Human muscle reserve cells (MuRCs) represent a quiescent myogenic population generated in vitro and an attractive model for studying muscle stem cell biology. This protocol outlines a robust and reproducible method to generate in vitro, quiescent human MuRCs from primary myoblasts (MBs) obtained from muscle biopsies of healthy donors undergoing orthopedic surgery. Muscle tissue is enzymatically dissociated and cultured in a defined growth medium (GM) to establish pure MB populations, which are isolated by flow cytometry sorting using $CD56^+/CD146^+/CD82^+$ markers. Human MBs can be expanded for up to seven passages (approximately 30 cell divisions). Upon reaching 80% confluency, human MBs are induced to differentiate by exposure to a serum-free differentiation medium (DM) for 48–72 h. Following brief trypsinization, multinucleated myotubes are removed and mononucleated MuRCs are isolated through a 20 μm filtration step. This approach yields a high number of quiescent human MuRCs, suitable for investigating muscle stem cell quiescence, activation, and function. The protocol provides a valuable platform to study the mechanisms underlying muscle regeneration and to develop therapeutic strategies that target muscle repair and disease.

Keywords Human skeletal muscle, Myoblasts, Muscle reserve cells, Pax7, Quiescence, Reactivation, EdU labeling, Skeletal muscle regeneration

1 Introduction

Skeletal muscle regeneration depends on a population of muscle stem cells (MuSCs) capable of entering and exiting quiescence [1–4]. Following injury, MuSCs exit quiescence, proliferate, and differentiate to repair damaged fibers, while a subset undergoes self-renewal to replenish the stem cell pool [5, 6]. Intramuscular transplantation of MuSCs has long been explored as a possible therapeutic approach for severe muscle trauma and muscular dystrophies. However, clinical trials using myoblasts (MBs) or mesoangioblasts have yielded limited success [7], largely because MuSCs rapidly lose regenerative

The authors "Diego Michel and Eléonore Parisel" have equally contributed to this chapter.

Nathalie Didier and Caroline E. Brun (eds.), *Muscle Stem Cells*, Methods in Molecular Biology, vol. 3093,
https://doi.org/10.1007/978-1-0716-5643-3_9,

potential when cultured ex vivo under conventional conditions [8, 9]. This decline is driven by premature activation from quiescence, underscoring the importance of maintaining a quiescent/dormant state in order to preserve MuSC function [10–13].

Muscle reserve cells (MuRCs) represent a distinct mononucleated population that remains undifferentiated during terminal myogenic differentiation in vitro and exhibits stem cell-like properties. Human MuRCs share functional similarities with human MuSCs, including the ability to re-enter the cell cycle and contribute to muscle repair [14, 15]. Understanding human MuRC biology is important for elucidating mechanisms of muscle homeostasis and for developing regenerative therapies for muscle-wasting conditions. Previously, we demonstrated that human MuRCs generated in vitro represent a population of quiescent $Pax7^{+}$ MuSCs arrested in a reversible G_0 state. These cells exhibit enhanced regenerative capacity and contribute to the $Pax7^{+}$ MuSC pool following transplantation into injured muscles [16]. Recent work has also revealed heterogeneity within the human MuRC population, identifying a $Pax7^{High}$ subpopulation in a deeper quiescent state with lower metabolic activity and reduced commitment to differentiation [17]. Furthermore, we showed that autofluorescence serves as a promising, label-free biomarker for identifying and isolating viable, functionally distinct subsets of human MuRCs [18].

The theoretical basis of this method lies in the observation that human MuRCs arise during myoblast differentiation in vitro, where a subset of cells escapes fusion and regains a quiescent phenotype. This quiescent state is defined by low metabolic activity and increased Pax7 expression, positioning human MuRCs as a good model for studying MuSC quiescence and reactivation. By replicating this process in vitro, human MuRCs can be isolated and examined under controlled conditions.

The major steps include:

1. Isolation of human primary myoblasts from muscle biopsies using enzymatic dissociation and flow cytometry cell sorting ($CD56^{+}/CD146^{+}/CD82^{+}$ markers).
2. Expansion of human myoblasts under defined conditions.
3. Induction of myogenic differentiation in serum-free medium for 48 h.
4. Removal of multinucleated myotubes by short trypsinization and filtration to isolate human MuRCs.

This protocol provides a reproducible platform for investigating the molecular signature and function of quiescent and/or reactivated human MuRCs.

2 Materials

2.1 Equipment

1. Microscope.
2. A heating water bath with magnetic stirring.
3. Centrifuge.
4. Precision scale.
5. Flow cytometry cell sorter (FACS Aria Fusion, Becton Dickinson).
6. Centrifuge.
7. Water bath.
8. −80°C storage freezer.
9. Liquid nitrogen cryostorage tank.
10. Widefield Axiocam Fluo microscope (Zeiss) through a 20× objective (EC Plan-Apochromat 20×/0.8) with an image acquisition software (Zeiss ZEN 3.11).
11. Images analysis software (e.g., Fiji).

2.2 Disposable Material

1. A sterile dissociation box containing the following: one scissors, one forceps, one magnetic bar.
2. A 200 mL dissociation plastic bottle.
3. 60 mm Petri dishes.
4. 10 cm Petri dishes.
5. 70 μm cell strainer.
6. 40 μm cell strainer.
7. 50 mL conical centrifuge tubes.
8. 5 mL polystyrene round-bottom tubes.
9. 15 mL conical centrifuge tubes.
10. 1.5 mL microcentrifuge tubes.
11. Cryotubes.
12. 15 mL conical centrifuge tubes.
13. Ice and ice bucket.
14. Progressive freezing box (e.g., Mr. Frosty freezing container, ThermoFisher Scientific).
15. 20 μm cell strainer (Miltenyi Biotec).
16. 5 mL conical centrifuge tube.
17. 15 mm Sterile coverslips (Epredia).
18. Forceps.
19. 12-well plate (FalconTM).
20. Microscope slides (Epredia).

2.3 Reagents and Buffers

1. 0.05% Trypsin–ethylenediaminetetraacetic acid (EDTA) solution.
2. Phosphate-buffered saline (PBS).
3. Growth medium (GM): 0.5 mg/mL bovine serum albumin (BSA), 0.5 mg/mL Fetuin, 1 mM Creatine, 0.04 mg/mL insulin, 0.39 μg/mL Dexamethasone, 5 μg/mL Gentamycin, 100 μg/mL Pyruvate, 50 μg/mL Uridine, 10 ng/mL Epidermal growth factor (EGF), 15% fetal calf serum (FCS) in Ham's F10.
4. Trypan Blue solution (Sigma-Aldrich).
5. Antibodies for FACS: mouse IgG1-κ anti-human CD56-AF488 (Bio-Rad), mouse IgG1-κ anti-human CD82-PE (Biolegend), mouse IgG1-κ anti-human CD146-PECy7 (BD Pharmigen), mouse IgG1-κ AF488 (BD Pharmigen), mouse IgG1-κ PE (Biolegend), mouse IgG1-κ PECy7 (BD Pharmigen).
6. Dimethyl sulfoxide (DMSO).
7. Differentiation medium (DM): 0.50 mg/mL BSA, 1 mM creatine, 10 μg/mL gentamycin, 100 μg/mL pyruvate, 50 μg/mL uridine, 10 μg/mL insulin in DMEM.
8. Paraformaldehyde (PFA) 4% in PBS.
9. Blocking buffer: PBS, 5% goat serum and 0.3% Triton X 100.
10. Antibody buffer: PBS, 5% BSA and 0.3% Triton X 100.
11. ProLong™ Glass Antifade Mountant with NucBlue™ (Invitrogen).
12. Primary antibodies for IF: rabbit IgG anti-human MyoD (Cell signaling) at 1/100, mouse IgG1 anti-human Pax7 (DSHB) at 1/100; Anti-MEF2C antibody (D80C1), rabbit monoclonal (Cell Signaling Technologies) at 1/500.
13. Click-iT™ Plus EdU Cell Proliferation Kit for Imaging (Invitrogen).
14. Secondary antibodies: Goat anti-rabbit IgG (H+L) AF546; goat anti-mouse IgG (H+L) AF647.

3 Methods

3.1 Isolation and Characterization of Human Muscle Reserve Cells

3.1.1 Dissociation of Human Muscle Biopsies

All procedures are performed at room temperature unless indicated otherwise.

The timeline of Sects. 3.1.1, 3.1.2 and 3.1.4 is presented in Fig. 1.

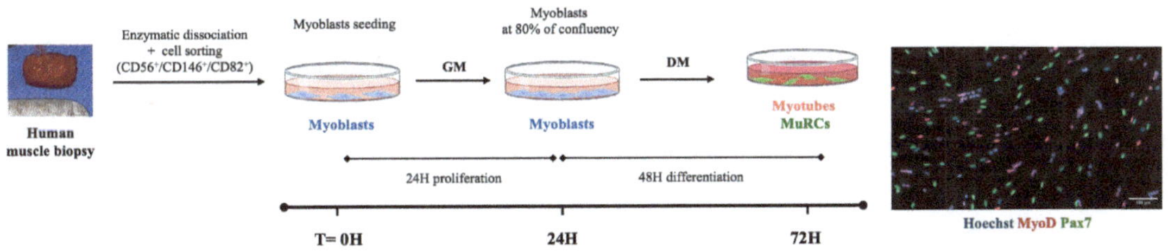

Fig. 1 Schematic representation of the in vitro model for human MuRC generation. The left panel illustrates the timeline of myoblast proliferation and differentiation, while the right panel shows a representative immunofluorescence image of human myotubes and human MuRCs. Cells were fixed, permeabilized, and stained with Hoechst (blue), Pax7 (green), and MyoD (red). Scale bar: 100 μm

1. Warm PBS, 0.05% Trypsin-EDTA, fetal calf serum (FCS), and growth medium (GM) to 37°C in a water bath.
2. Transfer the muscle sample into a 60 mm sterile Petri dish and close the lid.
3. Weigh the muscle sample with a precision scale (*see* **Note 1**).
4. Rinse the muscle biopsy with 5 mL of PBS and transfer it to a new sterile 60 mm Petri dish. Repeat this process three times. During each wash, carefully remove connective and adipose tissue using scissors and forceps.
5. Transfer the muscle biopsy into a new sterile 60 mm Petri dish and add 5 mL of 0.05% Trypsin-EDTA.
6. Using scissors, cut and mince the tissue into small pieces suitable for aspiration with a 10 mL sterile pipette.
7. Transfer minced muscle into a 200 mL dissociation plastic bottle containing a sterile magnetic bar (*see* **Note 2**).
8. Complete to 90 mL the final volume of 0.05% Trypsin-EDTA in the dissociation bottle.
9. Incubate at 37°C in a water bath for 60 min under mild agitation (20% speed).
10. Stop the enzymatic reaction by adding 10 mL of FCS to the dissociation bottle. Pipette up and down a few times to homogenize the mixture (*see* **Note 3**).
11. Place a 70 μm cell strainer on top of each of two 50 mL conical centrifuge tubes and transfer the supernatant containing the dissociated cells onto the cell strainer (*see* **Note 4**).
12. Centrifuge the tubes at 200 × *g* for 5 min.
13. Discard the supernatant.
14. Resuspend the first cell pellet in 10 mL of GM, then transfer it to the second conical tube and resuspend the second pellet.
15. Place a 40 μm cell strainer on top of one 50 mL conical centrifuge tube.

16. Transfer the supernatant containing the dissociated cells onto a 40 μm cell strainer to remove debris and larger fragments.
17. Centrifuge the tubes at 200 × *g* for 5 min.
18. Discard supernatant and resuspend the cells with 1 mL of GM.
19. Determine cell count after a 1:1 dilution in Trypan Blue solution.
20. Seed 3×10^5 cells into a 10 cm Petri dish containing 10 mL of GM, then incubate at 37°C in a humidified atmosphere with 5% CO_2.
21. Expand cells in GM for 5–7 days at 37°C in a humidified incubator with 5% CO_2, ensuring confluence does not exceed 80%, prior to flow cytometry sorting.
22. Replace half of the GM 3 days after initial seeding.

3.1.2 Isolation of Human Primary Myoblasts and Cell Sorting

1. Prewarm PBS, 0.05% Trypsin-EDTA, and GM to 37°C in a water bath.
2. Remove the 10 cm Petri dish from the incubator, aspirate the medium, and rinse the cells with 3 mL of PBS.
3. Aspirate the PBS and add 3 mL of 0.05% Trypsin-EDTA. Incubate the 10 cm Petri dish in a humidified incubator at 37°C with 5% CO_2 for 5 min.
4. Stop the enzymatic reaction by adding 5 mL of GM to the Petri dish. Pipette the mixture up and down several times to ensure thorough homogenization.
5. Transfer cell mixture into a 15 mL conical centrifuge tube and centrifuge at 240 × *g* for 3 min.
6. Aspirate the supernatant, resuspend the cells in 500 μL of GM and perform cell counting.
7. In a 5 mL tube, add 100,000 cells, adjust the volume to 500 μL with PBS, and add 1 μL of the following antibodies: mouse IgG1-κ AF488, mouse IgG1-κ PE and mouse IgG1-κ PECy7.
8. Transfer the remaining cells into a second 5 mL polystyrene round-bottom tube and add the following antibodies (for 3×10^6 cells): mouse anti-CD56-AF488 (3 μL), mouse anti-CD82-PE (1 μL), and mouse anti-CD146-PECy7 (3 μL). Incubate both tubes for 40 min on ice, protected from light.
9. Add 1 mL of PBS and centrifuge the cells at 240 × *g* for 3 min. Discard the supernatant, resuspend the pellet in 500 μL of PBS by pipetting up and down, and repeat this step once.
10. Centrifuge the cells at 240 × *g* for 3 min and discard the supernatant. Resuspend in 1 mL of GM per 5×10^6 cells.

Prepare a 5 mL conical tube containing 1 mL of GM to collect cells after cell sorting.

11. Sort CD56+/CD82+/CD146+ cells (i.e., myoblasts) using a flow cytometry cell sorter and collect cells in the 5 mL conical tube.
12. Centrifuge the 5 mL tube at 240 × *g* for 3 min and discard the supernatant to remove residual FACS buffer. Resuspend the cells in 1 mL of GM and perform cell counting.
13. Seed 5×10^5 cells into a 10 cm Petri plate containing 10 mL of GM and incubate at 37°C in a humidified atmosphere with 5% CO_2 up to 80% confluency (*see* **Note 5**).

3.1.3 Expansion and Cryopreservation of Human MBs

1. Human primary myoblasts can be expanded for up to seven passages (approximately 30 cell divisions). Passage cells before they reach 80% confluence to avoid premature differentiation. For this purpose, rinse myoblasts with 3 mL PBS. Aspirate and add 3 mL of 0.05% Trypsin-EDTA and incubate for 4 min at 37°C, 5% CO_2. Stop the reaction by adding 5 mL of GM (*see* **Note 6**). Perform gentle pipetting (up and down) to dissociate remaining cell clumps. Collect cells into a 15 mL conical tube and centrifuge at 240 × *g* for 3 min. Resuspend the pellet in 1 mL of GM. Count cells and seed 3×10^5 cells in a 10 cm dish containing 10 mL of GM.
2. Cryopreservation of human primary myoblasts: Suspend myoblasts at a concentration of 1×10^6 cells/mL in 1 mL GM supplemented with 10% DMSO in cryovials. Keep the cells on ice for 10 min. Freeze at −80°C using a progressive freezing box (approximately −1°C/min). For long-term storage, transfer cryovials to a liquid nitrogen tank (*see* **Note 7**).

3.1.4 Myogenic Differentiation and Isolation of Human Muscle Reserve Cells (MuRCs)

1. Myogenic differentiation is initiated when myoblasts reach approximately 80% confluence. Aspirate GM and replace it with 10 mL of prewarmed DM (*see* **Note 8**).
2. Maintain cells in an incubator at 37°C, 5% CO_2 for 48–72 h (*see* **Note 9**).
3. Rinse the 10 cm Petri dish with 3 mL PBS. Aspirate and add 3 mL of 0.05% Trypsin-EDTA, incubate at 37°C, 5% CO_2 for 1 min.
4. Check under the microscope for myotube detachment. If still attached, return to the incubator for 30 s and repeat until detached.
5. Myotube collection: Gently aspirate the supernatant containing myotubes using a 1 mL pipette and transfer it to 1.5 mL microcentrifuge tubes for downstream analyses (e.g., western blot) or discard (*see* **Note 10**).
6. Add 500 μL of 0.05% Trypsin-EDTA and incubate for 3 min at 37°C, 5% CO_2 to detach adherent MuRCs. Neutralize trypsin

by adding 5 mL GM. Perform gentle pipetting (up and down) to dissociate remaining clumps.

7. Filter cell suspension through the 20 μm strainer into a 15 mL conical tube. Centrifuge at 200 × *g* for 3 min (*see* **Note 11**).
8. Discard the supernatant and resuspend MuRCs in the appropriate medium based on the planned experiment. Proceed to cell counting.

3.2 Immunofluorescence-Based Detection of Myotubes and MuRCs

1. Using sterile forceps, place a sterile coverslip in a sterile 12-well plate.
2. Add 1 mL of GM and gently press the coverslip with the tip of the forceps to remove trapped air.
3. Seed 1.5×10^5 cells per well and incubate at 37°C, 5% CO_2 for 24 h.
4. When cells reach ~80% confluence, replace GM with 1 mL of DM and incubate at 37°C, 5% CO_2 for 48 h.
5. Wash gently with 500 μL of pre-warmed PBS once and remove immediately to prevent myotube detachment.
6. Fix cells with 500 μL of 4% PFA for 15 min.
7. Wash three times with 500 μL PBS, 5 min each.
8. Block with 250 μL blocking buffer for 30 min.
9. Aspirate and incubate with 250 μL antibody buffer containing the primary antibodies (mouse anti-human Pax7 and rabbit anti-human MyoD). Store at 4°C overnight.
10. Wash 3 times with 500 μL PBS, 5 min each.
11. Block again with 250 μL blocking buffer for 5 min.
12. Incubate with 250 μL antibody buffer containing a goat anti-rabbit AF546 (1/1000) and a goat anti-mouse AF647 (1/1000) for 75 min (*see* **Note 12**).
13. Wash three times with 500 μL PBS, 5 min each.
14. Mount coverslip on a glass slide using ProLong™ Glass Antifade Mountant with NucBlue™ (*see* **Note 13**).
15. Store slides at 4°C until imaging with a widefield Axiocam Fluo microscope using a 20× objective. Acquire a 6 × 6 image grid using Zeiss ZEN 3.11 software and analyze using Fiji software [19].

3.3 EdU-Based Cell Tracking During In Vitro Reactivation of Human MuRCs

One representative result is presented in Fig. 2.

1. Using sterile forceps, place a sterile coverslip in a sterile 12-well plate.
2. Add 1 mL of GM and gently press the coverslip with its tip to remove trapped air.

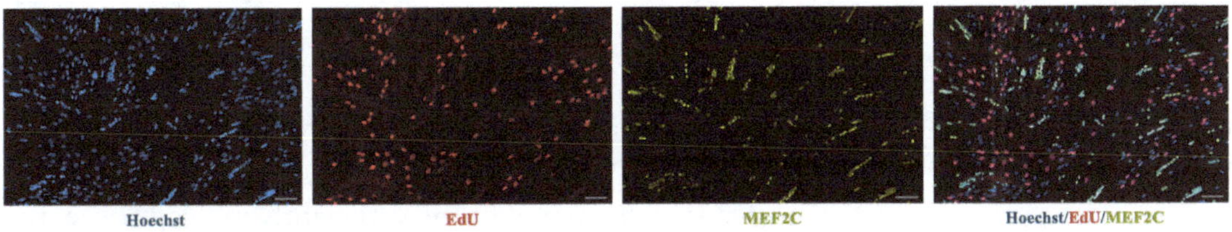

Fig. 2 Confluent monolayer of human myoblasts was switched to differentiation medium (DM) for 48 h. Human MuRCs were then reactivated in growth medium (GM) supplemented with EdU for 24 h. Shown are representative immunofluorescence images of human cells (myotubes + MuRC) after 48 h in DM and then reactivated for 24 h in GM, stained for EdU, Hoechst, and MEF2C. Scale bar: 100 μm

3. Seed 1.5×10^5 cells per well and incubate at 37°C, 5% CO_2 for 24 h.
4. When cells reach ~80% confluence, replace GM with 1 mL of DM and incubate at 37°C, 5% CO_2 for 48 h.
5. Replace DM with GM supplemented with EdU (1:1000) to reactivate MuRCs. Incubate at 37°C, 5% CO_2 for 6–24 h.
6. After 6–24 h, remove medium and wash gently with 500 μL of prewarmed PBS once and remove immediately to avoid myotube detachment.
7. Fix cells with 500 μL of 4% PFA for 15 min.
8. Wash three times with 500 μL PBS, 5 min each.
9. Block with 250 μL blocking buffer for 20 min.
10. Aspirate and incubate with 250 μL Click-iT® Plus reaction cocktail for 30 min (*see* **Note 12**).
11. Wash three times with 500 μL PBS, 5 min each.
12. Block with 250 μL blocking buffer for 30 min.
13. Aspirate and incubate with 250 μL antibody buffer containing the primary antibody (rabbit anti human MEF2C). Store at 4°C overnight.
14. Wash three times with 500 μL PBS, 5 min each.
15. Block with 250 μL blocking buffer for 5 min.
16. Incubate with 250 μL antibody buffer containing a goat anti-rabbit AF546 (1/1000) for 75 min.
17. Wash three times with 500 μL PBS, 5 min each.
18. Mount coverslip on a glass slide using ProLong™ Glass Antifade Mountant with NucBlue™ (*see* **Note 13**).
19. Store slides at 4°C until imaging with a widefield Axiocam Fluo microscope using a 20× objective. Acquire a 6 × 6 image grid using Zeiss ZEN 3.11 software and analyze using Fiji software [19].

4 Notes

1. Up to 3 g of muscle ensures optimal yield without requiring additional reagents.
2. Transfer a small portion of the sufficiently minced tissue, then add fresh trypsin to the Petri dish and continue cutting the remaining pieces into smaller fragments.
3. Any remaining connective tissue typically forms clumps during trypsinization, and these aggregates should be removed.
4. Pipette the 100 mL cell suspension up and down over the 70 μm cell strainer until it passes through. If the cell strainer becomes clogged, replace it with a new one.
5. Replace half of the GM every 2 days with 5 mL of GM prewarmed to 37°C to prevent premature myogenic differentiation and nutrient depletion.
6. To facilitate detachment, gently tap the side of the Petri dish without splashing liquid onto the lid.
7. After thawing from –80°C or liquid nitrogen, cells typically require one passage to recover from cryopreservation and expand. It is recommended to cryopreserve cells at an early passage (before passage 3).
8. Do not wash cells with PBS between the GM to DM switch.
9. You can start the myotubes and MuRC isolation protocol after 48 h in DM.
10. Myotubes appear as white filamentous structures in the dish.
11. Using a 20 μm strainer allows for the elimination of small remaining myotubes during MuRC isolation.
12. From this step onward, protect samples from light to prevent fluorochrome bleaching.
13. Rinse the coverslip once in a container of ddH_2O, then briefly touch the edge of the coverslip with absorbent paper to remove excess fluid.

Acknowledgments

This work was supported by the Fondation Suisse de Recherche sur les Maladies Musculaires. Ethics Approval Human muscle biopsies were obtained under the approval of the Commission Cantonale d'Éthique de la Recherche of the Canton of Geneva, Switzerland (approved project title: "Myogenic stem cells and improvement of muscle regeneration"; approval number: PB_2016-01793; approval date: 18 October 2018). Informed written consent was obtained from all study participants in accordance with the guidelines and regulations of the Swiss Health Authorities.

Competing Interests The authors declare no competing interests.

References

1. Lepper C, Partridge TA, Fan C-M (2011) An absolute requirement for Pax7-positive satellite cells in acute injury-induced skeletal muscle regeneration. Dev Camb Engl 138:3639–3646
2. Murphy MM, Lawson JA, Mathew SJ, Hutcheson DA, Kardon G (2011) Satellite cells, connective tissue fibroblasts and their interactions are crucial for muscle regeneration. Dev Camb Engl 138:3625–3637
3. Sambasivan R, Yao R, Kissenpfennig A, Van Wittenberghe L, Paldi A, Gayraud-Morel B, Guenou H, Malissen B, Tajbakhsh S, Galy A (2011) Pax7-expressing satellite cells are indispensable for adult skeletal muscle regeneration. Dev Camb Engl 138:3647–3656
4. Finnerty CC, McKenna CF, Cambias LA, Brightwell CR, Prasai A, Wang Y, El Ayadi A, Herndon DN, Suman OE, Fry CS (2017) Inducible satellite cell depletion attenuates skeletal muscle regrowth following a scald-burn injury. J Physiol 595:6687–6701
5. Relaix F, Bencze M, Borok MJ, Der Vartanian A, Gattazzo F, Mademtzoglou D, Perez-Diaz S, Prola A, Reyes-Fernandez PC, Rotini A, Taglietti V (2021) Perspectives on skeletal muscle stem cells. Nat Commun 12: 692
6. Ancel S, Stuelsatz P, Feige JN (2021) Muscle stem cell quiescence: controlling stemness by staying asleep. Trends Cell Biol 31:556–568
7. Skuk D, Tremblay JP (2015) Cell therapy in muscular dystrophies: many promises in mice and dogs, few facts in patients. Expert Opin Biol Ther 15(9):1307–19
8. Montarras D, Morgan J, Collins C, Relaix F, Zaffran S, Cumano A, Partridge T, Buckingham M (2005) Direct isolation of satellite cells for skeletal muscle regeneration. Science 309:2064–2067
9. Machado L, Esteves de Lima J, Fabre O, Proux C, Legendre R, Szegedi A, Varet H, Ingerslev LR, Barrès R, Relaix F, Mourikis P (2017) In situ fixation redefines quiescence and early activation of skeletal muscle stem cells. Cell Rep 21:1982–1993
10. Gilbert PM, Havenstrite KL, Magnusson KEG, Sacco A, Leonardi NA, Kraft P, Nguyen NK, Thrun S, Lutolf MP, Blau HM (2010) Substrate elasticity regulates skeletal muscle stem cell self-renewal in culture. Science 329:1078–1081
11. Judson RN, Quarta M, Oudhoff MJ, Soliman H, Yi L, Chang CK, Loi G, Vander Werff R, Cait A, Hamer M, Blonigan J, Paine P, Doan LTN, Groppa E, He W, Su L, Zhang RH, Xu P, Eisner C, Low M, Barta I, Lewis C-AB, Zaph C, Karimi MM, Rando TA, Rossi FM (2018) Inhibition of methyltransferase Setd7 allows the in vitro expansion of myogenic stem cells with improved therapeutic potential. Cell Stem Cell 22:177–190.e7
12. Sampath SC, Sampath SC, Ho ATV, Corbel SY, Millstone JD, Lamb J, Walker J, Kinzel B, Schmedt C, Blau HM (2018) Induction of muscle stem cell quiescence by the secreted niche factor Oncostatin M. Nat Commun 9: 1531
13. Arjona M, Goshayeshi A, Rodriguez-Mateo C, Brett JO, Both P, Ishak H, Rando TA (2022) Tubastatin A maintains adult skeletal muscle stem cells in a quiescent state ex vivo and improves their engraftment ability in vivo. Stem Cell Rep 17:82–95
14. Yoshida N, Yoshida S, Koishi K, Masuda K, Nabeshima Y (1998) Cell heterogeneity upon myogenic differentiation: down-regulation of MyoD and Myf-5 generates "reserve cells". J Cell Sci 111 6(Pt):769–79
15. Kitzmann M, Bonnieu A, Duret C, Vernus B, Barro M, Laoudj-Chenivesse D, Verdi JM, Carnac G (2006) Inhibition of Notch signaling induces myotube hypertrophy by recruiting a subpopulation of reserve cells. J Cell Physiol 208:538–48
16. Laumonier T, Bermont F, Hoffmeyer P, Kindler V, Menetrey J (2017) Human myogenic reserve cells are quiescent stem cells that contribute to muscle regeneration after intramuscular transplantation in immunodeficient mice. Sci Rep 7: 3462
17. Bouche A, Borner B, Richard C, Grand Y, Hannouche D, Laumonier T (2023) In vitro-generated human muscle reserve cells are heterogeneous for Pax7 with distinct molecular states and metabolic profiles. Stem Cell Res Ther 14:243

18. Bouche A, Michel D, Castets P, Hannouche D, Laumonier T (2025) Intracellular autofluore cence enables the isolation of viable, functional human muscle reserve cells with distinct Pax7 levels and stem cell states. Stem Cell Res Ther 17(1):12

19. Schindelin J, Arganda-Carreras I, Frise E, Kaynig V, Longair M, Pietzsch T, Preibisch S, Rueden C, Saalfeld S, Schmid B, Tinevez J-Y, White DJ, Hartenstein V, Eliceiri K, Tomancak P, Cardona A (2012) Fiji: an open-source platform for biological-image analysis. Nat Methods 9:676–682

Check for updates

Chapter 10

Modeling Skeletal Muscle Stem Cell Fate In Vitro

Muriel Sébastien, Emilie Christin and Vincent Gache

Abstract

Skeletal muscle stem cells (MuSCs) are necessary for muscle development and regeneration. Upon external cues, they activate from their quiescent state to self-renew and/or differentiate. The study of MuSCs is challenging as they tend to differentiate when isolated from their microenvironment. Here, we describe an in vitro model that mimics a skeletal muscle environment, allowing MuSC preservation together with myofiber maturation.

Keywords Skeletal muscle, In vitro, Niche modeling, Myofibers, Matrigel, MuSCs, Pax7

1 Introduction

Skeletal Muscle Stem Cells (MuSCs) are muscle-resident cells, originally discovered by Alexander Mauro [1]. They are considered the principal source of myogenic precursor cells, hence playing major roles in skeletal muscle regeneration [2]. While MuSCs are generally in a quiescent state in homeostatic conditions, they can be activated upon muscle damage or in diseases [3–5]. The activation of MuSCs and their entry into the cell cycle result in proliferation for self-renewal and/or differentiation and fusion to form *de novo* mature myofibers [6]. Interestingly, the MuSC microenvironment is highly relevant to this fate. Indeed, quiescent MuSCs are positioned in specific niches between myofibers and the basal lamina, where their stem cell properties are sustained by regulatory signals [7]. These signals arise from direct contact with other cells from the niche (e.g., skeletal muscle fibers, fibroblasts, endothelial cells, macrophages, etc.), as well as from secreted soluble factors and extracellular matrix molecules [8]. By integrating all these signals, MuSCs will remain in a quiescent state (Pax7-positive cells) or will activate and express successive transcription factors that will transform them into mononuclear myocytes (MyoD, MyoG), and later on into multinucleated myofibers [9]. *In vitro* modeling of mature myofibers or of their regenerative capacity

Nathalie Didier and Caroline E. Brun (eds.), *Muscle Stem Cells*, Methods in Molecular Biology, vol. 3093,
https://doi.org/10.1007/978-1-0716-5643-3_10,

is difficult using adult skeletal muscles, as MuSCs are sparse in these tissues and hard to culture once taken out of their environment [10]. The technique we present here combines the best of 2D skeletal muscle modeling by associating the high MuSC yield and good regenerative capacity of neonatal cells, with strong signaling from the extracellular matrix (Matrigel), altogether nicely mimicking the muscle environment [11]. In this work, we demonstrate through in vitro modeling of the myogenesis/regenerative process that, as muscle fibers form and mature, MuSCs (Pax7-positive cells) are preserved and physically interact with both immature (myotubes, expressing MyoG) and mature myofibers (with peripheral myonuclei and complete myofibrillogenesis). Overall, this work describes a new and original technique efficiently reproducing the MuSC niche and allowing the study of their fate in vitro.

2 Materials

All solutions and media used before fixation must be prepared under strictly sterile conditions. Here, we used wild-type C57Bl6/J newborns at post-natal day 5 (*see* **Note 1**).

2.1 Dissection

1. Dissection tools: one pair of scissors for animal and tissue mincing, one pair of small, pointed scissors (12 cm), and fine-tip straight forceps for precise sampling.
2. 70% Ethanol (v/v) in H_2O: for disinfection.
3. Small plastic dishes: for easier dissection.
4. Cold Phosphate-Buffer Saline (PBS): to collect and preserve tissues during sampling.

2.2 Digestion and Myoblasts Enrichment

1. Digestion mix: Weigh 175 mg dispase and 25 mg collagenase, and dilute in 50 mL of PBS. Mix thoroughly and filter-sterilize through a 0.22 μm filter. Prepare 8 mL aliquots and store at −20°C.
2. Pre-plating media: Dulbecco's Modified Eagle Medium (DMEM) supplemented with 4.5 g/L glucose, L-glutamine and sodium pyruvate, mixed with fetal bovine serum [FBS, 10% (v/v) final], and Penicillin/Streptomycin [P/S, 1% (v/v) final].
3. 37°C horizontal shaking incubator.
4. Cell culture incubator at 37°C and 5% CO_2.
5. Bench-top centrifuge (fitting 15 mL tubes) able to reach 500 g.
6. 15 mL centrifuge tubes.
7. Cell strainer with 40 μm mesh.

2.3 Differentiation and Maturation In Vitro

1. Proliferation media: Iscove's Modified Dulbecco's Medium (IMDM), mixed with FBS 20% (v/v) final, chicken egg extract [CEE, 1% (v/v) final], gentamycin 0.5% (v/v) final, P/S 1% (v/v) final, and 2.5 μL bFGF per 50 mL of media.
2. Differentiation media: IMDM, mixed with horse serum [HS, 2% (v/v) final] and P/S 1% (v/v) final.
3. Matrigel 1:100 solution: Dilute 250 μL Matrigel (growth factor reduced; Corning, or equivalent) in 25 mL ice-cold IMDM (*see* **Note 2**). Store at 4°C and use within 2 weeks.
4. Matrigel 1:2 solution: Dilute 250 μL Matrigel (growth factor reduced; Corning, or equivalent) in 250 μL cold IMDM. Make right before use, keep on ice and do not store or reuse leftovers.

2.4 Fixation and Immunofluorescent Staining

1. Phosphate-buffer saline (PBS).
2. Rinsing solution PBS-Triton 0.1% (v/v): Dilute 1 mL of a 10% (v/v) Triton X-100 solution in 100 mL of PBS. Store at 4°C up to a couple of months.
3. Fixative solution paraformaldehyde 4%-sucrose 4% (w/v): weigh 4 g of sucrose, add 12.5 mL of 32% paraformaldehyde, and complete to 100 mL with PBS. Aliquot and store at −20°C.
4. Permeabilization solution PBS-Triton 0.5% (v/v): Dilute 5 mL of a 10% (v/v) Triton X-100 solution in 100 mL of PBS. Store at 4°C up to a couple of months.
5. Blocking solution PBS-BSA 1% (w/v): weigh 0.5 g of bovine serum albumin (BSA), and dissolve gently in 50 mL PBS. Store at 4°C up to a couple of months.
6. Primary antibodies used: mouse IgG1 anti-Pax7 (DSHB, Hybridoma supernatant, 1:1); mouse IgG1 anti-Myogenin (5FD, Santacruz Biotechnologies, 1:500); mouse IgG1 anti-α -actinin (EA-53, Millipore-Sigma, 1:500).
7. Secondary antibody used: Alexa Fluor 488-conjugated Donkey anti-Mouse IgG (1:500).
8. Dyes used: DAPI (diluted according to the manufacturer's manual), Alexa Fluor 555-conjugated Phalloidin (1:250).

3 Methods

3.1 Preparations

1. Prepare the dissection bench (*see* **Note 3**).
2. Clean dissection instruments with 70% ethanol.
3. Place sterile PBS in a dish on ice (Fig. 1A).

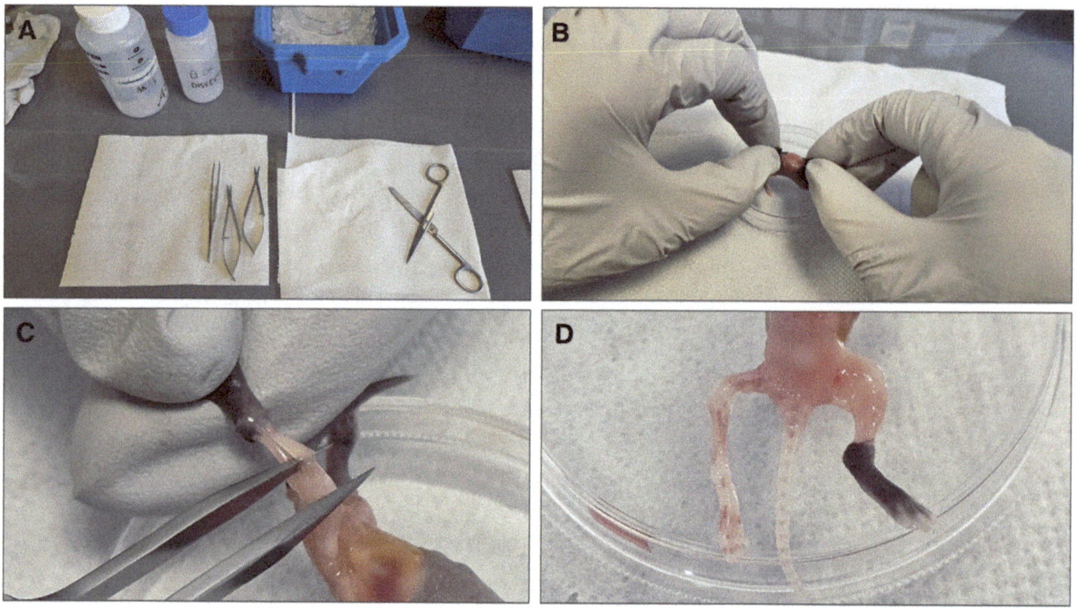

Fig. 1 Sampling muscles from the hindlimbs of post-natal mice. **A,** Representative image of the dissection bench set-up. Be sure to have clean tools, a dish with sterile PBS on ice, and 70% ethanol. **B,** To access the mouse hindlimbs, the skin of the back of the animal is teared down, then pulled down over the limbs. **C,** To sample the *tibialis anterior* muscle, the forceps are inserted in between the bones and muscle while holding the paw on the side. Once the muscle is grabbed using the forceps, the tendons at both ends can be cut. **D,** Picture of the hindlimbs of an animal, where the left paw has been fully sampled (*tibialis anterior*, *quadriceps*, *gastrocnemius*, *gluteus*), while the right paw is yet untouched

4. Thaw the required volume of digestion mix (1.5 mL/pup) at 37°C in the incubator.

3.2 Dissection

1. Proceed with mice euthanasia. Spray 70% ethanol on the belly and hindlimbs, then prepare one animal at a time.
2. Place the mouse on an upside-down clean dish. Grab the skin on the back of the animal with both hands, tear the skin, and pull it downwards to uncover the hindlimbs (Fig. 1B). Remove any visible fat (white tissues) from the limbs before collecting muscle tissues (*see* **Note 4**). Place all muscle tissues collected in cold PBS.
3. Place the mouse on its back and hold its paw to the side, insert the tweezers in between the *tibialis anterior* (TA) muscle and the bones below to separate them. Grab the middle of the TA with the forceps and cut both tendons at the ankle and the knee with the fine scissors (Fig. 1C).
4. Pinch the *quadriceps* along its length with the forceps, then cut around the forceps and close to the bones with the scissors.

5. Place the mouse facing down. Grab muscle tissue behind the knee with the forceps, and cut along the bones, from the ankle to the tail, taking most muscle groups from the hindlimb (Fig. 1D).
6. Repeat Steps 3–5 with the second limb, then proceed with all the animals from the litter with the same genotype.
7. Take the dish with tissues in PBS under a Biological Safety Cabinet (BSC) and go on with the next steps under sterile conditions.

3.3 Digestion and Myoblasts Enrichment

1. Dispose of as much PBS from the dish as possible, then tilt the dish and gather tissue pieces at the bottom of the dish, against the side.
2. Mince the tissue thoroughly for approximately 5 min with the scissors (*see* **Note 5**) to obtain a homogeneous consistency (Fig. 2A). The remaining liquid must appear cloudy.
3. Add the appropriate amounts of digestion mix (1.5 mL/pup) to the minced tissue and collect in 15 mL centrifuge tubes

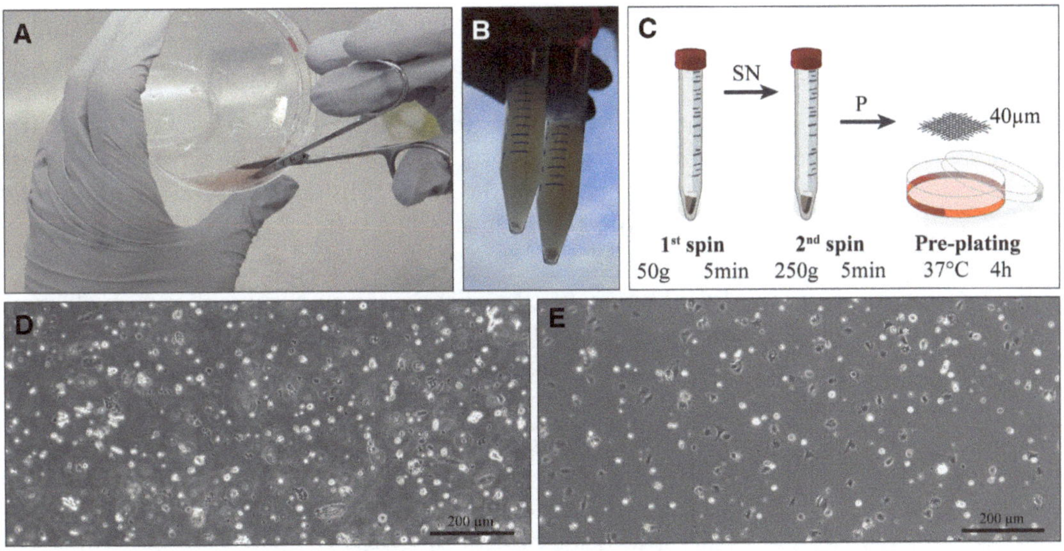

Fig. 2 Digestion and enrichment in myoblasts. **A,** Mincing of the tissue is done for approximately 5 min under sterile conditions. Holding the scissors against the side of the dish with the thumb and middle finger can facilitate the procedure. The tissue pieces should be of homogeneous sizes at the end. **B,** Picture of the general aspect of the solution at the end of the digestion. Some denser materials might stay at the bottom of the tubes, but no tissue clumps should remain. **C,** Schematics of the three steps to take after stopping the digestion: 1-a first spin at low speed, where the supernatants (SN) are kept, 2-a second spin at high speed, where the pellets (P) are kept, 3-filtering (40 μm) and preplating of the cells for 4 h at 37°C to allow attachment/removal of the fibroblasts in the cell suspension. **D,** Representative image of the preplating dish right after seeding of the cell suspension. **E,** Representative image of the preplating dish after 4 h at 37°C, 5% CO_2, as expected, many fibroblasts have attached to the dish, while myoblasts are still in suspension and can be collected

(*see* **Note 6**). If needed, pass the full volume through a serological pipette a couple of times (*see* **Note 7**).

4. Warm up the preplating media to 37°C.
5. Place the digestion tubes for 1 h in a horizontal shaking incubator set at 37°C, 240 rpm. Vigorously shake the tubes halfway through the digestion. At the end of the incubation, the solution must appear homogeneously cloudy without visible aggregates (Fig. 2B, *see* **Note 8**).
6. Stop the reaction by diluting the digestion mix by half in preplating media (i.e., 5 mL digestion mix + 5 mL preplating media).
7. Centrifuge the tubes for 5 min at 50 × *g*. Collect the supernatants in a new 15 mL tube and discard the pelleted debris (Fig. 2C and *see* **Note 9**).
8. Centrifuge the supernatants for 5 min at 250 × *g*. Keep the pellets this time and resuspend them in preplating media (*see* **Note 10**).
9. Pass the suspension through a previously wetted 40 μm mesh filter, and seed into uncoated dishes (Fig. 2C, D). Leave the dishes settle in the cell culture incubator (37°C, 5% CO_2) for a total of 4 h to allow fibroblast attachment to the dishes (*see* **Note 11**, Fig. 2E).

3.4 Differentiation and Maturation In Vitro

1. Coat the relevant number of dishes with a 1:100 solution of Matrigel (*see* **Note 12**). Incubate at room temperature at least 45 min. Rinse the coated dishes with IMDM right before seeding the cells. During this time, warm-up the proliferation media to 37°C.
2. Take the preplating dishes out of the incubator. Collect the media in a pipet and flow it over the surface of the dish to gather all unattached cells to the bottom. Transfer the cell suspension into a 15 mL tube.
3. Centrifuge the tube for 10 min at 250 × *g* (*see* **Note 13**). Resuspend the pellet in warm proliferation media and seed on Matrigel-coated dishes at the desired amounts (*see* **Note 14**). Place the dishes in the incubator (37°C, 5% CO_2) and leave alone 24 h before changing the media to remove dead cells and debris (Fig. 3A).
4. Once confluency is reached and myotubes start to form (approx. 72 h after seeding; *see* **Note 15**, Fig. 3B), replace the proliferation media with warmed-up differentiation media. Let differentiate for about 2 days (Fig. 3C).
5. At 2 days of differentiation, to avoid myotubes detachment, prepare a 1/2 solution of Matrigel and keep it on ice.

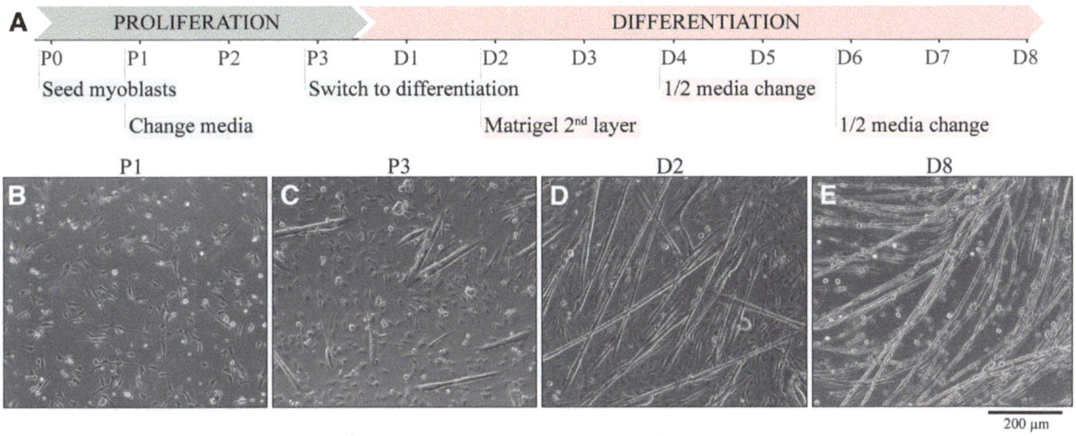

Fig. 3 Differentiation timeline and key steps. **A,** Overview of the expansion and differentiation stages per day starting from cell seeding (P0, proliferation 0) and going to 8 days of differentiation (D8), with key steps highlighted. The timing for these key steps may vary; see relevant Notes. **B,** Transmission image of the myoblasts 24 h after seeding in proliferation media (P1). **C,** Transmission image of the myoblasts 3 days after seeding (P3) in proliferation media. Some myotubes are starting to form; thus, the media can be switched to differentiation. **D,** Transmission image of the myotubes after 2 days in differentiation media (D2). To avoid myotube detachment past this point, a concentrated layer of Matrigel is added on top of the cells. **E,** Transmission image of the myotubes after 8 days in differentiation media (D8). Most myoblasts have fused, and striated patterns can be seen on some of the myofibers

6. Empty the dishes of the differentiation media, and add enough Matrigel to cover the cells (approximately 200 μL/well for a four-well ibidi μ-slide) in the dishes. Place the dishes back in the incubator for 45 min to achieve good Matrigel gelation (*see* **Note 16**).
7. Dispense warmed-up differentiation media on top of the Matrigel and place the cells back in the incubator. Change half of the media every other day.
8. Mature myofibers (with striated patterns) are present in the culture from around 5 days in differentiation media (Fig. 3D), and they can be maintained *in vitro* for at least 2 weeks. Pax7-positive cells are also preserved at least for 8 days in the differentiation media.

3.5 Fixation and Immunofluorescent Staining

1. Warm-up PBS and the fixative solution (PFA-Sucrose) to 37°C.
2. Take the dishes out of the incubator, empty them, and rinse the cells with warm PBS.
3. Add enough warmed-up fixative solution to cover the cells (depending on the containment), then place the dishes at 37°C for 20 min (*see* **Note 17**).

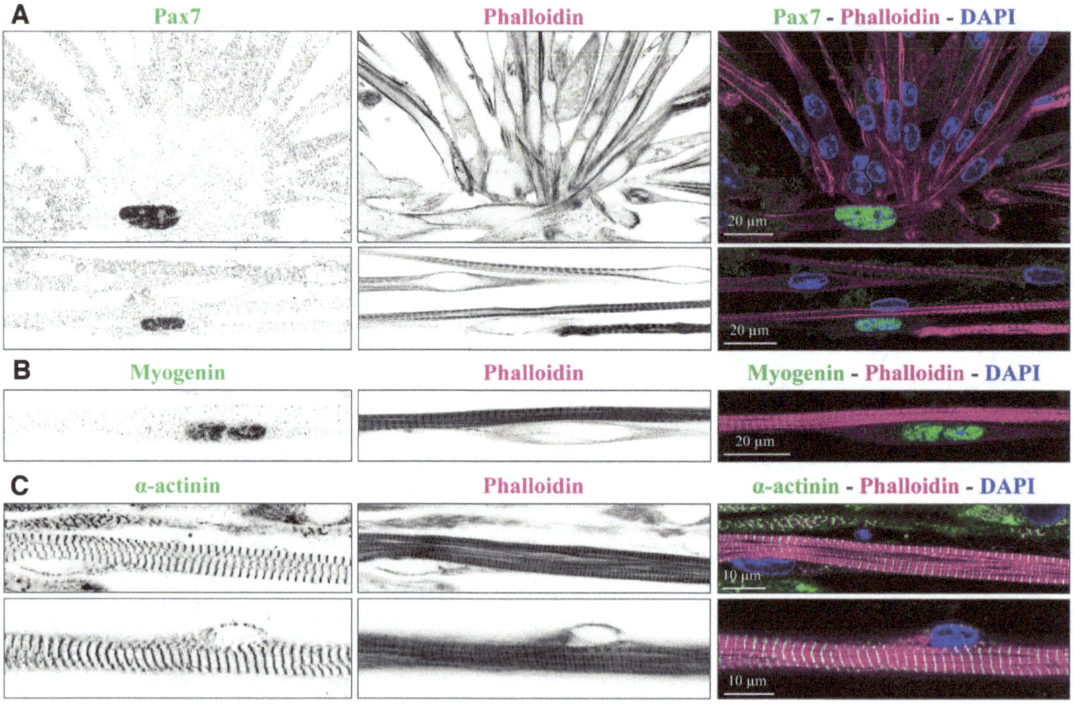

Fig. 4 Characterization of the culture at 8 days of differentiation. **A,** Immunofluorescent staining showing the presence of MuSCs (Pax7-positive cells) in between forming myofibers (striated Phalloidin stain). **B,** Immunofluorescent staining of showing early differentiation of MuSCs (Myogenin-positive cells), next to a myofiber. **C,** Immunofluorescent staining showing mature myofibers with α-actinin and phalloidin intercalated striations, and peripheral nuclei (DAPI)

4. Remove the fixative solution and rinse the cells three times with PBS (*see* **Note 18**).
5. Remove PBS to add permeabilization buffer. Incubate 5 min at room temperature.
6. Remove permeabilization buffer to add blocking buffer. Incubate at least 30 min at room temperature.
7. Replace blocking buffer by the primary antibody solutions (in blocking buffer). Here, we used antibodies against Pax7, Myogenin (MyoG), and α-actinin to observe several key steps of the lineage. Incubate overnight at 4°C in the dark (*see* **Note 19**).
8. Remove the primary antibody solutions and rinse three times with PBS-Triton 0.1%.
9. Replace PBS-Triton 0.1% with the secondary antibody solutions and conjugated dyes (in blocking buffer). Incubate at least 2 h at room temperature in the dark (*see* **Note 19**).

10. Remove the secondary antibody solutions and rinse once with PBS-Triton 0.1% and twice with PBS.
11. Replace the PBS by a 1× DAPI solution in PBS and incubate 5–10 min at room temperature in the dark.
12. Replace the DAPI by PBS and store at 4°C until imaging (*see* **Note 20**).
13. Image with a fluorescent microscope. As observed in Fig. 4, this culture method yields both MuSCs (Pax7-positive cells) and differentiated cells (early: myogenin, late: striated α-actinin), allowing a general study of the various stages of skeletal muscle development or regeneration.

4 Notes

1. Adjust sampling day according to the proliferative capacity wanted; post-natal days 1–4 could be amplified more than post-natal days 5–8.
2. Matrigel must be thawed on ice or overnight at 4°C to avoid gelation. Use cold media and pipet tips to dilute and manipulate.
3. Better be two people dissecting in parallel, no more than ten pups at a time.
4. If the intestine is pierced during the procedure, the muscle tissues should preferably not be collected to avoid contamination of the whole litter. Alternatively, remove the intestine, and carefully spray ethanol on the limbs and tools before sampling separately.
5. Hold the scissors down with your thumb and major fingers, with the blade against the side of the dish to facilitate mincing.
6. Keep the volume of digestion mix with tissues below 10 mL inside the 15 mL tubes, to achieve proper digestion.
7. When the suspension does not pass through the pipette, consider mincing more thoroughly the next time. Alternatively, pass through a serological pipette, by aspirating the solution while tapping the tip of the pipette against the dish. As an excess of trituration might also reduce digestion yield, keep the number of pipet flow-throughs to a minimum.
8. Visually check digestion efficiency and adjust timing accordingly (approximately 5 min should be sufficient). White swirls (remaining fats) in the suspensions are common and should not lead to an increase in total digestion time.

9. When collecting the first supernatant, avoid as much as possible taking the fats at the top of the tubes.
10. Adjust the volume of preplating media needed to the number of pups prepared. We generally seed no more than the equivalent of eight pups per P100 dish. Set the number of dishes needed accordingly.
11. A 4 h preplating classically yields many myoblasts, ensuring proper modeling of the skeletal muscle environment. However, this timing can be adapted according to the needs of the experiment: a shorter preplating time will result in a denser, but more heterogeneous population, while a longer preplating time might reduce the number of cells in the final population but achieve a better homogeneity.
12. We generally seed the equivalent of one pup per 4 cm^2 (two pups per four-well μ-slides, ibidi). That allows 2 days of amplification before reaching confluency and spontaneous differentiation, but these amounts can obviously be adjusted for faster or slower experiment timescale.
13. As the cell suspension contains a good number of blood cells, the pellet will certainly appear reddish at this point. The survival rate of red blood cells in culture being quite low, no extra steps were performed to remove them.
14. As an alternative to direct seeding, the cell pellet could also be resuspended in freezing media [FBS 90%—dimethyl sulfoxide (DMSO) 10%] and cryopreserved at −80°C at this stage for future use.
15. Differentiation onset strongly depends on seeding densities: if the culture is denser than described in this study, spontaneous differentiation may occur within 24–48 h post-seed. Conversely, if the culture is sparse, the cells might need up to a full week to reach confluency and start to fuse into myotubes. The timing for following steps, notably the Matrigel second layer, will therefore depend on the seeding densities.
16. The time dedicated to Matrigel settling may determine the gel stiffness; therefore, increasing or decreasing the incubation time could be optimized to impede or stimulate differentiation.
17. The sucrose in the fixative solution allows removal of the Matrigel layer. Any gel leftovers after fixation may prevent good antibody sensitivity during immunolabeling.
18. At this point, the cells can also be stored in PBS at 4°C for further use. For long-term storage, add 0.01% sodium azide to PBS, and store in a humid chamber to avoid evaporation. The cells must not dry out to be used.

19. For better results, centrifuge the antibody solutions for 10 min at 10,000*g*, prior to distribution on the cells. Incubation times can also be prolonged and performed with gentle shaking (10–15 rpm).
20. Although it is not required, the samples can be better preserved in the long term if mounted with an antifade mounting medium.

Acknowledgments

M.S. is supported by a Marie Skłodowska-Curie postdoctoral fellowship. V.G. received funding from the MyoNeurALP2 strategic pole (Association française contre les Myopathies, AFM-Téléthon) and from the Agence Nationale de la Recherche (ANR, NuCytAct). We would also like to acknowledge Dr. Caroline Brun for fruitful discussions on stemness and their ideal environment.

Competing Interests The authors declare no competing interests.

References

1. Mauro A (1961) Satellite cell of skeletal muscle fibers. J Cell Biol 9:493–495
2. Dumont NA, Bentzinger CF, Sincennes M, Rudnicki MA (2021) Comprehensive physiology. Compr Physiol 5:1027–1059
3. Garry, GA, Antony, M, Garry, DJ, 2016. Cardiotoxin induced injury and skeletal muscle regeneration. Springer, New York, pp. 61–71
4. Wallace GQ, McNally EM (2009) Mechanisms of muscle degeneration, regeneration, and repair in the muscular dystrophies. Annu Rev Physiol 71:37–57
5. Carlson BM (2003) Muscle regeneration in amphibians and mammals: passing the torch. Dev Dyn 226:167–181
6. Chargé SB, Rudnicki MA (2004) Cellular and molecular regulation of muscle regeneration. Physiol Rev 84:209–238
7. Krauss RS, Kann AP (2023) Muscle stem cells get a new look: dynamic cellular projections as sensors of the stem cell niche. BioEssays 45:e2200249
8. Evano B, Tajbakhsh S (2018) Skeletal muscle stem cells in comfort and stress. NPJ Regen Med 3:24
9. Buckingham M, Rigby PWJ (2014) Gene regulatory networks and transcriptional mechanisms that control myogenesis. Dev Cell 28: 225–238
10. Chal J, Pourquié O (2017) Making muscle: skeletal myogenesis in vivo and in vitro. Development 144:2104–2122
11. Pimentel MR, Falcone S, Cadot B, Gomes ER (2017) In vitro differentiation of mature myofibers for live imaging. J Vis Exp: JoVE (119) e55141

Check for updates

Chapter 11

Automated Quantitative Analysis of Myogenesis in Isolated Mouse and Rat Muscle Stem Cells

Pauline Garcia, Zachary Gurlekian, Rebecca Desaulniers and Nicolas A. Dumont

Abstract

Understanding the mechanisms governing myogenesis is essential for studying skeletal muscle development, regeneration, and disease. Here, we present a detailed and automated protocol for the quantitative analysis of myogenesis in isolated muscle stem cells (MuSCs) from both mice and rats. Mouse MuSCs are isolated from skeletal muscle tissue using magnetic-activated cell sorting (MACS), while rat MuSCs are purified by fluorescence-activated cell sorting (FACS), ensuring high purity and viability of stem cell populations in both models. Following isolation, MuSCs are cultured under defined proliferative or differentiating conditions and subsequently stained by immunofluorescence for key myogenic markers. To ensure unbiased and high-quality analysis, automated quantitative analysis is done using QuPath, enabling objective measurement of proliferation (Pax7, Ki67), differentiation (Myogenin), and fusion indices (Myosin Heavy Chain). Overall, this method describes a robust and fast automated analysis of myogenesis on mouse and rat using automated quantification.

Keywords Muscle stem cells, Myogenesis, Automated quantification, Image analysis, Immunofluorescence

1 Introduction

Myogenesis in both mice and rats is a complex and tightly regulated process by which skeletal muscle fibers form, grow, and mature, beginning during embryonic development and continuing into postnatal life. In both species, the process originates from mesodermal progenitor cells located in the somites, specifically within the dermomyotome, where key regulatory genes such as Pax3 and Pax7 initiate the specification of myogenic precursor cells [1–3]. These progenitors subsequently express the myogenic regulatory factors (MRFs) Myf5 and MYOD [4–6], committing them to the myogenic lineage. In mice, primary myogenesis occurs between embryonic days E10.5 and E12.5, when mononucleated myoblasts fuse to form the first multinucleated primary myotubes, which serve as scaffolds

Nathalie Didier and Caroline E. Brun (eds.), *Muscle Stem Cells*, Methods in Molecular Biology, vol. 3093,
https://doi.org/10.1007/978-1-0716-5643-3_11,

for later muscle fibers [7, 8]. This is followed by secondary myogenesis between E14.5 and E17.5, characterized by the formation of additional myotubes around the primaries through the fusion of fetal myoblasts [8, 9]. In rats, the same molecular and cellular events occur, but the timing is slightly delayed due to their longer gestational period, with primary myogenesis occurring around E12–E14 and secondary myogenesis between E16–E20 [10].

As differentiation progresses, activated myoblasts express Myogenin and MRF4, which drive terminal differentiation and myotube maturation [11–13]. Structural proteins such as myosin heavy chain (MyHC) appear at this stage, marking the assembly of contractile apparatus and the formation of functional myofibers [14]. By late gestation, muscle stem cells (MuSCs), which are responsible for postnatal growth and regeneration, begin to establish themselves beneath the basal lamina of developing fibers (they are also termed satellite cells based on this anatomical location).

After birth, both mice and rats experience significant muscle growth, driven by the continued fusion of MuSCs and the enlargement of existing fibers [10, 15]. During this postnatal period, mouse MuSCs remain active for the first 3–6 weeks, contributing additional nuclei to growing myofibers, after which they enter a quiescent state, becoming reactivated only in response to muscle injury or physiological stress. While mouse and rat myogenesis follow the same sequence of molecular and morphological events, the rat exhibits a slightly slower developmental timeline and generally larger muscle fiber size, reflecting species-specific differences in growth rate and body size rather than in fundamental biological mechanisms [16].

The process of myogenesis in mice and rats relies on the same tightly regulated network of myogenic regulatory transcription factors; however, the timing and amplitude of gene expression events can vary between species, influencing the balance between activation, self-renewal, and differentiation *in vivo*. Comparative studies using established immortalized myogenic cell lines, namely C2C12 in mice and L6 in rats, have demonstrated that both express PAX7, MYOD, and MYOGENIN, yet display functional and electrophysiological differences during early differentiation stages [17]. Furthermore, rat MuSCs show a higher proliferative and differentiation capacity when cultured on extracellular matrix (ECM) components derived from rat tissue compared to standard media [18]. A similar phenomenon is observed in mouse MuSCs, though triggered by distinct ECM-associated factors, suggesting that the response to regenerative cues and microenvironmental interactions can diverge between species [18–20]. Sarcopenia and age-related muscle wasting also begin at different ages in these rodents: mice generally maintain muscle function until approximately 24 months of age, whereas rats exhibit a marked decline around 20 months, reflecting species-specific differences in transcriptional and functional aging [16]. Transcriptomic profiling of skeletal muscle from both species revealed a broader set of age-associated genes in rats, suggesting a greater temporal and molecular impact of aging and an onset of sarcopenia that more closely parallels the human condition [21].

Importantly, although mouse models have historically dominated MuSC research, the use of rats offers several distinct advantages, including a larger muscle mass that facilitates stem cell isolation and certain physiological responses that more closely resemble human muscle biology. Consequently, rat models have recently been adopted for the study of sarcopenia, Duchenne muscular dystrophy, muscle regeneration, and preterm birth–associated muscle development [22, 23].

The analysis of myogenesis in vitro provides a powerful and controlled framework to study the molecular and cellular events underlying MuSC proliferation, differentiation, and fusion. When cultured under defined growth and differentiation conditions, isolated MuSCs or myoblast cell lines recapitulate a simplified sequence of myogenesis observed in vivo, enabling the quantification of lineage progression and functional maturation. Here, different strategies are used for MuSCs isolation in order to culture them in vitro. In mice, MuSCs were enriched using magnetic-activated cell sorting (MACS), which utilizes antibody-conjugated magnetic beads to perform a negative selection to exclude fibro-adipogenic progenitors, vascular and immune cells, and therefore collect only MuSCs. This approach provides rapid and gentle bulk enrichment, preserving cell viability and integrity. In contrast, MuSCs from rats were purified by fluorescence-activated cell sorting (FACS), in which fluorescently labeled cells were identified and separated based on negative and positive markers for MuSCs. Cells were stained with alpha7-integrin to mark the MuSCs, CD31, and CD45 to exclude endothelial and hematopoietic cells, respectively, and 7AAD for live cells/dead cells discrimination [24]. FACS enables precise gating and higher-purity recovery of MuSCs, and is advantageous given the relatively limited availability of rat-specific magnetic reagents. The use of MACS for mice and FACS for rats thus reflects both practical and biological considerations, balancing throughput and purity with the constraints of each species.

Traditionally, the analysis of myogenesis *in vitro* is assessed using immunofluorescence staining for key myogenic markers such as PAX7 (MuSC identity), Ki67 (proliferation) [25, 26], MYOD (activation and commitment), MYOG (terminal differentiation), and MyHC (myotube formation). Quantitative parameters such as the proportion of PAX7/MYOD-positive cells, the differentiation index (percentage of MYOG- or MyHC-positive cells) and the fusion index (the proportion of nuclei that are incorporated into multinucleated myotubes) are used to evaluate the efficiency of myogenic progression.

Despite the central role of myogenesis in skeletal muscle regeneration and disease, quantitative analysis of this process in cultured MuSCs has traditionally relied on manual assessment of micrographs. Conventional approaches, such as visual identification of myotubes and hand-counting of nuclei, are labor-intensive, time-consuming, and inherently subjective. These procedures not only restrict the number of experimental conditions that can be analyzed but also introduce operator bias and inter-observer variability, which compromise reproducibility and limit the sensitivity of comparative studies across species such as mice and rats. Automated analysis offers a powerful solution to these limitations by integrating high-content image acquisition with computational pipelines that can rapidly and objectively extract quantitative metrics from large image datasets [27, 28]. In the context of myogenesis, automation enables precise segmentation of individual nuclei and myotubes and unbiased detection of myogenic markers. Different open-source platforms, such as CellProfiler [29–31], ImageJ [32–34], and machine-learning-based segmentation algorithms, can be used to develop workflows that can be applied across different species and experimental conditions. By eliminating the need for manual scoring and applying uniform analysis criteria, automated pipelines increase throughput, and enhance reproducibility. Here, we show how to use the freely available QuPath and FIJI/ImageJ softwares to perform a reproducible work and methodology across the different scientific questions and species to quantify and analyze myogenesis in isolated mouse and rat MuSCs.

2 Materials

The following material is required for the experiments described in this chapter.

2.1 General Equipment

1. Standard humidified culture incubator (37°C, 5% CO_2)
2. Tissue culture laminar flow hood
3. Centrifuge 5804/5804R or other refrigerated benchtop centrifuge
4. Miltenyi Biotec GentleMACS Octo Dissociator
5. Water bath at 37°C
6. Heat block at 96°C
7. Guillotine (minimum 2.5×2.5 inch opening)
8. Dissection tools: Fine-tipped forceps, Blunt forceps, Micro-dissecting scissors
9. Widefield inverted epifluorescent Microscope (e.g., EVOS M5000)
10. BD FACSAria™ III Cell Sorter or other suitable cytometer

2.2 Euthanasia

1. Pentobarbital
2. Isoflurane USP

2.3 Reagents

1. Collagenase B
2. Dispase II
3. Red blood cell (RBC) lysis buffer
4. Ultrapure water
5. Bovine serum albumin (BSA)
6. Ethylenediaminetetraacetic acid (EDTA) (0.5 M), pH 8.0, RNase-free

2.4 Plastics

1. Serological pipettes 5 mL
2. Serological pipettes 10 mL
3. Cell strainers 100 μm
4. Syringe filter 0.2 μm
5. 100 mm TC-treated culture dish
6. Falcon® 12-well Clear Flat Bottom TC-treated Multiwell Cell Culture Plate
7. Falcon® 24-well Clear Flat Bottom TC-treated Multiwell Cell Culture Plate
8. LS column
9. gentleMACS™ tube C
10. 50 mL syringe
11. 40 μm cell strainer
12. 30 μm filter
13. Falcon™ 50 mL High Clarity Conical Centrifuge Tubes
14. Corning® Falcon® Centrifuge Tubes
15. Round polystyrene FACS tube with cap, 5 mL

2.5 FACS Antibodies

1. FITC alpha7-integrin.
2. PE-Cy7 CD31.
3. PE-Cy7 CD45.
4. 7-AAD. Other suitable viability markers such as DAPI may also be used, provided they are compatible with the antibody fluorochromes used for FACS.
5. Compensation beads.

2.6 MACS and Immunostaining Antibodies

1. MACS antibodies: Satellite Cell Isolation Kit.
2. Primary antibodies for immunostaining: Mice and rats: Anti-PAX7 antibody (AB_528428, DSHB), Anti-MYOD

antibody (SC-377460, Santa Cruz), Anti-MYOGENIN antibody (ab1835, Abcam), Anti-MyHC antibody (MF20; AB_2147781, DSHB); Mice: Anti-Ki67 antibody (14-5698-82, ThermoFisher); Rat: Anti-Ki67 antibody (MA5-14520, Thermofisher).

3. Secondary antibodies for immunostaining: Donkey anti-Mouse IgG1 Alexa Fluor™ 488, Donkey anti-Rabbit IgG Secondary Antibody, Alexa Fluor™, Donkey anti-Mouse IgG2b Secondary Antibody, Alexa Fluor™ 555, Donkey anti-Rat IgG Secondary Antibody, Alexa Fluor™ 555.
4. Hoechst 33342 (H3570, ThermoFisher).

2.7 Cell Culture

1. Ham's F-10 Medium 1× with L-Glutamine and Phenol Red 500 mL. The bottles are stored at 4°C.
2. Dulbecco's Modified Eagle Medium (DMEM) low glucose, pyruvate. The bottles are stored at 4°C.
3. Fetal bovine serum (FBS; standard, not heat-inactivated). Bottles are stored at −80°C for long term; once thawed and aliquoted, stored at −20°C.
4. Horse serum. Original bottles are stored at −80°C for long term; once thawed and aliquoted, stored at −20°C.
5. Phosphate-buffer saline 1×. The bottles are stored at 4°C.
6. HBSS, calcium, magnesium, no phenol red. The bottles are stored at 4°C.
7. Trypsin/EDTA. Original bottles are stored at −80°C for long term; once thawed and aliquoted, stored at −20°C
8. Red blood cell (RBC) lysis solution. The bottles are stored at 4°C.
9. UltraPure™ DNase/RNase-Free Distilled Water.
10. Mouse standard growth medium: Ham's F-10 Medium, 20% FBS, 1% Penicillin–Streptomycin, 50 μg/mL basic fibroblast growth factor (bFGF).
11. Rat standard growth medium: DMEM low glucose, 20% FBS, 1% Penicillin–Streptomycin, 10 μg/mL recombinant bFGF.
12. MACS buffer: PBS (1× pH 7.2), 0.5% BSA, 2 mM EDTA.
13. Mouse differentiation medium: 48.5% DMEM low glucose with pyruvate, 48.5% Ham's F-10 Medium, 2% horse serum, 1% Penicillin–Streptomycin.
14. Rat differentiation medium: DMEM low glucose with pyruvate, 5% horse serum, 1% Penicillin–Streptomycin.
15. Matrigel for coating tissue culture dishes. Store 500 μL aliquots at −20°C.

16. Rat Tail Collagen I for coating tissue culture dishes. Type I collagen coating solution is prepared by diluting a 3 mg/mL sterile collagen I stock solution in 0.1 M acetic acid to a final concentration of 0.1 mg/mL.

2.8 Reagents and Solutions for Immunostaining

1. Fixative solution: 2% paraformaldehyde (PFA), sodium phosphate buffer, PBS (1×).
2. Permeabilization buffer: 0.2% Triton X-100, 0.1 M Glycine, PBS (1×).
3. Blocking buffer: 2% BSA, PBS (1×).

2.9 Softwares

1. FIJI/ImageJ [35]
2. QuPath [36]

3 Methods

3.1 Day Before the Experiment

1. Prepare Collagenase/Dispase solution: Remove 200 mL of Ham's F-10 from a fresh sterile bottle and set aside for later use. Add Collagenase B (2.5 g) and Dispase II (1 g) powders into the bottle with 300 mL Ham's F-10, making sure to collect all the powder. Thoroughly dissolve the powders by shaking the bottle. Filter through 0.2 μm syringe filters using a 50 mL lock syringe, aliquot, and store at −20°C.
2. For mice: Prepare the MACS buffer by combining 0.5% BSA and 2 mM EDTA in PBS at pH 7.2. Prepare 50 mL of the MACS buffer for each mouse to be processed, ensuring the solution is filtered before use to avoid contamination. Store the MACS buffer at 4°C.
3. For mice: Prepare the Type I collagen coating solution by diluting a 3 mg/mL sterile stock solution in 0.1 M acetic acid to reach a final concentration of 0.1 mg/mL. Store the solution at 4°C.
4. Coat the satellite cell culture plates by adding 5 mL of the Type I collagen coating solution to each dish.
5. Incubate the plates for 15 min at room temperature to allow for proper surface attachment.
6. Retrieve the collagen solution from the dishes and store it in a sterile tube, as this solution may be reused up to five times.
7. Allow the coated dishes to dry for 15 min under a laminar flow hood before use.

8. For rats: Prepare FACS buffer, 2% of FBS in PBS 1×. Filter through a 0.22 μm syringe filter. Prepare approximately one 50 mL tube for every two rats. Store the solution at 4°C.

3.2 On the Day of the Experiment

1. Heat 5 mL of Collagenase B/Dispase II per mouse or rats in a water bath at 37°C for 20 min.
2. For rats: Prewarm FACS buffer and HBSS in a heating bath at 37°C for at least 20 min. Set the centrifuge to 15°C.
3. For rats: Coat 12-well plates with Matrigel (500 μL in 5 mL DMEM low glucose with 1% Penicillin–Streptomycin) and incubate at 37°C (*see* **Note 1**).

3.3 For Mice and Rats: Sample Preparation

1. Anesthetize the mice and rats using pentobarbital or isoflurane according to institutional regulations and approved ethics protocols.
2. Perform cervical dislocation for mice or decapitation for rats when the animals are unconscious.
3. Remove the skin of the hindlimb with scissors to expose the muscles.
4. Transfer the muscles into GentleMACS™ C tubes containing 5 mL of prewarmed mix of Collagenase/Dispase. For rats, place the dissected muscles from one leg into a single GentleMACS™ C tube. For mice, muscles from both legs may be combined and processed in one GentleMACS™ C tube.
5. Mince the tissue with scissors; no large pieces should remain.
6. Place the tubes upside down on the GentleMACS™ dissociator, attach heating sleeves, and run the appropriate skeletal muscle dissociation program (approximately 25 min) (*see* **Note 2**).

    a. Temperature on
    b. Spin 60 rpm, 3 s
    c. Spin −30 rpm, 9 s
    d. Loop ×6
    e. Spin 360 rpm, 5 s
    f. Spin −360 rpm, 5 s
    g. End loop
    h. Spin −30 rpm 12 s
    i. Temperature off
    j. End

The following steps are to be carried out in a sterile tissue culture hood.

3.4 For Mice (MACS): Cell Suspension

1. After 25 min of digestion/grinding, add 10 mL of MACS buffer to the tubes (rinse well what remains on the cap), perform some up and down using a 5 mL pipette, and transfer each muscle digestion in its own 50 mL conical tube through a 100 μm cell stainer (*see* **Note 3**).
2. Centrifuge for 10 min at 300 × *g*.
3. Discard the supernatant and add 5 mL of lysis buffer, prepared by diluting the 10X RBC lysis buffer stock solution 1:10 in ultrapure water, then incubate for 30 s to remove red blood cells and dead cells.
4. Add 10 mL of MACS buffer to the sample to wash and neutralize the Lysis Buffer.
5. Centrifuge the suspension for 10 min at 300 × *g*.

3.5 MuSCs Isolation Using MACS

1. Discard the supernatant.
2. Resuspend the pellet with 250 μL of MACS buffer.
3. Add 50 μL of Satellite Cell Isolation Kit. Resuspend by pipetting gently up and down and incubate for 15 min on ice.
4. Add 500 μL of MACS buffer to each sample.
5. Place the LS column in the magnetic field of a MACS Separator.
6. Activate the LS column by adding 3 mL of MACS buffer.
7. Add the sample to the column. Collect the continuous flow containing unlabeled cells, representing enriched satellite cells in a Falcon 15 mL.
8. Wash the column with 2 × 1 mL of MACS buffer.
9. Centrifuge 10 min at 300 × *g*.
10. Keep the pellet and resuspend in 1 mL of growth medium (*see* **Note 4**).

3.6 For Rats (FACS): Cell Suspension

1. Add 10 mL of prewarmed HBSS to stop the digestion.
2. Filter the suspension through a 40 μm cell strainer into a 50 mL tube (more than one strainer will be necessary for each rat). Wash the filter with 2 mL of HBSS.
3. Centrifuge at 600 × *g* for 5 min at 15°C.
4. Recover the supernatant and place in a second 50 mL tube, repeat centrifugation to collect additional cells (600 × *g*, 5 min). During this step, keep the first tube on ice.
5. Discard the majority of supernatant of the second tube. Note that the pellet may be unstable at this stage.
6. Resuspend the pellet and merge cells from both tubes in 5 mL of RBC lysis buffer. Incubate on ice for 5 min.

7. Add 10 mL HBSS, centrifuge at 600 × *g* for 5 min, and discard the supernatant.

3.7 Antibody Staining

1. Resuspend in 1 mL of FACS buffer and split cells in different falcon tubes for controls and compensation tubes by adding 35 μL for each tube [unstained, 7-AAD, Fluorescence Minus one (FMO) (*see* **Note 5**), single-stained (*see* **Note 6**)].
2. Prepare an unstained control tube containing 35 μL of cells to establish the baseline level of autofluorescence.
3. Prepare a dead cell control for viability staining by heating 35 μL of cells at 95°C for 5 min, followed by immediate cooling on ice.
4. Prepare single-stained compensation tubes by combining 7 μ L of FITC-α7 integrin with one drop of compensation beads.
5. Combine 5 μL each of PE-Cy7-CD31 and PE-Cy7-CD45 with one drop of beads in a separate tube.
6. Prepare the viability control by adding 5 μL of 7-AAD to a mixture of 35 μL live cells and the 35 μL of prepared dead cells.
7. Perform an additional wash with HBSS, centrifuge at 600 × *g* for 5 min, and discard the supernatant.
8. Resuspend the pellet in 100 μL FACS buffer before adding antibodies.
9. Prepare each sample for staining by adding 7 μL of FITC-α7 integrin, 5 μL of PE-Cy7-CD31, 5 μL of PE-Cy7-CD45, and 5 μL of 7-AAD.
10. Mix gently by pipetting up and down.
11. Incubate 20 min on ice, protected from light.

3.8 Washing and Preparation for Sorting

1. Wash cells with 10 mL FACS buffer.
2. Centrifuge at 500 × *g* for 5 min, discard the supernatant, and resuspend in 300 μL of FACS buffer for samples, and 200 μL for controls.
3. Filter cells through a 30 μm filter into 15 mL tubes. Wash the filter with 100 μL FACS buffer.
4. For recovery, precoat collection tubes with 500 μL FBS-containing medium.

3.9 Cell Sorting

1. Sort cells using a BD FACSAria™ III or equivalent, ensuring settings are adjusted according to the results of the four control tubes.
2. Select the cell population using an SSC-A/FSC-A gating strategy.

3. Remove doublets by applying an FSC-W/FSC-H gate to ensure single-cell analysis.
4. Select live cells by gating for FSC-H versus 7-AAD-A negative populations.
5. Isolate the α7-integrin-positive and CD31/CD45 negative cells using a FITC-A / PE-Cy7-A gating strategy.
6. Collect the sorted cells into coated tubes and maintain on ice until plating.

3.10 For Mice and Rats: Cell Culture

1. Put the cells in 100 mm dishes (Collagen-coated for mice or Matrigel-coated for rats) containing 10 mL of heated growth medium (mouse or rat standard proliferation medium).
2. Incubate under standard culture conditions (37°C, 5% CO_2) and change the media the day after.
3. Let the cells grow during a few days by changing the medium every 2 days with appropriate growth media.
4. Passage the cells once they reach approximately 70% confluence by first removing the growth medium and washing the cell layer once with 1× PBS (*see* **Note 7**).
5. Add 2 mL of 1× Trypsin and incubate the plate at 37°C for 3 min to facilitate cell detachment.
6. Confirm that more than 90% of the cells have detached from the plate surface by visual inspection under a microscope.
7. Inactivate the trypsin by gently adding 3 mL of growth medium and use the liquid flow to ensure all cells are fully detached.
8. Collect the cell suspension, transfer it into a 15 mL conical tube, and centrifuge at 400 × *g* for 5 min at 4°C.
9. Aspirate the supernatant carefully and resuspend the resulting cell pellet in 1 mL of fresh growth medium.
10. Plate the cells at the appropriate density into 24-well plates for the desired stage of analysis (*see* **Note 8**).
11. Seed cells at 20,000 cells/cm^2 in Growth Medium on a coated plate using at least four wells per biological replicate for analysis of early-stage myogenesis (proliferation).
12. Fix the proliferation plates 24 h after plating and proceed to immunostaining.
13. Seed cells at 40,000 cells/cm^2 on a Matrigel-coated plate using at least two wells per biological replicate for analysis of later-stage myogenesis (myotube fusion) (*see* **Note 9**).
14. Allow the cells to attach for at least 24 h until 80% confluence is reached before starting the differentiation of the myoblasts.

15. Remove the growth media and wash the cells once with 1× PBS.
16. Add 1 mL of Differentiation Medium per well and incubate the cells for 24, 48, 72, or 96 h depending on the specific stage of differentiation targeted for analysis (*see* **Note 10**).

3.11 Immuno-staining (Mice and Rats)

1. Remove cell culture media and wash the plates once with 1× PBS.
2. Remove PBS and add 1 mL of 2% PFA; incubate for 10 min at 21°C to fix the cells and preserve protein structures.
3. Wash the cells three times with 1 mL of 1× PBS for 5 min each to remove residual PFA.
4. Add 1 mL of permeabilization buffer and incubate for 10 min at 21°C to allow antibodies access to intracellular targets.
5. Wash three times with 1 mL of 1× PBS for 5 min each.
6. Add 1 mL of blocking buffer and incubate for 1 h at 21°C to reduce nonspecific antibody binding.
7. Remove the blocking buffer and apply the primary antibody solution for proliferating cells, consisting of Anti-PAX7 antibody (supernatant) diluted 1:2, and Anti-Ki67 antibody diluted 1:500, or Anti-MYOD antibody diluted 1:500 in blocking buffer (*see* **Note 11**).
8. For differentiating cells, apply a primary antibody solution consisting of Anti-MYOG antibody diluted 1:500 and Anti-MyHC (MF20) antibody (supernatant) diluted 1:20 in blocking buffer.
9. Incubate overnight (12–16 h) at 4°C.
10. Remove the antibody-containing solution and wash three times with 1 mL of 1× PBS for 5 min each.
11. Add 1 mL of the appropriate secondary antibody solution and incubate for 1 h at room temperature in the dark: for MYOD, use donkey anti-mouse Alexa Fluor 488; for PAX7-Ki67, use donkey anti-mouse Alexa Fluor 488 and donkey anti-rat Alexa Fluor 555; and for MYOG-MyHC (MF20), use donkey anti-rabbit Alexa Fluor 488 and donkey anti-mouse IgG2b Alexa Fluor 555, all diluted 1:1000 in blocking buffer.
12. Wash three times with 1 mL of 1× PBS for 5 min each.
13. Add 1 mL of Hoechst 33342 (0.1 mg/mL in PBS 1×) or DAPI (1 μg/mL in PBS 1×) to stain nuclei; incubate for 5 min at room temperature in the dark.
14. Wash three times with 1 mL of 1× PBS for 5 min each.
15. Add 1 mL of 1× PBS for imaging.

3.12 Imaging Parameters

1. Detect the staining using an inverted widefield epifluorescence microscope by using the blue channel for Hoechst 33342 or DAPI (excitation ~355 nm, emission 461 nm), the green channel (488 nm) for Alexa Fluor 488-labeled antibodies such as PAX7, MYOD, or MYOG, and the red channel (555 nm) for Alexa Fluor 555-labeled antibodies such as Ki67 or MF20.
2. Setup focus and acquire approximately five images per condition using a 10× objective to ensure representative sampling for quantitative analysis (Fig. 1). Taking ten images at 20× objective will also yield an accurate analysis.

3.13 Automation: Image Prepping for Automatic Myogenesis Analysis

1. Ascertain whether images taken from the microscope are recognized as an RGB image or a multi-channeled image on QuPath. The "Brightness and contrast" window will have multiple channels, including the channels Red, Green, and Blue, if QuPath recognizes the image RGB image. The window will only have the channels saved from the microscope if the image is instead recognized as multi-channeled image. Different microscopes save images in different formats (Fig. 2A).
2. Open a representative image in QuPath and access the "Brightness/Contrast" tool to determine if the software recognizes the file as an RGB image or a multi-channel image.
3. If the images are saved as multi-channeled images, skip ahead to subsection **Manual analysis of myogenesis**.
4. Convert RGB images into multi-channel images for myogenesis analysis by opening ImageJ/Fiji and dragging all relevant single-channel images into the application window (*see* **Note 12**).

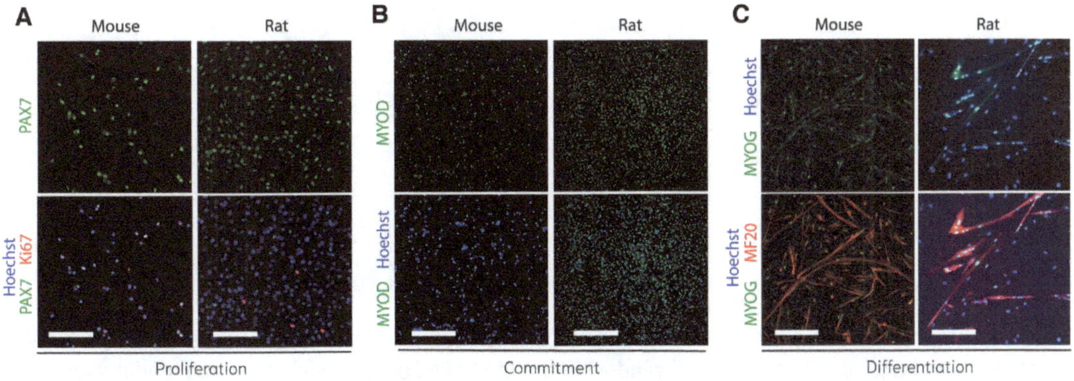

Fig. 1 **A,** Immunostaining against PAX7 (green) with or without Ki67 (red) and Hoechst (blue) to observe MuSC proliferation in mouse and rat. Scale Bar; 100 μm on the right and left. **B,** Immunostaining against MYOD (green) with or without Hoechst (blue) showing MuSC commitment in mouse and rat. Scale Bar; 150 μm on the right and 200 μm on the left. **C,** Immunostaining against MYOG (green), with or without MyHC/MF20 (red) and Hoechst (blue) to observe MuSC differentiation and fusion. Scale Bar; 100 μm on the right and the left

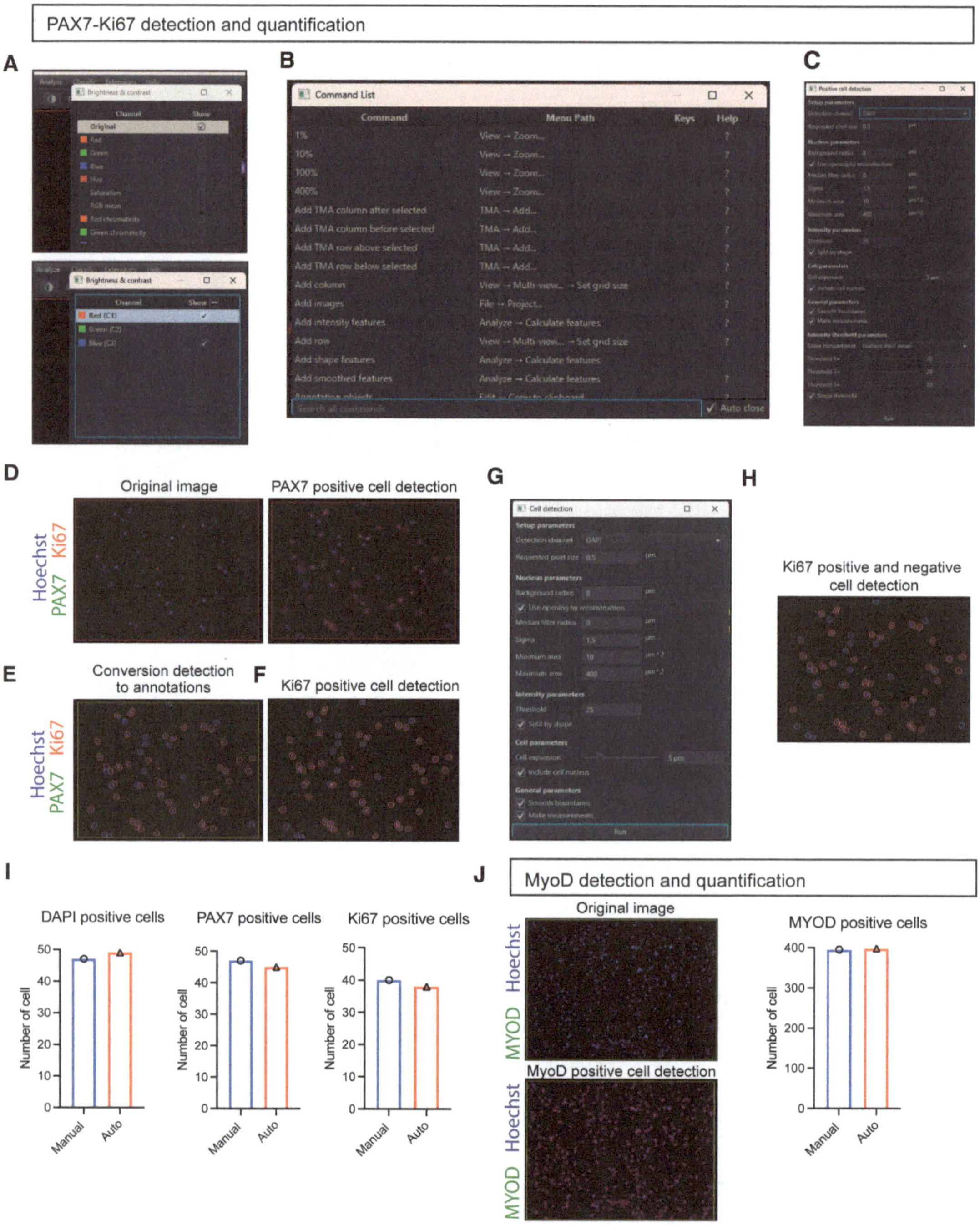

Fig. 2 **A,** "Brightness and contrast" window in Qupath for both RGB image (on top) and multi-channeled image (at the bottom). **B,** "Command list" window in QuPath to find commands' menu path and hotkeys or "to access" commands' window. **C,** "Positive cell detection" window in QuPath to run the "Positive cell detection" command. The parameters shown are used for PAX7 detection subsection in **PAX7-Ki67 costaining analysis**. **D,** Original image (left) and image showing results from PAX7 "Positive cell detection" command with the showcased parameters in PAX7-Ki67 costained cells (right). **E,** Results after running the "Detection-annotation conversion script" in the GitHub link after PAX7 positive cell detection in PAX7-Ki67 costained cells. **F,** Results

5. Access the "Merge Channels" tool under the "Image" tab and select the "Color" option to begin the alignment.
6. Assign each single-channel image to its corresponding color channel according to the fluorophores used.
7. Enable the "Create composite" option to ensure the channels remain independent for later analysis.
8. Save the resulting composite image in a compatible format for use in QuPath or subsequent processing.

3.14 Manual Analysis of Myogenesis: PAX7-Ki67 Analysis

1. Open QuPath software, then dragan empty folder inside QuPath window to create a project. The shortcut Ctrl + L opens a command list that provides quick access to any command whose name you know (Fig. 2B).
2. Drag and drop all composites/multi-channeled images of PAX7-Ki67 costained cells into the newly created project.
3. Select the option "Fluorescence" on the "Set image type" setting once the "Import images to project" window pops up.
4. Open the "Brightness and contrast" window and rename the channels by double-clicking on each channel. Make sure all images have the same channel names. A proposed name would be the antibody used for staining. In this case, it would be PAX7, Ki67, and Hoechst (or DAPI), for the green, red, and blue channels, respectively.
5. Set pixel size of images by accessing the "Image" tab, then double-clicking "Pixel width" and "Pixel height" (*see* **Note 13**).
6. Add a rectangular annotation that surrounds the whole image.
7. Access the "Positive cell detection" window under the "Analyze," then selecting "Cell detection" (Fig. 2C).
8. Set "Detection channel" to Hoechst (or DAPI) channel, then select the option "Nucleus: PAX7 channel mean" within the offered choices under the "Score compartment" parameter.

Fig. 2 (continued) after running "Positive cell detection" command for Ki67 marker in the "Positive" annotation in PAX7-Ki67 co-stained cells. **G,** "Cell detection" window in QuPath to run the "Cell detection" command. The parameters shown are used to demonstrate cell detection in "Negative" annotation in subsection **PAX7-Ki67 costaining analysis** (**H**) Results after running "Cell detection" command in "Negative" annotation in PAX7-Ki67 costained cells with the showcased parameters. **I,** Barplots comparing manual and automatic counting for total cell detection (Hoechst), PAX7 positive cells, and PAX7-Ki67 positive cells (**J**) The image on the right shows results after running "Positive cell detection" for MYOD in MYOD-stained cells. Barplot comparing manual and automatic counting of MYOD-positive cells

9. Adjust the parameters in the "Positive cell detection" window—including minimum/maximum area and intensity thresholds—until the command accurately identifies the target positive cells (Fig. 2D).
10. Preview the detection results under the "Hierarchy" tab to verify that the segmentation and classification align with the visual data.
11. Finalize these parameters on the initial image, as QuPath will automatically save them to the "Workflow" for standardized application across the remaining images in the dataset.
12. Convert detection from PAX7 positive cell detection into annotations (Fig. 2E). QuPath doesn't have the features to directly accomplish Step 12. The script "Detection-Annotation converter.groovy", which can be found in the GitHub link presented in Sect. 3.17, will be needed to perform those steps. Drag and drop the script in the "Script editor" under the "Automate" tab.
13. Delete the rectangular annotation by left-clicking it under the "Annotations" tab, then clicking on "Delete." All annotations are listed in alphabetical order. Therefore, all annotations classified "Negative" will appear in the annotations classified "Positive."
14. Select all "Negative" annotations by left-clicking the first "Negative" annotation on the list, then Shift + Left-clicking on the last "Negative" annotation on the list. All "Negative" annotations should be highlighted.
15. Merge all "Negative" annotations by right-clicking the highlighted annotations, moving the cursor over "Edit multiple," then selecting "Merge selected."
16. Repeat Steps 14 and 15 to merge all "Positive" annotations.
17. Select the merged "Positive" annotation, then access the "Positive cell detection" window again.
18. Set "Detection channel" to Hoechst (or DAPI) channel and "Score compartment" to Nucleus: Ki67 channel mean, then run the task.
19. Adjust the parameters (i.e., minimum and maximum area, and intensity threshold) until the command aptly recognizes positive cells (Fig. 2F). The Ki67 positive cell detection command's parameters are saved on the Workflow (see below in Sect. 3.17), making it possible to reuse the same parameters for another positive cell detection.
20. Select the merged "Negative" annotation, then access the "Cell detection" window under the "Analyze," then selecting "Cell detection" (Fig. 2G)

21. Set "Detection channel" to Hoechst (or DAPI) channel, there is no "Score compartment" parameter for the "Cell detection" command.
22. Adjust the parameters (i.e., minimum and maximum area, and intensity threshold) until the command aptly recognizes cells (Fig. 2H). Detection results can be previewed under the "Hierarchy" tab. When all "Negative" or "Positive" annotations are merged, QuPath no longer recognizes each individual cell. While total PAX7-positive cells can be quantified from the "Positive cell detection" command, PAX7-negative cells need to be quantified with the "Cell detection" command. The parameters of the cell detection will be saved on the Workflow (see below in Sect. 3.17), making it possible to reuse the same parameters for another cell detection.
23. Save the measurements file (*see* **Note 14**).
24. Export the results for PAX7-Ki67 cell counting through the "Export measurements" window under the "Measure" tab.
25. Choose which images to export measurements from, then select "Annotations" as the "Export type." Click on the "Output file" parameter to set a place to save the measurements and name the measurements file.
26. Export the data file.
27. Open an empty Excel file, then drag and drop the.csv files containing the measurements for PAX7-Ki67 costaining cell counting to perform quantitative analysis (Fig. 2I) (*see* **Note 15**).

3.15 Manual Analysis of Myogenesis: MYOD Analysis

1. Repeat Steps 1–9 from the subsection **PAX7-Ki67 analysis** using composites/multi-channeled images of MYOD-stained images (Fig. 2J). Set "Detection channel" to "Hoechst (DAPI)" channel, then select the option "Nucleus: MYOD channel mean" within the offered choices under the "Score compartment" parameter (*see* **Note 16**).
2. Repeat Steps 22-26 from the subsection **PAX7-Ki67 analysis** to export measurements.
3. Open an empty Excel file, then drag and drop the.csv file containing the measurements for MYOD staining cell counting to perform quantitative analysis.

3.16 Manual Analysis of Myogenesis: MF20-MYOG Analysis

1. Open QuPath software, then drag a new empty folder inside to create a project.
2. Drag and drop a small number (Five images recommended) of composites/multi-channeled images of MF20-MYOG costained cells into the newly created training project (*see* **Note 17**).

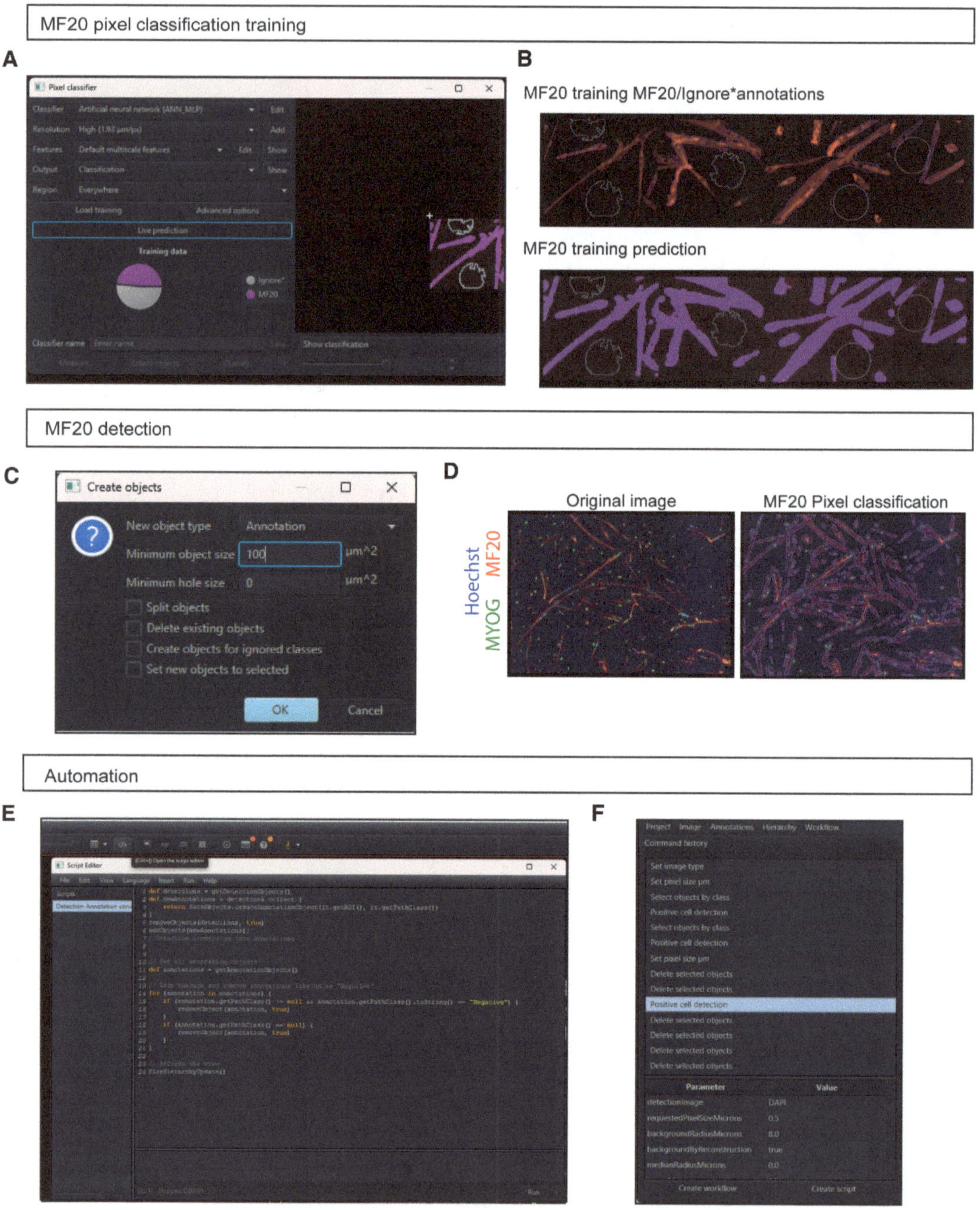

Fig. 3 A, "Pixel classifier" window in QuPath to run the "Train pixel classifier" command with an ongoing live prediction of the MF20 pixel classifier training. **B,** The image on top shows the training image with "MF20" annotations (purple) and "Ignore*" annotations (gray), while the bottom image shows the results of the live prediction of the MF20 pixel classifier training in the training image of MF20-MYOG costained cells. **C,** "Create object" window in QuPath to run the "Load pixel classifier" command. The parameters shown are used to demonstrate MF20 pixel classifier loading in subsection **MF20-MYOG costaining analysis**. **D,** Original image (left) and image showing results from loading the MF20 pixel classifier "Positive cell detection" with the showcased parameters in MF20-MYOG costained cells (right). **E,** "Script Editor" window in QuPath. Scripts

3. Repeat Steps 3–5 from the subsection **PAX7-Ki67 analysis**.
4. Under the "Annotations" tab, select the default "Region*" classification, then check the "Auto set" box.
5. Within all of the imported images, add a rectangular annotation on sections representative to myotube presence (*see* **Note 18**).
6. Create a training image by going under the "Classify" tab, then selecting "Training images," and finally "Create training image."
7. Under the "Classification" parameter, select the one corresponding to the rectangular annotations newly created.
8. Keep default settings for the rest of the parameters, then click "OK."
   a. A new image will be created by putting side by side all annotated sections.
9. Set the training image type to "Fluorescence."
   a. Channel names and pixel size should be transferred over from the original pictures.
10. Open the "Pixel classifier" window by going under the "Classify" tab, then selecting "Pixel classification," and finally "Train pixel classifier" (Fig. 3A).
11. Under the "Resolution" parameter, select "Very high (downsample = 0.62 μm/px)." Alternatively, "High (downsample = 1.23 μm/px)" can be used. The distance/pixel ratio changes depending on the pixel size. If you see "Very high (downsample = 2.00)" or "High (downsample = 4.00)", that means the pixel size hasn't been set.
12. Navigate to the "Features" parameter and click "Edit" to open the "Select features" window.
13. Select "MF20" as the sole channel for analysis to ensure the classifier focuses exclusively on the myotube staining.
14. Set the scales to "1.0" and, if holes are visible within the highlighted MF20-stained myotube during the training preview, add a "4.0" scale to fill those gaps.
15. Enable the "Gaussian", "Laplacian of Gaussian", and "Weighted deviation" features to optimize the edge detection and texture recognition of the myotubes.

Fig. 3 (continued) can be written directly in this window, or .groovy files can be dropped in this window to load pre-written scripts. **F**, Display of the "Workflow" tab in QuPath. A list of all commands ran in a selected image is shown under the "Command history" tab. When selecting a command, its parameters and script are shown below the "Command history" tab. Clicking on "Create script" will open a "Screen Editor" window with all of the commands ran in the selected image written as a script

16. Under the "Annotations" tab, create the classification "MF20".
17. Create "MF20" annotations on stained myotube regions and "Ignore*" annotations on unstained background regions using the "Auto set" option and the "Magic brush" tool.
18. Uncheck the Hoechst and MYOG channels in the "Brightness/Contrast" window before creating the training annotations to ensure these stains do not interfere with the MF20 classification. Exercise caution when using the "Magic brush" tool, as it is highly sensitive and may inadvertently select background areas if the intensity thresholds are not precisely defined.
19. Once a few annotations are made, click on "Live prediction" in the "Pixel classifier" window.
20. Complete the training until the pixel classifier aptly detects MF20-stained regions (Fig. 3B). The live prediction will show the results of the current training, showing what the training recognizes as MF20 staining and what it recognizes as background.
21. Adjust training by adding annotations or editing existing ones until live prediction can appropriately recognize myotubes. For best results, make sure the distribution of annotations MF20/Ignore* is balanced. Start by annotating the MF20 regions, then use the Ignore* annotations to refine the training. When training is done incorrectly, the training will annotate loads of lone pixels when it is trying to recognize MF20. It is better to delete all annotations and restart annotating MF20 and Ignore* regions when this happens.
22. Name the pixel classifier, then save it.
23. Optionally, before starting a new QuPath project for MF20/MYOG analysis, use the pixel classifier on the full images used for training to see how it looks. Steps 25–27 describe the process to apply a pixel classifier. If the training looks good, skip to Step 26.
24. Go back to the training image to edit the pixel classifier if issues are found in the training.
25. Repeat Steps 10–17 while keeping the original annotations if needed. Simply add more "MF20" and "Ignore*" annotations to the existing annotations.
26. Create a new QuPath project for MF20-MYOG costained cells analysis.
27. Repeat Steps 1–6 from the subsection **PAX7-Ki67 analysis** using composites/multi-channeled images of MF20-MYOG costained images.

28. Access the MF20 pixel classifier in the QuPath pixel classifier training project's file. In the "classifiers" folder, there is a "pixel_classifiers" folder.
29. Copy the "pixel_classifiers" folder, then paste it inside the "classifiers" file of the QuPath MF20-MYOG analysis project's file.
30. Apply pixel classifier by going under the "Classify" tab, then selecting "Pixel classification," then finally "Load pixel classifier."
31. Select the previously created pixel classifier, then select the "Everywhere" option under the "Region" parameter.
32. Click on the "Create objects" option and select "Full image" to open the corresponding configuration window (Fig. 3C).
33. Set the "New object type" parameter to "Annotation" to ensure the classified pixels are converted into manageable shapes.
34. Define a "Minimum object size" of 100 μm^2 to filter out small artifacts or nonspecific staining noise (*see* **Note 19**).
35. Check the "Delete existing objects" box if you are reapplying a new pixel classifier to ensure previous data do not overlap with the new analysis.
36. From this point, MF20-stained regions (myotubes) are detected by the pixel classifier (Fig. 3D). MF20-positive cells can be detected by running a "Cell detection" on "Hoechst" channel, where cells will be identified in both "MF20" annotation and the rectangular annotation. This will make it possible to count MF20-positive cells and calculate the fusion index.
37. To identify MF20-MYOG costained positive cells, run a "Positive cell detection" using "Hoechst" as the detection channel and "Nucleus: MYOG mean" under the "Score compartment" parameter as shown previously in PAX7-Ki67 and MYOD analyses. This will, similarly to "Cell detection," show MYOG-positive cells in both "MF20" and rectangular annotations.
38. To simply count MYOG-positive cells, follow Steps 1–9 in the subsection **PAX7-Ki67 analysis** using MYOG to run the "Positive cell detection" command.

3.17 Automatic Analysis of Myogenesis

1. The scripts "Automation PAX7-Ki67.groovy", "Automation MYOD.groovy", and "Automation MF20-MYOG.groovy" can be used to automate the processes shown above. These scripts can be found in the GitHub link following this paragraph. However, some adjustments are necessary to properly analyze the images, since channel threshold might change

depending on staining and microscope conditions. As mentioned in the subsection **MF20-MYOG analysis**, various analyses can be used after loading the MF20 pixel classifier. Therefore, the "Automation MF20-MYOG.groovy" is incomplete and needs a script of analysis to be complete. GitHub link: https://github.com/ZacharyGur/Automatic-myogenesis-analysis.git

2. Open the "Script Editor" window under the "Automate" tab (Fig. 3E).
3. Drag and drop the appropriate script for analysis.
4. Change Pixel size and Channel names within the script accordingly.
5. Access the "Workflow" tab and click on "Create script" (Fig. 3F).
6. Replace lines in analysis script containing QuPath commands (i.e., "Cell detection," "Positive cell detection," and "Load pixel classifier") with Workflow script containing the specific parameters of commands ran on personal project. The Workflow will contain the script of commands ran in the QuPath project. For example, when running the "Positive cell detection" command on Step 8 of **PAX7-Ki67 analysis**, a line of script will appear in the Workflow. Copying that line and pasting it in the analysis script will let the script perform the same "Positive cell detection."
7. Click on the three vertical dots on the bottom right of the "Script Editor" window, then "Run for projects" to open the "Select project images" window.
8. Choose which images need to be analyzed or click on the double arrows to select all the images, then run the script by clicking "OK."
9. Wait while the script goes through all the images.
10. When the batch processing ends, the whole project will be saved and exported to the location of choice.
11. Drag and drop the desired.csv file into an empty Excel file to access measurements to perform quantitative analysis.

4 Notes

1. To prepare Matrigel, put it on ice and place a box of sterile pipet tips and tubes to cool alongside. Verify lot number for protein concentration, dilute with cold DMEM to reach a concentration of 5 mg/mL using precooled pipet tips and

tubes. Store in 500 μL aliquots at −20°C. Do not refreeze any remaining diluted Matrigel once it has been thawed.

2. A GentleMACS-free alternative for obtaining viable MuSCs is to perform a mechanical and enzymatic digestion by cutting the dissected muscles in small pieces and incubating them at 37°C in a digestion solution containing collagenase II (0.2%–0.5%) during 1 h, and then perform a second digestion with collagenase II (0.2%–0.5%) and dispase II (2–4 U/mL) during 30 min. Pipet the samples up and down a few times every 15 min to facilitate digestion. The resulting cell suspension is then filtered sequentially through 70 and 40 μm strainers to remove debris, centrifuged, and resuspended in growth medium.
3. Add MACS buffer if the sample is too viscous to ensure proper filtration.
4. To increase the purity of MuSC isolation, a positive selection step can be performed after the negative selection by using α 7-integrin magnetic beads. These beads bind specifically to cells expressing α7-integrin on their surface (i.e., MuSCs), thereby removing the remaining non MuSC populations. However, this additional step may result in a lower overall yield.
5. FMOs are used to determine true signal and to account for spectral overlap by combining all fluorochromes used, except one.
6. Use beads for single stains as this accounts for independent fluorochrome signal. Use cells for 7-AAD to determine viability, and cells for the unstained control to detect autofluorescence and background signal.
7. Following MACS, in case of fibroblasts contamination, the myogenic fraction can be enriched by performing a first passage using a shorter incubation time with trypsin (2 min instead of 5 min) as myogenic cells will detach faster than fibroblasts. Alternatively, preplating can be performed by incubating the cells in a noncoated dish for 3 h to let the fibroblasts attach to the dish and then transfer the medium containing the floating myogenic cells in a collagen-coated dish.
8. Immunostainings are performed using two technical replicates (two wells) per biological sample. Since MYOD and PAX7 cannot be included in the same well due to their shared isotype (mouse IgG1), four wells per biological sample are required for the proliferation condition.
9. Because myotubes detach more easily than myoblasts due to changes in their adhesion properties during myogenesis,

Matrigel is used to provide a more supportive adhesive surface for the myotubes.

10. During differentiation, the culture contains mostly mononucleated myocytes at 24 h, small multinucleated myotubes at 48 h, and more mature myotubes by 72 h. Because mature myotubes have weaker adhesive properties, most of them will start to detach after 96 h.
11. As PAX7 and MYOD are both mouse IgG1 antibodies, using them in the same well would lead to cross-reactivity during detection. Therefore, they cannot be applied simultaneously in a single well.
12. When there are many composite images to create, the process of individually merging all single-channeled images of a picture is tedious. It is possible to semiautomate the merging process through ImageJ/Fiji with the macro "ImageJ-Fiji semiautomatic single-channeled images merger macro.ijm", which can be found in the GitHub link presented in the Sect. 3.17. To use macro (.ijm file), drag and drop it in the ImageJ/Fiji window to open the macro window.
13. Images taken at 10× from the Evos M5000 will have pixel width and height of 0.603 μm, while images taken at 20× from the same microscope will have a pixel size of 0.308 μm. The pixel size will change depending on the microscope and the objective used.
14. It's important to save your QuPath project regularly, since an analysis' results from an unsaved image won't be exported.
15. On the Excel file, the "Negative" annotation will show a value of 0 for both "Num Negative" and "Num Positive" columns since the "Cell detection" command does not output data pertinent to those columns. It is therefore safe to ignore those columns for "Negative" annotations.
16. MYOD analysis is very similar to the first part of PAX7-Ki67 analysis, since only the positive cell detection command is needed. The only difference being the channels used to run the "Positive cell detection" command.
17. For MF20-MYOG analysis, contrarily to the previous analyses, this first QuPath project will serve as a training project for the pixel classifier that will be used. It is recommended to keep a different folder for any pixel or object classifier training to make it easier to edit them in the future when needed.
18. The size of the annotation is not important, what is important is that the training is done on a smaller sample than the whole project. The training will be done faster if done on five smaller rectangular annotations than five full images.

19. The suggested "Minimum object size" is based on our experience, but both object size and hole size can be adjusted as needed for optimal results.

Funding

P.G. is supported by a fellowship from Muscular Dystrophy Canada and Canadian Institutes of Health Research (CIHR). N.A.D. is supported by a Canada Research Chair (CRC-2024-00025), as well as research grants from Natural Sciences and Engineering Research Council of Canada (NSERC; #2018-05979) and CIHR (#PJT-186202, PJT-186092 and PJT-186202).

Competing Interests The authors have no competing interests to declare.

References

1. Seale P, Sabourin LA, Girgis-Gabardo A et al (2000) Pax7 is required for the specification of myogenic satellite cells. Cell 102:777–786
2. Williams BA, Ordahl CP (1994) Pax-3 expression in segmental mesoderm marks early stages in myogenic cell specification. Development 120:785–796
3. Relaix F, Rocancourt D, Mansouri A et al (2005) A Pax3/Pax7-dependent population of skeletal muscle progenitor cells. Nature 435:948–953
4. Zammit PS (2017) Function of the myogenic regulatory factors Myf5, MyoD, myogenin and MRF4 in skeletal muscle, satellite cells and regenerative myogenesis. Semin Cell Dev Biol 72:19–32
5. Rudnicki MA, Schnegelsberg PN, Stead RH et al (1993) MyoD or Myf-5 is required for the formation of skeletal muscle. Cell 75:1351–1359
6. Kablar B, Krastel K, Ying C et al (1997) MyoD and Myf-5 differentially regulate the development of limb versus trunk skeletal muscle. Development 124:4729–4738
7. Chal J, Pourquié O (2017) Making muscle: skeletal myogenesis in vivo and in vitro. Development 144:2104–2122
8. Hutcheson DA, Zhao J, Merrell A et al (2009) Embryonic and fetal limb myogenic cells are derived from developmentally distinct progenitors and have different requirements for beta-catenin. Genes Dev 23:997–1013
9. Messina G, Cossu G (2009) The origin of embryonic and fetal myoblasts: a role of Pax3 and Pax7. Genes Dev 23:902–905
10. Narusawa M (1981) Change in number of skeletal muscle fibers during postnatal growth of the rat. Jpn J Phys Fit Sports Med 30:103–113
11. Rhodes SJ, Konieczny SF (1989) Identification of MRF4: a new member of the muscle regulatory factor gene family. Genes Dev 3:2050–2061
12. Hasty P, Bradley A, Morris JH et al (1993) Muscle deficiency and neonatal death in mice with a targeted mutation in the myogenin gene. Nature 364:501–506
13. Kassar-Duchossoy L, Gayraud-Morel B, Gomès D et al (2004) Mrf4 determines skeletal muscle identity in Myf5: myod double-mutant mice. Nature 431:466–471
14. Schiaffino S, Rossi AC, Smerdu V et al (2015) Developmental myosins: expression patterns and functional significance. Skelet Muscle 5: 22
15. White RB, Biérinx A-S, Gnocchi VF et al (2010) Dynamics of muscle fibre growth during postnatal mouse development. BMC Dev Biol 10:21
16. Börsch A, Ham DJ, Mittal N et al (2021) Molecular and phenotypic analysis of rodent models reveals conserved and species-specific modulators of human sarcopenia. Commun Biol 4:194
17. Portiér GL, Benders AG, Oosterhof A et al (1999) Differentiation markers of mouse C2C12 and rat L6 myogenic cell lines and the effect of the differentiation medium. In Vitro Cell Dev Biol Anim 35:219–227
18. Stern MM, Myers RL, Hammam N et al (2009) The influence of extracellular matrix derived from skeletal muscle tissue on the proliferation

and differentiation of myogenic progenitor cells ex vivo. Biomaterials 30:2393–2399
19. Zhang W, Liu Y, Zhang H (2021) Extracellular matrix: an important regulator of cell functions and skeletal muscle development. Cell Biosci 11:65
20. Palade J, Pal A, Rawls A et al (2019) Molecular analysis of muscle progenitor cells on extracellular matrix coatings and hydrogels. Acta Biomater 97:296–309
21. Shavlakadze T, Kyaduru S, Siggins A et al (2023) Age-related gene expression signatures from limb skeletal muscles and the diaphragm in mice and rats reveal common and species-specific changes. Skelet Muscle 13: 11
22. Deprez A, Molina T, Cagnone G et al (2025) Preterm birth conditions alter muscle stem cells and their niche, causing lasting impairments in muscle regeneration and function. J Cachexia Sarcopenia Muscle 16:e70058
23. Relaix F, Lafuste P, Taglietti V et al (2025) The rise of rat models for Duchenne muscular dystrophy and therapeutic evaluations. Skelet Muscle 15: 26
24. Beauchamp JR, Heslop L, Yu DS et al (2000) Expression of CD34 and Myf5 defines the majority of quiescent adult skeletal muscle satellite cells. J Cell Biol 151:1221–1234
25. Gerdes J, Lemke H, Baisch H et al (1984) Cell cycle analysis of a cell proliferation-associated human nuclear antigen defined by the monoclonal antibody Ki-67. J Immunol 133:1710–1715
26. Gerdes J, Schwab U, Lemke H et al (1983) Production of a mouse monoclonal antibody reactive with a human nuclear antigen associated with cell proliferation. Int J Cancer 31:13–20
27. Al-Khazraji BK, Medeiros PJ, Novielli NM et al (2011) An automated cell-counting algorithm for fluorescently-stained cells in migration assays. Biol Proced Online 13: 9
28. Rundo L, Tangherloni A, Tyson DR et al (2020) ACDC: automated cell detection and counting for time-lapse fluorescence microscopy. Appl Sci (Basel) 10:6187
29. Lau YS, Xu L, Gao Y et al (2018) Automated muscle histopathology analysis using CellProfiler. Skelet Muscle 8: 32
30. Brown A, Morris B, Kamau JK et al (2023) Automated image analysis pipeline development to monitor disease progression in muscular dystrophy using cell profiler. Arch Microbiol Immunol 7:178–187
31. Laghi V, Ricci V, De Santa F et al (2022) A user-friendly approach for routine histopathological and morphometric analysis of skeletal muscle using CellProfiler software. Diagnostics (Basel) 12:561
32. Desgeorges T, Lyot S, Lyon S et al (2019) Open-CSAM, a new tool for semi-automated analysis of myofiber cross-sectional area in regenerating adult skeletal muscle. Skelet Muscle 9: 2
33. Mayeuf-Louchart A, Hardy D, Thorel Q et al (2018) MuscleJ: a high-content analysis method to study skeletal muscle with a new Fiji tool. Skelet Muscle 8: 25
34. Reyes-Fernandez PC, Periou B, Decrouy X et al (2019) Automated image-analysis method for the quantification of fiber morphometry and fiber type population in human skeletal muscle. Skelet Muscle 9: 15
35. Schindelin J, Arganda-Carreras I, Frise E et al (2012) Fiji: an open-source platform for biological-image analysis. Nat Methods 9:676–682
36. Bankhead P, Loughrey MB, Fernández JA et al (2017) QuPath: open source software for digital pathology image analysis. Sci Rep 7: 16878

Check for updates

Chapter 12

Directed Differentiation of Human Pluripotent Stem Cells into Skeletal Muscle Progenitor Cells

Dieu-Huong Hoang, Nimisha Gupta, Sam Torres, Michael R Hicks and April D Pyle

Abstract

Human pluripotent stem cells (hPSCs) hold the potential to differentiate into skeletal muscle progenitor cells (SMPCs), which provide a human model for muscle disease modeling, drug screening, and a promising cell source for future stem cell-based muscle therapies. Directed differentiation is a transgene-free approach to guide hPSCs through the developmental path and differentiate them into embryonic-like SMPCs. However, the directed differentiation of functional SMPCs from hPSCs can be challenging due to cell line variations leading to low efficiency and reproducibility. In this section, we introduce a robust protocol that has been validated in multiple hPSC lines, providing cues on how to ensure efficient SMPC differentiation routinely. We describe the directed differentiation and isolation of SMPCs, using well-characterized markers ERBB3 and NGFR. Previous work from our lab has generated transcriptomic profiles of hPSC-derived SMPCs and functionally characterized the differentiated SMPCs in vitro and in vivo using different mouse models, and in this chapter, we will provide a detailed method for generating SMPCs used in these and additional studies.

Keywords Human pluripotent stem cells, Directed differentiation, PAX7, Skeletal muscle progenitor cells, Myotubes

1 Introduction

Since their discovery in the late 1990s, human pluripotent stem cell (hPSC) technology has rapidly grown and revolutionized human biological research [1]. HPSCs include embryonic stem cells (hESCs) and induced pluripotent stem cells (hiPSCs), which are somatic cells reprogrammed back into an embryonic-like state. HPSCs can differentiate into skeletal muscle progenitor cells (SMPCs), which recapitulate the main functions of primary muscle stem cells: differentiate and fuse to make myotubes in vitro, and generate myofibers in vivo when engrafted into skeletal muscle of immunocompromised mice. Disease modeling using hPSCs-derived SMPCs have become a valuable tool, overcoming the major limitation related to the restricted accessibility

Nathalie Didier and Caroline E. Brun (eds.), *Muscle Stem Cells*, Methods in Molecular Biology, vol. 3093,
https://doi.org/10.1007/978-1-0716-5643-3_12,

of primary cells. hPSC-derived SMPCs have been employed to study different aspects of muscle myopathies, from modelling myonuclear abnormalities in laminopathies [2], to mimicking muscle atrophy and impaired oxidative stress responses in facioscapulohumeral muscular dystrophy [3]. The ability of hPSC-derived SMPCs to recapitulate genetic disease hallmarks with high fidelity makes it a powerful platform for drug screening. Indeed, a drug screen using an hPSC model of dysferlinopathy was performed and identified Nocodazole, a compound with potential to increase dysferlin protein levels. Nocodazole improves membrane resealing function in dysferlinopathy, becoming a potential therapeutic solution for the disease [4]. Moreover, the high proliferative capacity of hPSCs is frequently associated with an abundant and renewable cell source, making them particularly attractive for regenerative medicine. Several studies from our lab and others have shown that hPSC-derived SMPC engraftment into immunocompromised DMD-null NSG mice, a model of Duchenne muscular dystrophy, can efficiently restore dystrophin expression in myofibers [5]. Engraftment of hPSC-derived SMPCs has also been shown to contribute to an improvement of muscle function, manifested by increasing contractility, and fatigue tolerance in DMD-null NSG mice [6, 7]. Although muscle cell therapy based on hPSC-derived SMPCs requires further optimization due to the persistent issue of cell maturation, they still hold significant promise for future therapeutic applications. Indeed, recent work has shown that in vivo maturation does occur over time [8–10].

Two main approaches have been developed to generate SMPCs from hPSCs. The first consists of using a transgene to transcriptionally reprogram the hPSCs, to overexpress the major regulators of myogenesis, such as Pax3 [11], Pax7 [7, 12], or MyoD [13–15]. These methods yield a high number of pure myogenic progenitors in a short amount of time. However, the direct reprogramming skips the developmental maturation steps, resulting in relatively immature SMPCs [16, 17]. In contrast, the directed differentiation approach aims to guide hPSCs through the developmental path by adding signaling molecules in the culture medium at specific time points throughout the directed differentiation timeline. This approach mirrors human development in the dish, resulting in SMPCs that are transcriptionally aligned with human developing weeks 7–12 SMPCs [18, 19]. In the embryo, myogenic progenitor cells arise from the paraxial mesoderm. Several groups developed directed differentiation protocols by first activating WNT/β-catenin signaling to induce differentiation of hPSCs into paraxial mesoderm [18, 20–23]. Precise modulation of the WNT pathway is critical for the specification of mesodermal lineages, and consequently, formation of Pax7+ SMPCs by the end of the protocol. Paraxial mesoderm cells will next be differentiated into somitic cells with the modulation of BMP and TGFβ signaling, then into dermomyotome by modulating WNT/β

-catenin signaling. The final transition steps of the dermomyotome towards Pax7 + SMPCs were promoted by growth factors (HGF, IGF1, and/or bFGF) [19, 24].

Directed differentiation protocols can be challenging and face several difficulties, especially the reproducibility between different hPSC lines. Here, we describe a stepwise directed differentiation protocol that has been validated in greater than ten independent hPSC lines and robustly generates well-characterized SMPCs [25], which have demonstrated transplantation feasibility and functional potential in multiple mouse models [5, 21, 26, 27].

2 Materials

2.1 hPSC Line and hPSC Culture

1. hPSC H1
2. mTeSR Plus (STEMCELL Technologies, 100-0276)
3. Matrigel hESC-Qualified Matrix, LDEV-free (Corning, 354277)
4. DMEM/F12
5. Penicillin–Streptomycin
6. D-PBS
7. EDTA

2.2 Directed Differentiation of hPSC into SMPCs

1. TrypLE (Gibco, 12605010).
2. Seeding medium (Day 1): mTeSR Plus (STEMCELL Technologies, 100-0276), 0.5% penicillin–streptomycin, 2 μM Thiazovivin (STEMCELL Technologies, 72252).
3. Mesoderm induction medium (Days 2–3): E6 Media (Gibco, A1516401), 0.5% Penicillin–Streptomycin, 8 μM CHIR 99021 (Tocris, 4423).
4. Pax3 specification medium (Days 4–11): E6 media (Gibco, A1516401), 0.5% Penicillin–Streptomycin.
5. Pax3 expansion medium (Days 12–18): StemPro-34 SFM (base media and nutrient supplement) (Gibco, 10639011), 0.5% Penicillin–Streptomycin, 5 ng/mL bFGF, 1× GlutaMAX Supplement, 0.45 mM Monothioglycerol, 11 μg/mL Recombinant Human Transferrin.
6. Pax7 specification medium (Days 19–22): E6 media (Gibco, A1516401), 0.5% Penicillin–Streptomycin.
7. Pax7 expansion medium (Days 23–30): E6 media (Gibco, A1516401), 0.5% Penicillin–Streptomycin, 10 ng/mL IGF-1.
8. Myogenic differentiation medium (Days 31–34): DMEM/F12, 0.5% Penicillin–Streptomycin, 1.2% N2 supplement, 1%

Insuline Transferrin Selenium (ITS-G) (Gibco, 41400045), 10 ng/mL IGF-1.

9. Myogenic maturation medium (Days 35–42): DMEM/F12, 0.5% Penicillin–Streptomycin, 1.2% N2 supplement (Gibco, 17502048), 1% Insuline Transferrin Selenium (ITS-G), 10 ng/mL IGF-1, 3 μM SB-431542 (Tocris, 1614).

2.3 FACS Sorting and Expansion of ERBB3+ NGFR+ SMPCs

1. FACS Buffer: D-PBS, 0.5% Penicillin–Streptomycin, 0.4% bovine serum albumin, 2 mM EDTA.
2. Human TruStain FcX (Fc Receptor Blocking Solution) (Biolegend, 422302).
3. Antibodies: FITC anti-human CD57 Antibody (Biolegend, 359604), PE anti-human erbB3 Antibody (Biolegend, 324706), AlexaFluor 647 anti-human CD271 (BD Bioscience, 560326).
4. SMPCs Expansion medium: Myocult-SF Expansion Supplement (STEMCELL Technologies, 05982) diluted in DMEM low glucose, 20 ng/mL bFGF.

3 Methods

hPSCs are maintained using standard hPSC culture methods. When hPSC colonies are compact and uniform in size, a directed differentiation protocol can be carried out for 42 days. By the end of the directed differentiation, hPSC-derived SMPCs can be isolated using fluorescence-activated cell sorting (FACS).

3.1 Culture and Maintenance of hPSCs

hPSCs are maintained in a serum-free, feeder cell-free environment in Matrigel-coated six-well culture plate and cultured in mTeSR Plus medium. For routine maintenance, change cell medium with 2 mL of mTeSR Plus every day, and pass the cells with EDTA solution every 5–7 days, depending on the hPSC line. hPSCs are ready for passage when the majority of the colonies are compact and large with a dense center.

1. Coat a six-well plate with Matrigel hESC-Qualified Matrix following the product guidelines (*see* **Note 1**). Ensure even distribution of Matrigel solution and incubate at room temperature (15°C–25°C) for at least 1 h (*see* **Note 2**).
2. To pass the hPSCs, aspirate the medium from the well, and gently wash the cells with 2 mL of D-PBS.
3. Aspirate the D-PBS, add 1 mL of 0.5 mM EDTA, and ensure even distribution. Incubate the plate in a 37°C incubator for 4 min.

4. Aspirate the EDTA solution, flush the well with 3 mL of mTeSR Plus to detach hPSC colonies from the well, using 5 mL pipette.
5. To plate cells, aspirate the Matrigel solution from the newly coated plate. Add 2 mL of mTeSR Plus per well. From 3 mL of cell suspension, plate 30–100 μL of cells into each well, depending on the hPSC lines (*see* **Note 3**).

3.2 Directed Differentiation of hPSCs into SMPCs (Fig. 1)

3.2.1 Seeding of hPSCs: Day 1

1. hPSCs are ready to be plated for differentiation when colonies are compact and large. Ensure the hPSCs colonies are of good quality with no spontaneously differentiated colonies (*see* **Note 4**).
2. Change the hPSCs medium with Seeding media containing ROCK inhibitor for at least 1 h to reduce single-cell-induced apoptosis (*see* **Note 5**).
3. Aspirate the medium from the well, and gently wash the cells with 2 mL of D-PBS.

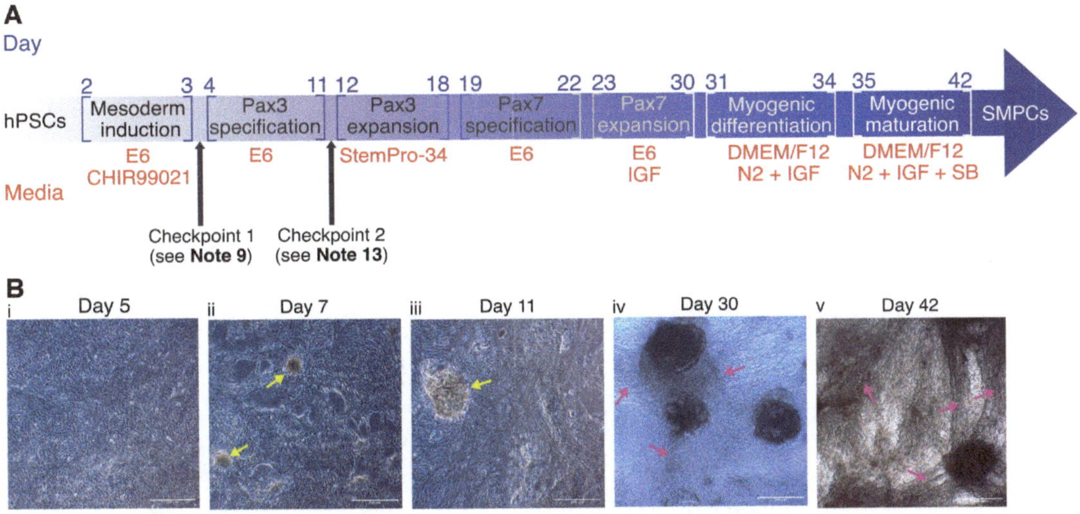

Fig. 1 Overview of directed differentiation protocol. **A,** Timeline of the directed differentiation protocol. The schematic diagram illustrates the key developmental stages, the main composition of the media, and checkpoints during the directed differentiation. **B,** Representative images of the morphology of the hPSC colony during key steps of the directed differentiation protocol. At Day 5 (**Bi**), cells begin to reach confluency and form 3D structures by Day 7 (**Bii**, yellow arrows). The 3D structures continue to grow and increase in complexity (**Biii**, yellow arrows). Myofibers start to form around Day 30 (**Biv**, magenta arrows). By the end of the protocol, myofibers mature, and contract (**Bv**, magenta arrows). Scale bar = 200 μm

4. Aspirate the D-PBS, add 0.5 mL of TrypLE per well of the six-well plate, and ensure even distribution. Incubate the plate in a 37°C incubator for 4 min.
5. Triturate using a P1000 μL pipette. Inactivate TrypLE with 1.5 mL of Seeding media (ratio 1:3 of TrypLE:Seeding media).
6. Collect the cell suspension in a 15 mL tube. Centrifuge the cell suspension at 300 G for 5 min.
7. Aspirate and discard the supernatant. Resuspend the cell pellet in 1 mL of Seeding medium and proceed to count the cells.
8. To plate 3×10^5 cells per well, prepare the cell solution with Seeding medium at a concentration of 1.5×10^5 cells/mL, and add 2 mL of cell solution per well in a Matrigel-coated six-well plate (*see* **Note 6**).
9. Allow seeded cells to recover in a 37°C incubator for 24 h.

3.2.2 Directed Differentiation

1. Paraxial mesoderm induction (Days 2–3): To induce paraxial mesodermal cell fate, treat hPSCs with 8 μM CHIR 99021 in E6 medium for 2 days. Change with 2 mL per well of Mesoderm induction medium, every 24 h. To ensure the precise concentration, dilute CHIR 99021 in E6 medium immediately before feeding (*see* **Notes 7–9**).
2. Pax3 cell specification (Days 4–11): Change with 2 mL per well of Pax3 specification medium, every 24 h (*see* **Notes 10–13**).
3. Pax3 cell expansion (Days 12–18): Thaw StemPro-34 Nutrient Supplement overnight on ice. Prepare Pax3 expansion medium and filter the medium using a vacuum filter system (pore size 0.22 μm). Completed Pax3 expansion media can be stored at 4°C for up to 2 weeks. Change with 2 mL per well of Pax3 expansion medium, every 24 h (*see* **Note 11**).
4. Pax7 cell specification (Days 19–22): Change with 2 mL per well of Pax7 specification medium, every 24 h (*see* **Note 11**).
5. Pax7 cell expansion (Days 23–30): Change with 2 mL per well of Pax7 expansion medium, every 24 h (*see* **Notes 11 and 14**).
6. Myogenic differentiation (Days 31–34): Change with 2 mL per well of Myogenic differentiation medium, every 24 h (*see* **Note 11**).
7. Myogenic maturation (Days 35–42): Change with 2 mL per well of Myogenic maturation medium, every 24 h (*see* **Notes 11 and 15**).

3.3 FACS Analysis and Sorting of ERBB3+ NGFR+ SMPCs (see Note 16 and Fig. 2)

3.3.1 Cell Dissociation

1. Prepare FACS Buffer and sterile filter using a vacuum filter system (pore size 0.22 μm).
2. Aspirate the medium from the well, and gently wash the cells with 2 mL of D-PBS (*see* **Note 17**).
3. Aspirate the D-PBS, add 1 mL of TrypLE per well of the six-well plate, and ensure even distribution. Incubate the plate in a 37°C incubator for 5–7 min.
4. Triturate using a P1000 μL pipette. Inactivate TrypLE with 3 mL of ice-cold FACS Buffer (ratio 1:3 of TrypLE:FACS Buffer) (*see* **Notes 18 and 19**).
5. Filter the cell suspension through a 70 μm, then a 40 μm cell strainer (*see* **Note 20**).
6. Centrifuge the tube containing cells at 500*g*, for 10 min, at 4°C.

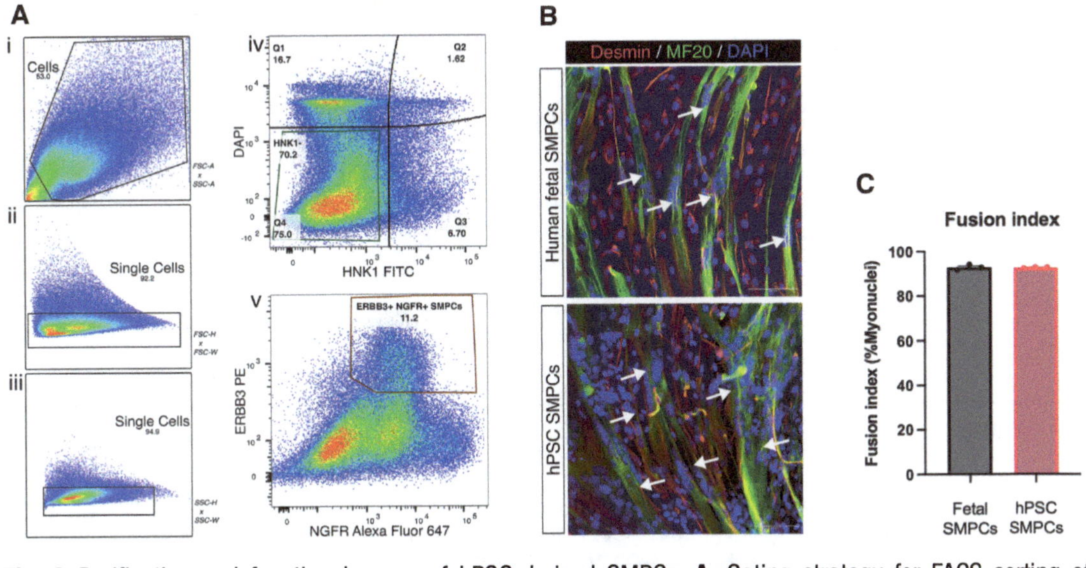

Fig. 2 Purification and functional assay of hPSC-derived SMPCs. **A,** Gating strategy for FACS sorting of hPSC-derived SMPCs after directed differentiation. Cells were dissociated using TrypLE and filtered through 70 μm and 40 μm filters to obtain mononucleated cells. Debris and doublets were excluded using FSC/SSC gates (**Ai, Aii, Aiii**). Mononucleated cells are labelled with HNK1, ERBB3, and NGFR. hPSC-derived SMPCs are HNK1$^-$ ERBB3$^+$ NGFR$^+$ (gates **Aiv**, then **Av**). **B,** Representative results of differentiation assay of human fetal SMPCs (CD31$^-$ CD45$^-$CD235a$^-$ CD56$^+$ CD82+, top), and hPSC-derived SMPCs (HNK1$^-$ ERBB3$^+$ NGFR$^+$, bottom). Similar to human fetal SMPCs, hPSC-derived SMPCs form multinucleated myotubes (white arrows) expressing myosin heavy chain proteins (MF20, green). Scale bar = 100 μm. **C,** Example of fusion index analysis showing comparable fusion potential of fetal and hPSC-derived SMPCs

7. Aspirate and discard the supernatant. Resuspend the cell pellet in 1 mL of FACS buffer and proceed to count the cells.

3.3.2 FACS Staining: Prepare Cells for FACS Sorting Using Standard Protocol

1. Prepare the cell suspension at a concentration of 1×10^7 cells/mL.
2. Block nonspecific binding of Fc receptors using Human TruStain FcX. Add 5 μL of Human TruStain FcX per 100 μL of staining volume. Mix and incubate at room temperature for 10 min. It is not necessary to wash cells between the blocking and staining steps.
3. Before adding the antibodies for labeling, aliquot cells for controls for FACS: single antibody controls and Fluorescence Minus One (FMO) controls (*see* **Notes 21 and 22**).
4. Stain the cell suspension with a cocktail of FITC anti-human CD57 antibody (1 μL/20 μL of staining volume), PE anti-human erbB3 antibody (1 μL/30 μL of staining volume), and AlexaFluor 647 anti-human CD271 antibody (1 μL/12.5 μL of staining volume) (*see* **Note 23**). Mix and incubate at 4°C for 20 min, protect from light.
5. Wash the antibodies using ten parts volume of FACS buffer. Centrifuge the tube containing cells at 500 *g*, for 10 min, at 4°C.
6. Resuspend the cell pellet at a concentration of 1×10^7 cells/mL. Transfer the cells into FACS tubes.
7. Immediately before sorting, add 1 μg/mL DAPI for viability staining.
8. hPSC-derived SMPCs are $HNK1^-$ $ERBB^+$ $NGFR^+$, and can be FACS sorted using a gating strategy as shown in Fig. 2.
9. $ERBB^+$ $NGFR^+$ SMPCs are maintained in culture in SMPCs expansion medium.

4 Notes

1. Use the lot number to check the dilution factor on Corning Quality Certificate Lookup (typically, the dilution factor is between 270 and 350 μL). Dilute one volume of the dilution factor of Matrigel to 25 mL of ice-cold DMEM/F12 to coat six-well plates (2 mL/well).
2. After 1 h at room temperature, the Matrigel-coated plate can be kept in the 37°C incubator for up to 14 days. Aspirate Matrigel solution only before use; do not leave the coated plate dry.

3. Avoid breaking the colonies into too small of clumps. Avoid excessive pipetting of the cell suspension. Use wide bore tip to keep relatively big clumps.
4. Quality of hPSCs is essential for directed differentiation. Cultures with abnormal and spontaneously differentiated colonies are challenging to differentiate into mesodermal cells. During routine maintenance of hPSCs, abnormal colonies can be manually scraped off using a P1000 μL pipette tip. To ensure efficient directed differentiation, pass hPSCs at least three passages, counting from thawing, before starting the differentiation protocol. Karyotyping should be performed regularly.
5. For this protocol, a commonly used ROCK inhibitor, Y-27632 (STEMCELL Technologies, 72308), functions equally well to reduce apoptosis. Y-27632 at a final concentration of 10 μM can be used in place of Thiazovivin.
6. Cell density is critical for successful directed differentiation. Different hPSC lines have different ideal seeding densities, which requires empirical optimization. When first starting the differentiation on an hPSC line, a range of 2×10^5–7×10^5 cells per six-well (equivalent to 2×10^4–7×10^4 cells/cm^2) can be used to determine optimal seeding density. For optimization purposes, 12-well and 24-well plates can be used to determine the ideal density of cells/cm^2. Strategies for optimization and the effect of cell density on the directed differentiation was reported in a previous study [27].
7. Different hPSC lines have different sensitivity to CHIR 99021 treatment. This step is critical for successful directed differentiation. Similar to cell density, it is strongly recommended to determine the optimal CHIR 99021 concentration for a new hPSC line, using a titration ranging from 5 to 15 μM. Strategies for optimization and the effect of CHIR 99021 concentration on the directed differentiation was reported in a previous study [27].
8. A small proportion of cell death is expected following CHIR 99021 treatment. However, after 2 days of CHIR 99021 treatment, at Day 4, cells are expected to reach >70% of confluence and have single cell morphology.
9. Paraxial mesoderm specification is critical for the formation of myogenic progenitors at the latter stage of this protocol. Following CHIR 99021 treatment, the transition into mesodermal cells can be verified as a control checkpoint of the protocol. At Day 4, expression of Brachyury (T), Mesogenin 1, and TBX6 can be assessed by qPCR or immunostaining and compared with the expression level at Day 0 [28], to confirm the successful paraxial mesoderm induction.

10. A high proportion of cell death is expected during Pax3 cell specification and does not interfere with the directed differentiation. Between Days 4 and 5, cells are expected to reach 100% confluence. Troubleshooting: Unable to obtain confluency at Day 5 of the differentiation is likely due to low seeding density or CHIR-induced cell death. Consider stopping the differentiation and starting over with a higher seeding density.
11. Rapidly growing cells during this step can lead to increased consumption of cell medium, resulting in low pH levels, indicated by yellow coloration of the medium. If the medium turns yellow rapidly, consider increasing the medium used for feeding up to 2.5–3 mL.
12. Starting from Day 7, cells will organize into 3D structures, which will support the differentiation into myogenic cells and the formation of myofibers at the latter stages of the protocol. The 3D structures start to form at Day 7 and continue to increase in size and complexity until Day 11, indicating that the cells are well specified and ready for the next steps of the protocol. Troubleshooting: Undetectable 3D structures at Day 10 are likely due to unsuccessful mesoderm specification during Days 2 and 3. Consider stopping the differentiation and starting over with a more optimal CHIR 99021 concentration.
13. Following Pax3 cell specification, the emergence of Pax3 cells can be verified as a control checkpoint of the protocol. At Day 12, expression of Pax3 and Six1 can be assessed by qPCR or immunostaining and compared with the expression level at Day 0 [27], to confirm the successful specification.
14. By Day 30, myotube formation can be observed. Troubleshooting: If myotube formation is delayed, consider prolonging the Pax7 expansion step to additional 1–4 days.
15. Spontaneous contractions of myofibers can be observed.
16. SMPC enrichment using FACS can be performed between Days 42 and 55, with similar efficiencies. After Day 42, cells can be maintained in myogenic maturation medium. However, spontaneous contractions can lead to detachment of the thick cell layer.
17. By the end of the directed differentiation, the cells form a thick heterogeneous cell layer. Be cautious not to aspirate the cell layer when removing medium. We recommend leaving out around 200 μL of liquid in the well to not disturb the cell layer, and increasing the number of D-PBS washes (two to three washes) to ensure efficiency of TrypLE subsequently.

18. After 42 days of differentiation, the thick cell layer formed complex ECM structures that are difficult to dissociate using TrypLE. We suggest complementing enzymatic dissociation with mechanical dissociation. Briefly, after 5 min of incubation in TrypLE, manually break the cell layer into small pieces using two P1000 pipette tips. After inactivation of TrypLE with FACS buffer, triturate the cell suspension using a series of 10 mL, then 5 mL pipettes, and a P1000 pipette. This will help to release the cells from the ECM.
19. Cell dissociation for FACS sorting affects cell viability. Usually, we observed 60%–70% viability after dissociation. To reduce cell death, minimize the time of TrypLE treatment, and avoid incubating the cells in TrypLE for over 10 min. After dissociation, keep the cells at 4°C throughout the preparation and FACS sorting to reduce cell death.
20. To reduce cell loss between filters, wash the tubes and filters multiple times using FACS buffer.
21. Single color controls can be stained with beads instead of cells (Invitrogen, 01222242).
22. For controls for FACS, we suggest using 20 μL of cell suspension (equivalent to 2×10^5 cells) per control.
23. The antibody dilutions described were titrated for our staining condition and instrument (BD FACSAria H). Antibody titration needs to be done in-house for different antibody clones or different FACS instruments.

Competing Interests

Dr. Pyle is co-founder of MyoGene Bio and Anew cell therapies.

References

1. Thomson JA, Itskovitz-Eldor J, Shapiro SS et al. (1998) Embryonic stem cell lines derived from human blastocysts. Science 282:1145–1147
2. Steele-Stallard HB, Pinton L, Sarcar S et al. (2018) Modeling skeletal muscle laminopathies using human induced pluripotent stem cells carrying pathogenic LMNA mutations. Front Physiol 9:1332
3. Caron L, Kher D, Lee KL et al. (2016) A human pluripotent stem cell model of facioscapulohumeral muscular dystrophy-affected skeletal muscles. Stem Cells Transl Med 5:1145–1161
4. Kokubu Y, Nagino T, Sasa K et al. (2019) Phenotypic drug screening for dysferlinopathy using patient-derived induced pluripotent stem cells. Stem Cells Transl Med 8:1017–1029
5. Hicks MR, Saleh KK, Clock B et al (2023) Regenerating human skeletal muscle forms an emerging niche in vivo to support PAX7 cells. Nat Cell Biol 25(12):1758–1773
6. Bourgeois Yoshioka CK, Takenaka-Ninagawa N, Goto M et al. (2024) Cell transplantation-mediated dystrophin supplementation efficacy in Duchenne muscular dystrophy mouse

motor function improvement demonstrated by enhanced skeletal muscle fatigue tolerance. Stem Cell Res Ther 15:313
7. Darabi R, Arpke RW, Irion S et al. (2012) Human ES- and iPS-derived myogenic progenitors restore Dystrophin and improve contractility upon transplantation in dystrophic mice. Cell Stem Cell 10:610–619
8. Crist SB, Azzag K, Kiley J et al (2024) The adult environment promotes the transcriptional maturation of human iPSC-derived muscle grafts. NPJ Regen Med 9:1–11
9. Abreu P, Garay BI, Nemkov T et al. (2024) Metabolic changes during in vivo maturation of psc-derived skeletal myogenic progenitors. Cells 13:76
10. Sun C, Kannan S, Choi IY et al. (2022) Human pluripotent stem cell-derived myogenic progenitors undergo maturation to quiescent satellite cells upon engraftment. Cell Stem Cell 29:610–619.e5
11. Darabi R, Santos FNC, Filareto A et al. (2011) Assessment of the myogenic stem cell compartment following transplantation of Pax3/ Pax7-induced embryonic stem cell-derived progenitors. Stem Cells 29:777–790
12. Kim J, Oliveira VKP, Yamamoto A, Perlingeiro RCR (2017) Generation of skeletal myogenic progenitors from human pluripotent stem cells using non-viral delivery of minicircle DNA. Stem Cell Res 23:87–94
13. Rao L, Tang W, Wei Y et al. (2012) Highly efficient derivation of skeletal myotubes from human embryonic stem cells. Stem Cell Rev and Rep 8:1109–1119
14. Albini S, Coutinho P, Malecova B et al. (2013) Epigenetic reprogramming of human embryonic stem cells into skeletal muscle cells and generation of contractile myospheres. Cell Rep 3:661–670
15. Goudenege S, Lebel C, Huot NB et al. (2012) Myoblasts derived from normal hESCs and dystrophic hiPSCs efficiently fuse with existing muscle fibers following transplantation. Mol Ther 20:2153–2167
16. Esteves de Lima J, Relaix F (2021) Master regulators of skeletal muscle lineage development and pluripotent stem cells differentiation. Cell Regen 10:31
17. Hamer MS, Rossi FMV (2025) Multitasking muscle: engineering iPSC-derived myogenic progenitors to do more. Front Cell Dev Biol 12:1526635
18. Xi H, Fujiwara W, Gonzalez K et al. (2017) In vivo human somitogenesis guides somite development from hPSCs. Cell Rep 18:1573–1585
19. Chien P, Xi H, Pyle AD (2022) Recapitulating human myogenesis *ex vivo* using human pluripotent stem cells. Exp Cell Res 411:112990
20. Shelton M, Metz J, Liu J et al. (2014) Derivation and expansion of PAX7-positive muscle progenitors from human and mouse embryonic stem cells. Stem Cell Rep 3:516–529
21. Hicks MR, Hiserodt J, Paras K et al. (2018) ERBB3 and NGFR mark a distinct skeletal muscle progenitor cell in human development and hPSCs. Nat Cell Biol 20:46–57
22. Chal J, Al Tanoury Z, Hestin M et al. (2016) Generation of human muscle fibers and satellite-like cells from human pluripotent stem cells in vitro. Nat Protoc 11:1833–1850
23. Bou Akar R, Lama C, Aubin D et al. (2024) Generation of highly pure pluripotent stem cell-derived myogenic progenitor cells and myotubes. Stem Cell Rep 19:84–99
24. Chal J, Oginuma M, Al Tanoury Z et al. (2015) Differentiation of pluripotent stem cells to muscle fiber to model Duchenne muscular dystrophy. Nat Biotechnol 33:962–969
25. Xi H, Langerman J, Sabri S et al. (2020) A human skeletal muscle atlas identifies the trajectories of stem and progenitor cells across development and from human pluripotent stem cells. Cell Stem Cell 27:158–176.e10
26. Stearns-Reider KM, Hicks MR, Hammond KG et al. (2023) Myoscaffolds reveal laminin scarring is detrimental for stem cell function while sarcospan induces compensatory fibrosis. NPJ Regen Med 8:16
27. Jaime OG, Arias J, Pavani S et al. (2023) SIX1+ PAX3+ identify a progenitor for myogenic lineage commitment from hPSCs. Development 150:dev201509
28. Shelton M, Kocharyan A, Liu J et al. (2016) Robust generation and expansion of skeletal muscle progenitors and myocytes from human pluripotent stem cells. Methods 101:73–84

Part IV

Metabolism and Mitochondrial Function

Chapter 13

Flow Cytometry Analysis of Mitochondrial Function in Muscle Stem Cells

Lucia Acampora, Melania Murolo, Shahragim Tajbakhsh and Barbara Gayraud-Morel

Abstract

Mitochondrial metabolism plays a central role in regulating muscle stem cell (MuSC) quiescence, activation, and differentiation, with dynamic metabolic transitions accompanying changes in cell fate and function. Flow cytometry–based analysis of mitochondrial function provides a rapid and accessible approach to characterize metabolic states at the single-cell level. Here, we present a detailed protocol to analyze mitochondrial activity, particularly by assessing mitochondrial membrane potential and mitochondrial mass in freshly isolated MuSCs and myoblasts using the fluorescent dyes TMRE and MitoTracker Deep Red in cytometry. This method is compatible with antibody-based identification of MuSCs and includes controls using mitochondrial uncouplers to validate $\Delta\Psi$m-dependent staining. This protocol enables reliable discrimination between quiescent, activated, and differentiating muscle cells and can be applied to study physiological and pathological conditions in different contexts, including injury, aging, and diseases.

Keywords Stem cells, Satellite cells, Skeletal muscle, Flow cytometry, Mitochondria, Metabolism, TMRE, MitoTracker

1 Introduction

Cell metabolism has become a central theme in biomedical research, as metabolic states reflect and shape cell functions and are frequently altered in diseases. This is particularly relevant in skeletal muscle, a tissue at the center of body metabolism regulation, with high energetic demand and remarkable regenerative capacity. In addition, metabolic activity and mitochondrial content contribute to the classification of slow and fast myofibres, thereby conferring distinct functional properties to skeletal muscle [1].

Adult muscle stem cells (MuSCs) are quiescent (G_0) under homeostatic conditions [2–4] and enter cell cycle upon stimulation by external cues to sustain tissue maintenance or to repair injured

The authors “Lucia Acampora and Melania Murolo” contributed equally.

Nathalie Didier and Caroline E. Brun (eds.), *Muscle Stem Cells*, Methods in Molecular Biology, vol. 3093,
https://doi.org/10.1007/978-1-0716-5643-3_13,

myofibers. Quiescence is an actively maintained state that ensures the persistence of a functional pool of stem cells and provides resilience against stress conditions [5–7]. It is defined by reversible cell–cycle exit and is characterized by a low metabolic activity, reduced protein translation, a small cell size and cytoplasmic content [8–11]. These parameters can be readily quantified by flow cytometry, a technique accessible to most laboratories, using freshly isolated MuSCs from resting or injured muscles, as well as cultured myoblasts.

Distinct metabolic pathways are associated with different muscle cell states. Fatty Acid β-oxidation (FAO) provides a major energy source during quiescence [10, 12], whereas glycolysis is strongly upregulated during activation and proliferation. Later, oxidative phosphorylation (OxPhos) becomes predominant upon differentiation [10, 12, 13]. In quiescent MuSCs, mitochondrial content and overall metabolic activity are low but increase upon activation and differentiation [9, 10, 14, 15]. However, during aging changes in mitochondrial dynamics lead to inefficient OxPhos and MuSC impairment [9].

At the center of cellular metabolism, mitochondria play a crucial role by producing most of the cellular ATP through OxPhos and lipid oxidation. Beyond energy production, mitochondria are involved in apoptosis, intracellular calcium regulation, and reactive oxygen species (ROS) production. Metabolic state transitions are accompanied by marked changes in mitochondrial activity, dynamics, and substrate usage, making mitochondrial readouts informative proxies of the functional status of cells. Mitochondrial function and cellular metabolic parameters can be quantified using various techniques, including Agilent Seahorse technology and Oroboros O2k, which provide mitochondrial real-time measurements of oxygen consumption rate (OCR) and extracellular acidification rate (ECAR) [9, 10, 16]. However, these approaches require specialized equipment and high numbers of cells or samples to generate robust results [17]. This requirement is particularly limiting for studies in which the amount of starting material is low, such as experiments involving stem cells of aged tissues.

Alternatively, mitochondrial function can be quickly evaluated using mitochondria-specific dyes, such as Tetramethylrhodamine ethyl ester (TMRE) and Tetramethylrhodamine methyl ester (TMRM) [18], that can be readily applied to live cells during flow cytometry or imaging. These dyes are hydrophobic cationic molecules that diffuse across the plasma and mitochondrial membranes and preferentially accumulate in negatively charged mitochondria, reporting mitochondrial membrane potential, and polarization. The more negative the mitochondrial membrane potential ($\Delta\Psi m$; often used as a proxy of mitochondrial energetic state), the greater the accumulation of the positively charged dye in the matrix. Mitochondrial membrane potential is generated during oxidative phosphorylation, in which electron transport through complexes I, III, and IV of the electron transport chain (ETC), reduces oxygen to water and pumps protons out from the mitochondrial matrix into the intermembrane space (IMS). This proton translocation establishes pH and electrochemical gradients across the inner mitochondrial membrane. At the end of the ETC, the ATP synthase (complex V) uses the $\Delta\Psi m$ to drive proton re-entry from the IMS to the matrix enabling ATP synthesis. $\Delta\Psi m$ is negative and typically ranges, for example, in neural cells, between -150 mV and -180 mV. Other dyes, such as MitoTrackers [19], differ in their chemical composition and may be sensitive or insensitive to mitochondrial membrane potential according to their molecular structure [20]. For example, MitoTracker Red and Red FM fail to stain depolarized mitochondria, whereas MitoTracker Green, Deep Red, and Deep Red FM are largely independent of membrane potential and instead covalently label mitochondrial proteins, making them suitable for assessing mitochondrial content (e.g., mitochondrial mass). Their properties reside in the ability of their chloromethyl groups to react with free thiol groups on cysteine residues of proteins located in the inner mitochondrial membrane. This feature provides the advantage of most MitoTracker dyes to be retained after fixation, in contrast to TMRE or MitoTracker Red, which have a low membrane binding affinity and are therefore incompatible with fixation.

A large range of dyes, each with distinct chemical properties and applications, allows the measurement of mitochondrial mass, membrane potential, reactive oxygen species (ROS) production, and apoptosis. Although the biochemical mechanisms underlying the function of these dyes have been thoroughly examined in previous reviews, only a limited number of studies have provided practical guidance for potential pitfalls [21–23].

Leveraging mitochondria-specific dyes and cytometry, we provide streamlined and rapid experimental strategies to capture and interpret mitochondrial activity changes in quiescent and activated myogenic cells. These approaches have been already largely used to characterize muscle cell states and activity under physiological and pathological conditions, including aging, fasting, injury, and quiescence [6, 8, 15, 24–26].

2 Materials

Mitochondrial dyes are commercially available from several suppliers and are proposed in multiple fluorophores, allowing optimization of color combination for flow cytometry. TMRE responds with high sensitivity to changes in mitochondrial membrane potential ($\Delta\Psi m$) and concentrates in mitochondria with relatively low binding to mitochondrial lipids and proteins. MitoTracker™ Deep Red FM is used to quantify mitochondrial mass and is insensitive to ΔΨm. Unlike TMRE, MitoTracker™ Deep Red FM has the advantage of being retained after paraformaldehyde fixation. In this protocol, TMRE (Ex 549 nm/Em 574 nm) and MitoTracker™ Deep Red FM (Ex 644 nm/Em 665 nm) detection is performed using yellow and red lasers, respectively. Although this protocol focuses on flow cytometry, these dyes can also be used for imaging mitochondrial morphology, which dynamically changes depending on cell state and activity.

2.1 Equipment and Materials

1. Dissection tools for muscle dissection.
2. Falcon® Round-Bottom Tubes with Cell Strainer Cap, 5 mL (Corning).
3. Matrigel® Basement Membrane Matrix, *LDEV-Free (Corning).
4. NucBlue™ Live ReadyProbes™ Reagent (Hoechst 33342, Invitrogen).
5. Propidium iodide (PI) (Calbiochem), 1 mg/mL aqueous solution.
6. TrypLE Express Enzyme (1×), no phenol red (Gibco).
7. Acquisitions were performed on BD FACSAria™ III cell sorter or BD FACSymphony A5, using BD FACSDiva™. Data were analyzed using FlowJo™ v10 software.

2.2 Preparing Stock Solutions

1. Collagenase solution: dissolve the Collagenase type 2 according to the unit number provided specifically by the manufacturer on each lot of Collagenase (Worthington). Prepare a 1000 units/mL solution in 10% Horse Serum/Ham-F10. For all hindlimb muscle, use 10 mL of 1000 units/mL of Collagenase type 2. Add DNAse (DNAse I, Sigma) at 10 μg/mL final concentration for injured muscle digestion.
2. Dispase solution: dissolve the Dispase according to the unit number provided by the manufacturer specifically on each lot of Dispase (Gibco). Dispase is used at the final concentration of 1.3 U/mL.
3. MitoTracker™ Deep Red FM (here referred to as MitoTracker, Invitrogen, M22426), 1 mM stock solution: Before resuspending the powder, bring the tube to room temperature. Each tube contains 50 μg of the lyophilized powder (MW: 543.58 g/mol) to dissolve in 92 μL of anhydrous dimethyl sulfoxide (DMSO, Sigma-Aldrich, D2653) to a final concentration of 1 mM (*see* **Note 1**). MitoTracker™ Deep Red FM solution appears blue. Store at −20°C in a box protected from light (up to 12 months). An intermediate 10 μM stock solution is recommended (add 10 μL of 1 mM to 990 μL of DMSO). MitoTracker™ Deep Red is used here at 70 nM

working concentration (*see* **Note 2**). Working concentration can vary between 25 and 500 nM but should be employed in a lower concentration range to prevent toxicity and allow correct equilibration across membranes.

4. Tetramethylrhodamine, Ethyl Ester, Perchlorate (TMRE, Invitrogen, T669) 10 mM stock solution: Prepare a highly concentrated 10 mM stock solution by dissolving 25 mg of TMRE (MW: 515 g/mol) powder in 4.85 mL of anhydrous DMSO. The TMRE solution appears red-orange. Store 1 mL aliquots at −20°C in a box protected from light.
   Prepare an intermediate concentrated TMRE 100 μM stock solution by diluting 50 μL of the 10 mM TMRE stock with 4.95 mL of anhydrous DMSO. Make 250 μL aliquots. Store in freezer at −20°C in a box protected from light (up to 12 months). Here, TMRE is used at 200 nM working concentration (*see* **Note 2**). The working concentration should be maintained in 15–500 nM range [22].
5. Carbonyl cyanide p-(trifluorothoxy) phenylhydrazone (FCCP, Sigma, C2920) 10 mM stock solution: Dissolve 10 mg (MW: 254.17 g/mol) in 3.90 mL of anhydrous DMSO. Make 50 μL aliquots. FCCP is used at a 10 μM final concentration. Store at −20°C protected from light.
   Warning: FCCP is a respiratory chain uncoupler and is toxic. Wear appropriate personal protective equipment when handling this compound.
6. DAPI 1 mg/mL stock solution: Resuspend 1 mg of 4′,6-diamidino-2-phenylindole (DAPI dihydrochloride, Sigma-Aldrich) powder in 1 mL of distilled water. An intermediate solution of 1 μg/mL is recommended. Use DAPI at a final concentration of 100 ng/mL. Store at 4°C protected from light.
7. Ham-F10: Dissolve one bottle of Ham's F10 Nutrient Mixture, with L-glutamine, without sodium bicarbonate (Ham-F10) (Sigma, N6635) in 1 L of distilled water. Add 1.2 g of sodium bicarbonate (Sigma, S5761) to balance the acidic pH. Mix well. Store at 4°C.
8. 10% horse serum/Ham-F10 (10% HS/F10): Inactive complement in horse serum by 30 min incubation in 56°C water bath. Add 50 mL of heat-inactivated horse serum to 450 mL of Ham-F10. Filter through a 0.22 μm filter if the medium will be stored for more than 1 day. Store medium at 4°C until use.
9. Staining buffer: 2% Horse Serum (HS, Gibco) in F10. Add 10 mL of 10% HS/F10 to 40 mL of Ham-F10 (*see* **Note 3**).
10. MuSC growth medium: DMEM/F-12, GlutaMAX™ Supplement (Gibco), 20% fetal bovine serum (FBS),

2% Ultroser-G (Pall Biosepra, 15950-017) and 1% Penicillin–Streptomycin.

11. Matrigel solution: For coating, Matrigel is used at a final concentration of 1 mg/mL.

An accurate concentration is provided by the manufacturer for each lot of Matrigel (typically around 10 mg/mL). Thaw the Matrigel on ice. At high concentration, Matrigel forms a gel and cannot be diluted properly. Prepare a 75 cm^2 flask on ice with the appropriate volume of ice-cold DMEM to obtain a final concentration of 1 mg/mL. Transfer 10 mL of Matrigel quickly into the cold DMEM. Aliquot the solution into 1.5 mL Eppendorf tubes kept on ice. Store the aliquots at −20°C. Do not reuse thawed aliquots. For cell culture coating, thaw Matrigel aliquots at room temperature. Transfer 1 mL (larger volumes may be required for surfaces greater than 10 cm^2) into a cell culture plate to cover the entire surface, then remove the excess solution with a micropipette and transfer into the next cell plate if needed. Incubate the coated plate at 37°C for 30 min. After incubation, add medium to the plate to prevent excessive drying until use (within the same day).

2.3 Antibody Mix for MuSCs

1. Antibodies are diluted in 10% HS/F10 and always kept on ice.
2. List of antibodies: Anti-CD31-eFluor 450 (1/75, Invitrogen, 48-0311-82); Anti-CD45-BV421 (1/75, BD Horizon, 563890); Anti-Sca1-Pacific Blue (1/75, BioLegend, 108120); Anti-Alpha7-integrin-Alexa 647 (1/1000, AbLab, 67-0010-05); CD106 (VCAM-1)-eFluor 660 (1/75, eBioscience, 50-1061-82); CD106 (VCAM-1)-PE (1/75, eBioscience, 12-1061-82); CD106 (VCAM-1)-biotin (1/75, BD, 553331); and Anti-CD34 - FITC (1/75, BD, 553733). Fluorophores are indicated as an example and can be modified by the user accordingly.

3 Methods

3.1 Isolation of Quiescent or Injured Muscle Cells

Mitochondrial staining described here was performed on freshly isolated MuSCs from uninjured or injured muscle, as well as in vitro cultured myoblasts. Freshly isolated MuSCs from resting muscle are referred to here as quiescent, although MuSCs subjected to enzymatic digestion and FACS exhibit transcriptomic changes compared with truly quiescent *in situ* cells, isolated with fixation-based protocols [27].

1. Conventionally, mononucleated cells from muscle are obtained after enzymatic digestion. The procedure used in this protocol

to digest the muscle uses Collagenase type 2 and dispase [28, 29]. Briefly, hindlimb muscles from uninjured or injured muscles are dissected and digested sequentially with collagenase and dispase diluted in 10% HS/F10, washed, mechanically dissociated by repeated passage through a 20½ gauge syringe and filtered with a 40 μm cell strainer (*see* **Note 4**). In general, pellet from full hindlimb cell suspension is resuspended with 500 μL of 2% HS/F10.

2. For injured muscle digestion, DNAse at 10 μg/mL final concentration is added to the collagenase solution during the incubation time. Injured samples have a high number of damaged cells that release DNA, which tends to aggregate and cause clogging during FACS analysis. Adding DNAse during the digestion eliminates extracellular DNA and facilitates the FACS procedure.
3. MuSC-derived myoblasts from *Tg:Pax7-nGFP* mouse (cultured for 4 days in growth medium) were collected by using TrypLE for 4 min at 37°C. Cells were collected in 10 mL of 10%HS/F10 and centrifuged for 10 min at 500 g at RT. The supernatant was discarded, and the cell pellet was resuspended in 2% HS/F10 for mitochondrial staining.

3.2 TMRE and MitoTracker Staining of Quiescent and Activated Pax7-nGFP+ cells for FACS Analysis

In this section, MuSCs were isolated from adult *Tg:Pax7-nGFP* mice [30], where GFP is a reporter of the transcription factor Pax7 transcriptional expression, a specific marker of skeletal MuSCs in muscle [31] (Fig. 1).

1. Centrifuge cell suspension for 10 min at 500 *g* at 4°C.
2. Aspirate supernatant by tilting tube to avoid pellet disruption.
3. Resuspend pellet in 500 μL of 2% HS/F10.
4. Prepare mitochondrial dye solutions. Dilute TMRE (for 200 nM working concentration, add 20 μL of 100 μM stock in 10 mL of serum-free Ham-F10) and MitoTracker (for 70 nM working concentration, add 70 μL of 10 μM intermediate stock solution in 10 mL of serum-free Ham-F10) (*see* **Note 5**).
5. Transfer cell suspension (~ 40 μL if using hindlimb cell suspension, *see* **Note 6**) in a 2 mL microtube for single or combined TMRE and MitoTracker staining.
6. Add 1 mL of single or combined TMRE and MitoTracker diluted solution into 2 mL tube containing the cell suspension. Mix well by gently pipetting up and down several times with a micropipette. To allow adequate gas exchange during incubation, pierce the Eppendorf tube cap with a needle.

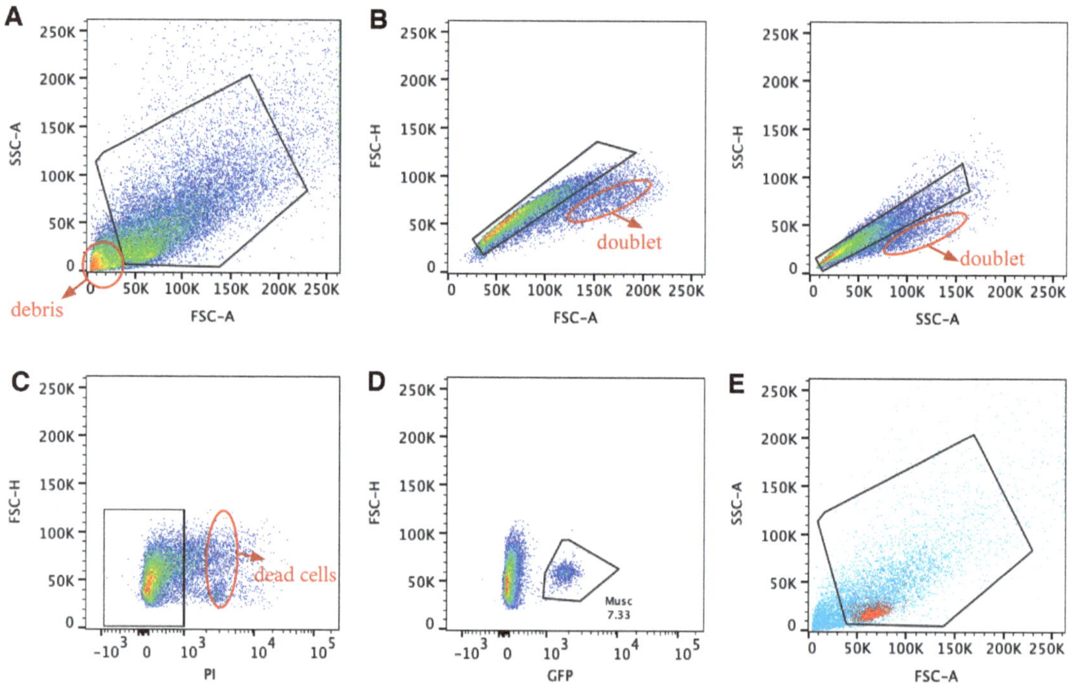

Fig. 1 Flow cytometry gating strategy to identify Pax7-nGFP+ MuSCs from a bulk of muscle cell suspension isolated from adult (here, 3-month-old) *Tg:Pax7-nGFP* mouse. **a,** The first gate is set on the main cell population based on cell size (Forward Scatter-A) and granularity (complexity, Side Scatter-A), excluding cell debris found on the bottom left corner. **b,** FSC-A *vs.* FSC-H and SSC-A *vs.* SSC-H are used to exclude doublets. **c,** Dead cells are excluded with viability dye, such as PI. PI can be excited with UV laser using the appropriate laser/filter configuration for the instrument (here, Ex. 355 nm/Em. 730 nm, in BD FACSAria™ III). **d,** Pax7-nGFP+ cells are detected on the FITC/GFP channel. GFP protein is excited at 488 nm (blue laser) and detected at 510 nm. **e,** Backgating the Pax7-nGFP+ population in the SSC-A and FSC-A dot plot shows that MuSCs have a small size and low granularity

7. Incubate at 37°C for 30 min.
8. Add 1 mL of 2% HS/F10 medium to wash the cells, gently resuspend by pipetting, and centrifuge 10 min at 500 *g* at RT.
9. Aspirate the supernatant gently.
10. Resuspend pellet in 300 μL of 2% HS/F10.
11. Maintain cells in the dark and on ice while preparing the FACS.
12. To prevent modification of the equilibrium established in mitochondria, perform the acquisition shortly after staining (*see* **Note 7**).
13. Transfer the cell suspension to a filter-cap FACS tube fitted with a 35-μm nylon mesh cell strainer.

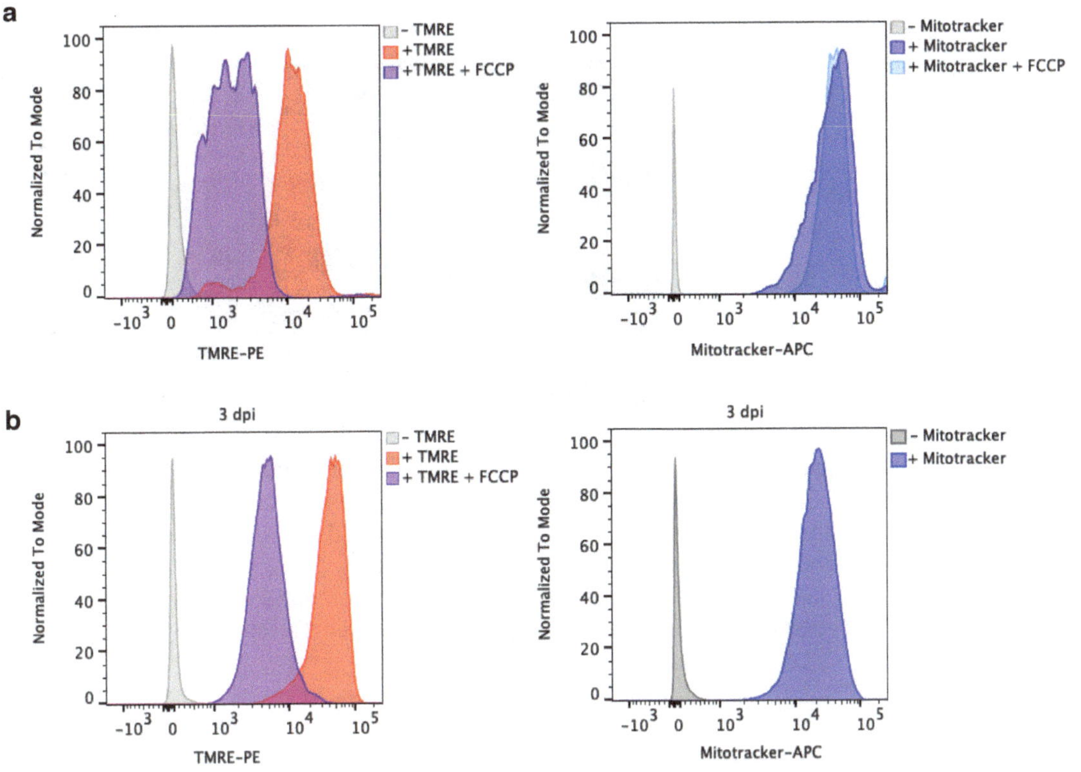

Fig. 2 Histograms for TMRE and MitoTracker fluorescence in quiescent and proliferative Pax7-nGFP+ cells. TMRE and MitoTracker stainings of quiescent MuSCs (**a**), and proliferative Pax7-nGFP+ cells isolated 3 days after TA muscle injury (3 dpi) by intramuscular injection of 30 μL of 1.2% $BaCl_2$ (**b**). Addition of 10 μM FCCP during TMRE/MitoTracker staining collapses ΔΨm, resulting in reduced TMRE fluorescence, without affecting MitoTracker signal in quiescent MuSCs (**a**) (*see* **Note 9**)

14. Five minutes before processing for analysis by FACS, add viability dye that allows identification of dead cells (DAPI or PI according to antibody panel and FACS configuration).
15. In general, aim to acquire at least 10,000 cells from the population of interest, if sufficient biological material is available (*see* **Note 6**).

Representative histograms of TMRE and MitoTracker measurements in quiescent MuSCs (Fig. 2a) and injury-induced myoblasts (Fig. 2b) from *Tg:Pax7-nGFP* mice are provided (*see* **Notes 8 and 9**).

3.3 TMRE and MitoTracker Staining of Pax7-nGFP+ MuSCs for Live-Cell Imaging

The analysis of mitochondria morphology is crucial for defining transitions in cell state and physiology across multiple contexts. Although flow cytometry provides precise quantitative measurements of mitochondrial mass and activity, live-cell imaging offers additional insights into MuSC metabolic remodeling. To this end, TMRE and

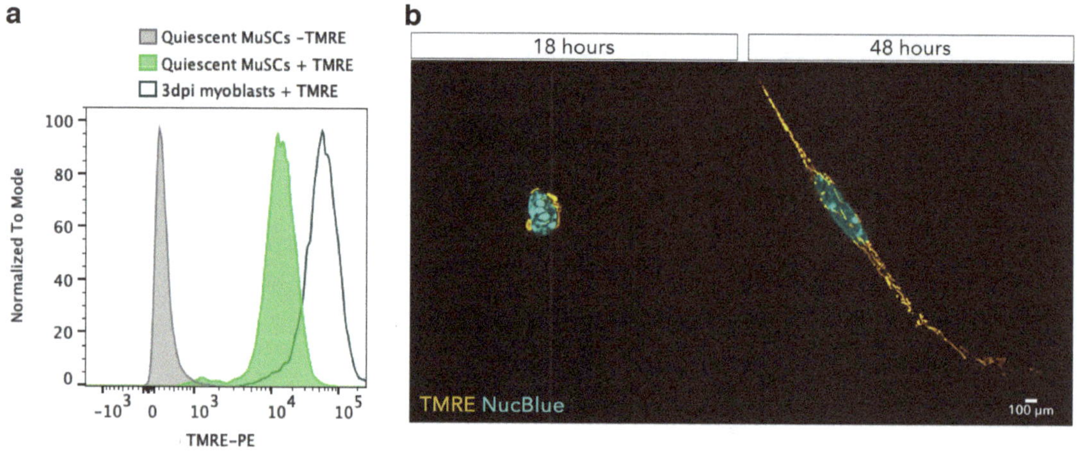

Fig. 3 TMRE staining in quiescent and activated myogenic cells analyzed through flow cytometry and live imaging. **a,** Quiescent MuSCs exhibit lower metabolic activity than 3 dpi myoblasts, as indicated by lower levels of TMRE fluorescence. **b,** Representative live-cell images of isolated Pax7-nGFP$^+$ MuSCs stained with TMRE and NucBlue at 18 and 48 h post-plating. TMRE and NucBlue were used to label mitochondria and nuclei, respectively. Cells were imaged using an Opera Phenix high-content microscope (PerkinElmer) equipped with a 63× water-immersion objective (NA 1.15) in confocal mode, with an incubation chamber providing environmental control at 37°C, 5% CO_2, and 3% O_2

MitoTracker can be used in combination to stain MuSCs in vitro, followed by live imaging (Fig. 3b).

1. Isolate MuSCs by FACS based on GFP intensity when using *Tg:Pax7-nGFP* mice, alternatively with an antibody cocktail. Sorted MuSCs can be collected into 1.5 mL tubes containing 300 μL of MuSC growth medium, and subsequently plated, or alternatively plated directly into 1 mg/mL Matrigel-coated culture plates at a density of 10,000 cells/cm^2 when using a cytometer equipped with an automated single-well plating system.
2. Gently centrifuge the plate at 50 g for 3 min at RT to facilitate cell adhesion.
3. Before live-cell imaging, prepare the staining solution by diluting TMRE to a final concentration of 200 nM and/or 70 nM MitoTracker Deep Red FM in serum-free prewarmed DMEM/F12.
4. Wash cultured cells once with serum-free prewarmed DMEM/F12.
5. Add the staining solution to the cells.
6. Incubate for 30 min at 37°C in a humidified incubator.
7. Wash cultured cells once with MuSC growth media.

8. Prepare the NucBlue solution adding two drops per mL of MuSC growth media.
9. Incubate for 20 min at RT.
10. Wash cultured cells once with MuSC growth media.
11. Image cells using a confocal high-content microscope. A 63× immersion objective is recommended to obtain high-quality images.

Differences in mitochondria activity can be observed between different cell states by cytometry, as shown here in quiescent and 3 days post-injury (dpi) muscle cells (Fig. 3a). In addition, mitochondria dynamics and activity can be visualized by imaging TMRE and MitoTracker staining in live cells (Fig. 3b) (*see* **Note 8**). Images show the paucity of mitochondria in quiescent MuSCs, contrasting with the extended network in myoblasts.

3.4 Combined TMRE or MitoTracker Staining with Antibody Labeling of Quiescent and Activated MuSCs

Genetically modified mice [30, 32] are useful to readily detect and isolate MuSCs based on the expression of a single fluorophore. However, experimental conditions require often myogenic cells to be identified by antibody staining. The combination of antibodies is based on elimination of non-myogenic populations (Lin^+), including anti-CD45 for the hematopoietic lineage, anti-CD31 for the endothelial cells, and SCA-1 for Fibro-Adipogenic Progenitors (FAPs). Double positive α7-integrin$^+$/CD34$^+$ or α7-integrin$^+$/VCAM$^+$ cell subset represents the quiescent MuSC population. Isolating MuSCs using double-positive markers will improve the accuracy of identifying a pure myogenic population. In myoblasts, VCAM expression is retained, whereas CD34 is absent.

1. Transfer muscle cell suspension (~80 μL if starting from hindlimb cell suspension, *see* **Note 6**) in a 2 mL microtube.
2. Prepare antibody mixes for all samples including appropriate controls:
   a. Unstained.
   b. Single stain for each fluorophore of the antibody panel. Single stains are essential to determine and correct the level of compensation between spectral overlap of different fluorophores (*see* **Note 10**).
   c. Fluorescence Minus One (FMO) control: a sample containing all antibodies except one, used to account for fluorescence spillover and to set FACS gates accurately.
   d. TMRE or MitoTracker only.
   e. Full antibody staining including: anti-CD45, anti-CD31, anti-SCA1 (Ly6A), α7-integrin and anti-CD34 or anti-VCAM.

3. Prepare the mix to perform staining in a 100 μL final volume, completing with 10% HS/F10 if needed. Once antibody mix is added to the cell suspension, gently pipette up and down at least ten times with micropipette to mix antibodies efficiently.
4. Incubate for 45 min on ice (*see* **Note 11**).
5. Wash by filling tube with 10% HS/F10. Gently pipette up and down at least ten times with micropipette to wash efficiently.
6. Centrifuge for 10 min at 500*g* at 4°C.
7. Remove supernatant by aspiration.
8. Resuspend in Ham-F10 without serum and proceed to TMRE or MitoTracker staining as described above (Sect. 3.2) (*see* **Note 12**).
9. Before loading for analysis by FACS, add live/death marker, such as DAPI to visualize dead cells using the UV laser.

Representative histograms of TMRE measurement in quiescent α7-integrin$^+$CD34$^+$ MuSCs from wild-type mice are provided (Fig. 4). To analyze myoblasts from injured muscle, we first validated the expression of VCAM in Pax7-nGFP$^+$ myogenic cells at 3 dpi (Fig. 5a). Next, TMRE was quantified in lin-VCAM$^+$ cells in 3dpi wild-type muscle (Fig. 5b). Mitochondrial dyes can be used to compare the metabolic status of different cell populations, as shown here between in vivo (3 dpi) and in vitro proliferating myoblasts (4 days post-plating) (Fig. 5c).

3.5 Modulation of Membrane Potential by FCCP

In polarized, metabolically active mitochondria, membrane potential depends on the proton gradient generated by the ETC. To validate that the signal quantified by flow cytometry reflects the mitochondrial ΔΨm, it is recommended to perturb the ETC using protonophore uncouplers such as FCCP (lipophilic molecule that dissipates the proton gradient). It allows proton diffusion across the inner membrane, resulting in uncoupling of electron transport from ATP production, and reduction of ΔΨm. Addition of FCCP forces maximal electron transport activity and collapses ΔΨm.

We recommend performing the following protocol:

1. Transfer muscle cell suspension (~40 μL if hindlimb cell suspension, *see* **Note 6**) in a 2 mL microtube, maintained on ice. Pierce the tube cap with a needle to allow adequate gas exchange during incubation.
2. Prepare a 10 μM FCCP and 200 nM TMRE working solution in Ham-F10 without serum (add 5 μL of 10 mM FCCP stock solution, and 10 μL of 100 μM solution to 5 mL of Ham-F10 without serum).
3. Add 1 mL of the mixed solution FCCP/TMRE to the cell suspension.

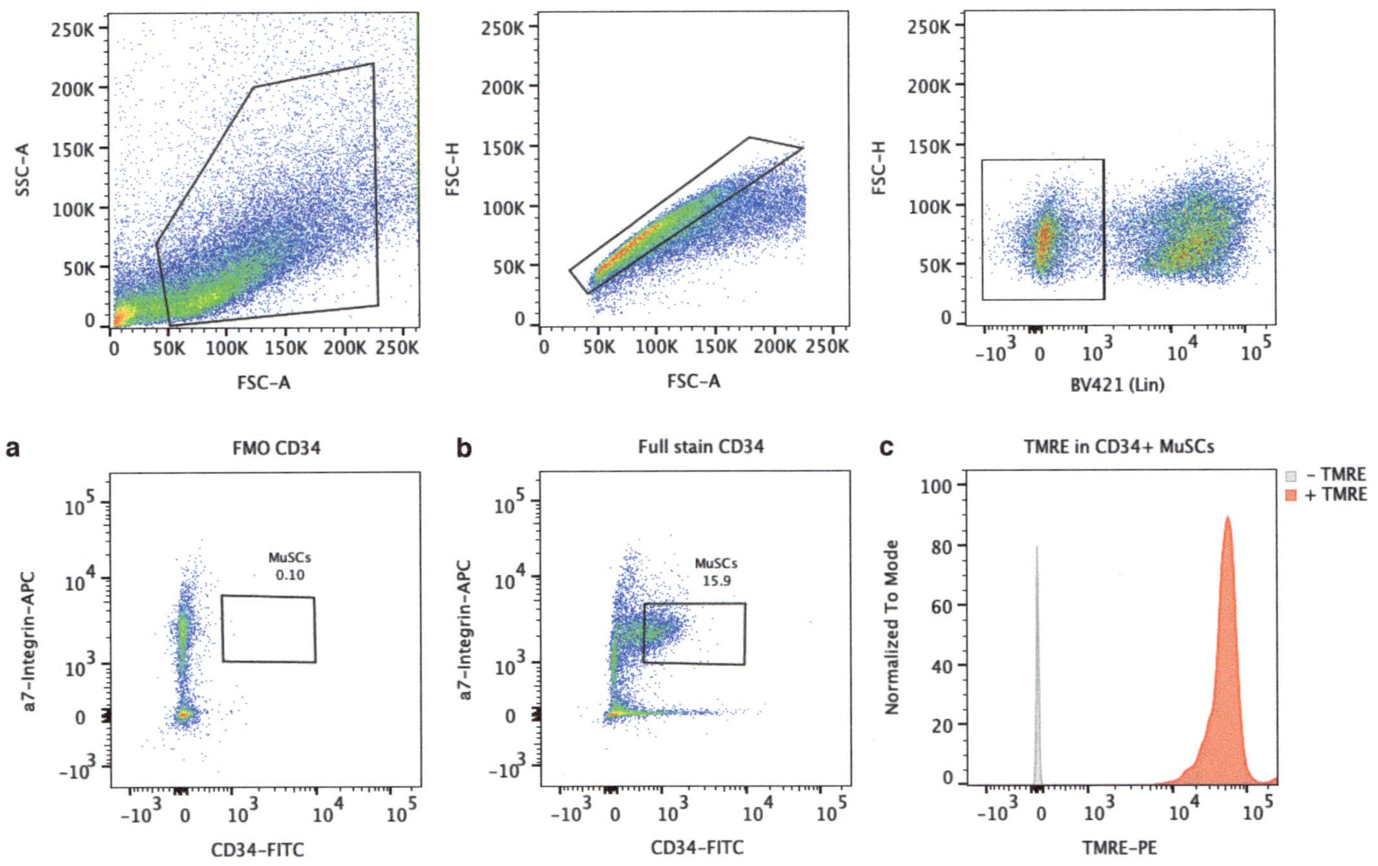

Fig. 4 TMRE staining of quiescent MuSCs using α7-integrin$^+$/CD34$^+$. MuSCs were prepared from a bulk quadriceps muscle from adult C57BL/6 N mice. Cells positive for CD45, CD31, or SCA-1 were excluded. **a,** Fluorescence-minus-one (FMO) control without CD34-FITC antibody. **b,** MuSC staining, including CD34-FITC antibody. MuSCs were identified as double positive for α7-integrin$^+$ and CD34$^+$. **c,** TMRE staining in MuSCs defined as α7-integrin$^+$/CD34$^+$ population

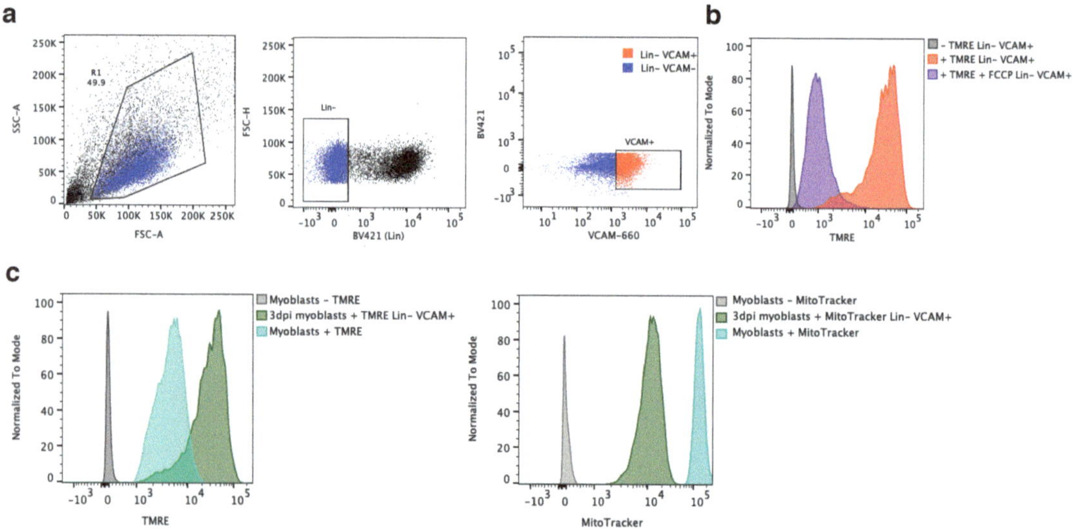

Fig. 5 TMRE and MitoTracker staining in VCAM+ myoblasts. **a,** Pax7-nGFP+ activated/proliferating myoblasts isolated from TA muscle of *Tg:Pax7-nGFP* mice 3 dpi, gated within the Lin- population and stained with VCAM-660. **b,** Myoblasts from 3 dpi wild-type-injured TA were gated as SCA1−CD45−CD31−(Lin-)VCAM+. VCAM+ myoblasts were stained with TMRE, in presence or absence of FCCP. **c,** Comparison of TMRE and MitoTracker fluorescence of *in vivo* (3 dpi) and 4 days *in vitro* cultured myoblasts. Anti VCAM-660 and anti VCAM-PE were combined with TMRE and MitoTracker, respectively, to detect myoblasts in injured muscle and perform mitochondrial analysis

4. Incubate at 37°C for 30 min.
5. Centrifuge 10 min at 500 *g* at RT.
6. Aspirate gently the supernatant.
7. To sustain the inhibition, FCCP must be maintained throughout the staining procedure and FACS acquisition. Use 300 µL of 2%HS/F10 with 10 µM FCCP to resuspend the cell pellet.
8. Maintain cells in the dark and on ice while preparing for FACS acquisition. Acquire the data in presence of the inhibitor.

Representative histograms of FCCP treatment with TMRE and MitoTracker measurements in quiescent MuSCs and activated/proliferative myoblasts in wild-type mice are provided. FCCP induces the depolarization of the inner membrane and abolishes the ΔΨm (Figs. 2 and 5b).

4 Notes

1. Lyophilized drugs should be dissolved in high-quality anhydrous DMSO; we suggest using a freshly opened DMSO vial (5 mL vials, Sigma D2653-5 × 5 mL) to prepare stock solutions.

2. Titration of final working concentration of dyes is recommended for each cell type and experimental setup. High dye concentration (μM range) should be avoided, as it may result in quenching, a process that reduces fluorescence due to high fluorophore density (where one molecule absorbs light emitted by another molecule). Concentration above 1 μM of TMRE may also interfere with ETC function [23].
3. The medium used in our collagenase/dispase digestion protocol is Ham-F10 supplemented with 2% horse serum, therefore, we use it as a diluent for our staining. However, other growth media can be used. Medium containing phenol red should be avoided to prevent spectral interference and artefacts in flow cytometry measurements.
4. Collagenase and dispase batches should be tested to validate their innocuity on exposed cell surface receptors, which could be interfering if using cell surface markers to identify myogenic cells. Favor high-purity collagenase preparations. Do not incubate samples over 2 h in enzymes.
5. TMRE staining can be performed in regular serum or BSA-containing medium. However, manufacturers recommend using MitoTracker in a serum-free medium to prevent oxidation of the dye by serum catalase.
6. In general, aim to acquire 10,000 cells from the population of interest for analysis. In our setup, we use 40 μL, corresponding to 1/15 of the total hindlimbs bulk preparation from a *Tg:Pax7-nGFP* mouse, for which no additional antibody staining is required to identify GFP-labeled MuSCs. If one or two additional antibody-staining steps are required, as with wild-type cells, double the amount of starting material to compensate for cell loss during centrifugation and pipetting. If the analysis is performed on a single mouse muscle or on aged muscles, MitoTracker–TMRE staining should be performed on the entire muscle cell preparation to avoid obtaining an insufficient number of cells for analysis. In this case, 1 mL of TMRE solution may be added directly to the cell pellet for staining.
7. Mitochondrial dyes do not necessarily need to be washed; however, manufacturers recommend washing to obtain a sharper fluorescent signal. As an option, to maintain mitochondrial equilibrium, washed cells can be resuspended in a lower concentration of TMRE (1:4 of the initial concentration).
8. Dyes exhibit cumulative toxicity over time; therefore, stained cells can be sorted for short-term imaging but cannot be recovered for long-term cell culture.
9. Histogram representations were generated using FlowJo™ v10 Software. They show the frequency of distribution of

gated cells (here, quiescent or activated Pax7-nGFP+ cells) *versus* the fluorophore fluorescence intensity (here, TMRE-PE or MitoTracker-647) in cells. The modal "Normalized To Count" mode in FlowJo™ was selected to compare samples with unequal event counts. This mode does not alter raw mean intensity data or statistical values but facilitates visual comparison between samples when overlaid on the same plot. Mean fluorescence intensity (MFI) values provided in FlowJo™ statistical table can be used to generate graphs.

10. This panel is designed to minimize fluorophore compensation, as experiments are often performed with limited biological material, such as analyzing a single muscle. However, depending on the fluorophores chosen, single-stain controls may still be required to perform accurate compensation.
11. Do not perform antibody staining concomitantly with TMRE staining. Antibody staining is typically performed on ice, this is not compatible with TMRE staining, which requires incubation at 37°C to ensure active mitochondria.
12. In our setup, MitoTracker Deep Red FM and the anti α7-integrin-647 have overlapping emission spectra, making their use incompatible in MuSCs. To overcome this problem, the anti α7-integrin antibody should be coupled with a different fluorophore compatible with the antibody panel designed to gate MuSCs, or another dye specific to mitochondrial mass should be used, such as MitoTracker Green (specificity and limitation of each dye should be carefully checked before use).

Acknowledgments

We acknowledge funding support from the Institut Pasteur, Agence Nationale de la Recherche (Laboratoire d'Excellence Revive, Investissement d'Avenir; ANR-10-LABX-73 to ST), European Research Council (Advanced Research Grant #101055234 to ST), La Fondation ARC pour la Recherche sur le Cancer and the Centre National de la Recherche Scientifique (PGA1 RF20190208619). This work has been supported by the Fondation ARC pour la recherche sur le cancer with Postdoctoral fellowships granted to MM (#ARCPOST-DOC2025010009305).

Competing Interests The authors declare no competing or financial interests.

References

1. Schiaffino S, Reggiani C (2011) Fiber types in mammalian skeletal muscles. Physiol Rev 91:1447–1531
2. de Morree A, Rando TA (2023) Regulation of adult stem cell quiescence and its functions in the maintenance of tissue integrity. Nat Rev Mol Cell Biol 24:334–354
3. Evano B, Tajbakhsh S (2018) Skeletal muscle stem cells in comfort and stress. NPJ Regen Med 3:24
4. Relaix F, Zammit PS (2012) Satellite cells are essential for skeletal muscle regeneration: the cell on the edge returns centre stage. Development 139:2845–2856
5. Benjamin DI, Both P, Benjamin JS et al. (2022) Fasting induces a highly resilient deep quiescent state in muscle stem cells via ketone body signaling. Cell Metab 34:902–918.e906
6. Latil M, Rocheteau P, Chatre L et al. (2012) Skeletal muscle stem cells adopt a dormant cell state post mortem and retain regenerative capacity. Nat Commun 3: 903
7. Vahidi Ferdousi L, Rocheteau P, Chayot R et al. (2014) More efficient repair of DNA double-strand breaks in skeletal muscle stem cells compared to their committed progeny. Stem Cell Res 13:492–507
8. Gayraud-Morel B, Le Bouteiller M, Commere PH et al (2018) Notchless defines a stage-specific requirement for ribosome biogenesis during lineage progression in adult skeletal myogenesis. Development 145(23):dev162636
9. Hong X, Isern J, Campanario S et al. (2022) Mitochondrial dynamics maintain muscle stem cell regenerative competence throughout adult life by regulating metabolism and mitophagy. Cell Stem Cell 29:1298–1314.e1210
10. Pala F, Di Girolamo D, Mella S et al (2018) Distinct metabolic states govern skeletal muscle stem cell fates during prenatal and postnatal myogenesis. J Cell Sci 131(14):jcs212977
11. Signer RA, Magee JA, Salic A et al. (2014) Haematopoietic stem cells require a highly regulated protein synthesis rate. Nature 509:49–54
12. Ryall JG, Dell'Orso S, Derfoul A et al. (2015) The NAD(+)-dependent SIRT1 deacetylase translates a metabolic switch into regulatory epigenetics in skeletal muscle stem cells. Cell Stem Cell 16:171–183
13. Bhattacharya D, Scime A (2020) Mitochondrial function in muscle stem cell fates. Front Cell Dev Biol 8:480
14. Dell'Orso S, Juan AH, Ko KD et al (2019) Single cell analysis of adult mouse skeletal muscle stem cells in homeostatic and regenerative conditions. Development 146(12):dev174177
15. Rodgers JT, King KY, Brett JO et al. (2014) mTORC1 controls the adaptive transition of quiescent stem cells from G0 to G(Alert). Nature 510:393–396
16. Hong X, Munoz-Canoves P (2023) Measuring oxygen consumption rate (OCR) and extracellular acidification rate (ECAR) in muscle stem cells using a seahorse analyzer: applicability for aging studies. Methods Mol Biol 2640:73–88
17. Jacques M, Kuang J, Bishop DJ et al. (2020) Mitochondrial respiration variability and simulations in human skeletal muscle: the Gene SMART study. FASEB J 34:2978–2986
18. Ehrenberg B, Montana V, Wei MD et al. (1988) Membrane potential can be determined in individual cells from the nernstian distribution of cationic dyes. Biophys J 53:785–794
19. Poot M, Zhang YZ, Kramer JA et al. (1996) Analysis of mitochondrial morphology and function with novel fixable fluorescent stains. J Histochem Cytochem 44:1363–1372
20. Neikirk K, Marshall AG, Kula B et al. (2023) MitoTracker: a useful tool in need of better alternatives. Eur J Cell Biol 102:151371
21. Cottet-Rousselle C, Ronot X, Leverve X et al. (2011) Cytometric assessment of mitochondria using fluorescent probes. Cytometry A 79:405–425
22. Tovar-Ferrero O, Rubio J, Zorzano A et al. (2025) Measuring mitochondrial membrane potential. EMBO J 44:7334–7345
23. Perry SW, Norman JP, Barbieri J et al. (2011) Mitochondrial membrane potential probes and the proton gradient: a practical usage guide. Biotechniques 50:98–115
24. Brun CE, Sincennes MC, Lin AYT et al. (2022) GLI3 regulates muscle stem cell entry into G(Alert) and self-renewal. Nat Commun 13: 3961
25. Rocheteau P, Gayraud-Morel B, Siegl-Cachedenier I et al. (2012) A subpopulation of adult skeletal muscle stem cells retains all template DNA strands after cell division. Cell 148:112–125
26. Rodgers JT, Schroeder MD, Ma C et al. (2017) HGFA is an injury-regulated systemic factor that induces the transition of stem cells into G(Alert). Cell Rep 19:479–486
27. Machado L, de Lima JE, Fabre O et al. (2017) In situ fixation redefines quiescence and early activation of skeletal muscle stem cells. Cell Rep 21:1982–1993
28. Liu L, Cheung TH, Charville GW et al. (2015) Isolation of skeletal muscle stem cells by fluorescence-activated cell sorting. Nat Protoc 10:1612–1624

29. Yue L, Wan R, Luan S et al. (2020) Dek modulates global intron retention during muscle stem cells quiescence exit. Dev Cell 53:661–676.e666
30. Sambasivan R, Yao R, Kissenpfennig A et al. (2011) Pax7-expressing satellite cells are indispensable for adult skeletal muscle regeneration. Development 138:3647–3656
31. Seale P, Sabourin LA, Girgis-Gabardo A et al. (2000) Pax7 is required for the specification of myogenic satellite cells. Cell 102:777–786
32. Murphy MM, Lawson JA, Mathew SJ et al. (2011) Satellite cells, connective tissue fibroblasts and their interactions are crucial for muscle regeneration. Development 138:3625–3637

Check for updates

Chapter 14

Measurement of Oxygen Consumption Rate of Primary Mouse Skeletal Muscle Stem Cells and Their Progeny on the Seahorse Analyzer

Audrey Saugues, Lucie Ancel, Ha My Ly, Lola E. R. Lessard, Rémi Mounier and Anita Kneppers

Abstract

To establish mature myofibers with a complex mitochondrial network, skeletal muscle stem cells (MuSCs) need to undergo large metabolic alterations. This goes beyond an increase in mitochondrial abundance during myogenesis and includes mitochondrial morphological adaptations by fusion and fission. Recent insights into the epigenetic, transcriptomic, and cell signaling alterations supported by mitochondrially generated intermediates and end products demonstrate how mitochondrial metabolism dictates MuSC behavior. The measurement of mitochondrial oxidative phosphorylation (OXPHOS), as an indicator of mitochondrial function under stressed states, therefore gives crucial insight into metabolic alterations that drive MuSC function. In this chapter, we describe the measurement of the oxygen consumption rate of primary mouse skeletal muscle stem cells and their progeny on the Seahorse analyzer.

Keywords Skeletal muscle, Muscle stem cells, Myotubes, Metabolism, Oxygen consumption rate, Cellular respiration, Seahorse technology

1 Introduction

Skeletal muscle tissue constitutes approximately 45% of the total body weight in healthy humans and is a major determinant of the basal metabolic rate. The metabolic activity is predominantly dictated by the myofibers, which are large syncytia that are made up of a highly organized contractile network fueled by mitochondria that intercalate between the myofibril bundles [1]. To establish mature myofibers with such complex mitochondrial network, skeletal muscle stem cells (MuSCs) need to undergo large metabolic alterations [2, 3]. Indeed, quiescent MuSCs are generally associated with a low metabolic need, that is predominantly supplied by mitochondrial fatty acid oxidation and oxidative phosphorylation (OXPHOS) [4–7]. Upon an activation stimulus, MuSCs undergo a relative shift towards anaerobic glycolysis

Nathalie Didier and Caroline E. Brun (eds.), *Muscle Stem Cells*, Methods in Molecular Biology, vol. 3093,
https://doi.org/10.1007/978-1-0716-5643-3_14,

that generates the metabolic intermediates and end products required for rapid biosynthesis of nucleotides, lipids, and proteins during MuSC growth and proliferation [6, 8, 9]. Importantly, glycolytic metabolism generates pyruvate, which in turn stimulates mitochondrial biogenesis and OXPHOS [6, 8, 10], contributing to a gradual increase in OXPHOS during the progression towards myogenic commitment [11, 12]. OXPHOS becomes predominant during differentiation, which may be driven by the combined actions of MyoD and Myogenin [9, 13–17].

Mitochondrial regulation during MuSC fate progression goes beyond mitochondrial abundance. Mitochondrial morphology, shaped by fusion and fission, plays a crucial role in MuSC metabolism and function. This was elegantly demonstrated using genetic models impairing mitochondrial fission or fusion, resulting in impaired metabolic flexibility and disrupted muscle regeneration [18, 19]. Restoration of impaired mitochondrial dynamics and mitophagy in MuSCs from aged mice also restored MuSC function [18, 19], demonstrating the causal role of mitochondrial metabolism in MuSC fate progression.

The epigenetic, transcriptomic, and cell signaling alterations supported by mitochondrially generated intermediates and end products that dictate MuSC behavior have been extensively reviewed by Sartorelli and Ciuffoli [3]. The measurement of mitochondrial OXPHOS, as an indicator of mitochondrial function under stressed states, therefore gives crucial insight into metabolic alterations that drive MuSC function. The Agilent Seahorse XF Cell Mito Stress Test is an endpoint measure of key parameters of mitochondrial function inferred by the direct measurement of the oxygen consumption rate (OCR) of cells. It is a plate-based live cell assay that monitors OCR in real time upon the addition of modulators of respiration. These modulators are Oligomycin, Carbonyl cyanide-4 (trifluoromethoxy) phenylhydrazone (FCCP), Rotenone, and Antimycin. After measurement of *basal respiration* (OCR in unperturbed state), oligomycin is added to inhibit ATP synthase, resulting in a drop of OCR reflecting *ATP-linked respiration* and resulting in an OCR that reflects the *mitochondrial proton leak* (Fig. 1). Subsequently,

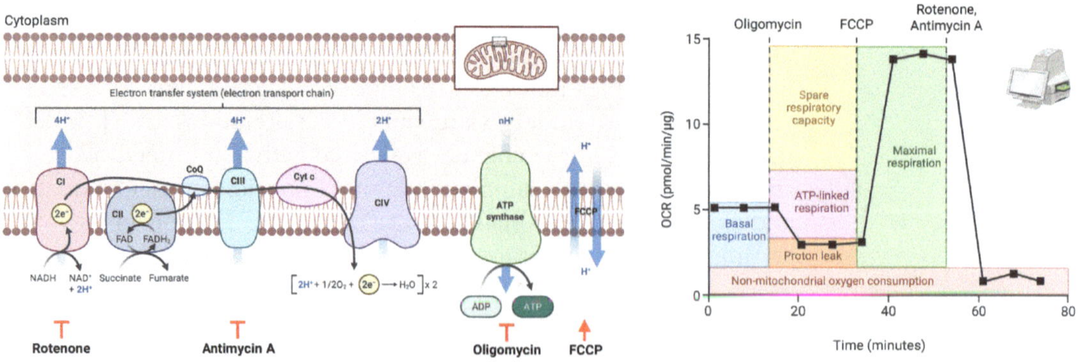

Fig. 1 Schematic of oxygen consumption measurement during mitochondrial stress test. This figure was generated using BioRender

mitochondrial respiration is uncoupled using FCCP that acts as a proton ionophore, a mobile carrier that shuttles protons across the inner mitochondrial membrane, resulting in an increase in OCR that reflects the *maximal respiration*. Finally, Rotenone and Antimycin are added to block complex I and III to inhibit the generation of a proton gradient, thereby blocking the mitochondrial OCR. The resulting OCR reflects *non-mitochondrial oxygen consumption*. From the difference between the basal respiration and the maximal respiration, we can extrapolate the *spare respiratory capacity*.

2 Materials

2.1 Seahorse Plate Coating and Cell Seeding

1. Matrigel® Basement Membrane Matrix (Corning, #354234)
2. 0.05% Trypsin–Ethylenediaminetetraacetic acid (EDTA) solution (Gibco)
3. Phosphate-Buffered Saline (PBS)
4. Growth medium: DMEM/F-12, GlutaMAX™ supplement (Gibco), 20% heat-inactivated fetal bovine serum, 2% Ultroser G (Pall Gelman Sciences), 1% penicillin–streptomycin
5. Differentiation medium: DMEM/F-12 GlutaMAX™ Supplement(Gibco),2%horseserum,1%penicillin–streptomycin
6. Seahorse XF Cell Mito Stress Test Kit (Agilent, #103015-100) containing single-use aliquots of oligomycin, FCCP and rotenone/antimycin
7. Seahorse XFe96/XF Pro Cell Culture Microplates (Agilent, #103794-100)

2.2 Seahorse Experiment Preparation

1. Sterile double-distilled water (ddH_2O)
2. XF Calibrant solution (Agilent #100840-000)
3. A humidified, non-CO_2 incubator
4. Seahorse XFe96/XF Pro Sensor Cartridges (Agilent, #103793-100)

2.3 Seahorse Analysis

1. Assay medium (pH 7.3–7.5): Seahorse medium (Agilent, #103575-100), 17 mM Glucose (prepared from powder; Sigma, #G8270), 2 mM GlutaMAX (Gibco, #35050061), 1 mM sodium pyruvate (Gibco, #11360070).
2. Oligomycin 100 μM stock: (vial from kit + 630 μL assay medium at 37°C).
3. FCCP 100 μM stock: (vial from kit + 720 μL assay medium at 37°C).
4. Rotenone/Antimycin A (Rot/A) 50 μM stock: (vial from kit + 540 μL assay medium at 37°C).

5. Agilent Seahorse XFe96 Analyzer or equivalent.
6. Seahorse Wave Pro Software.

2.4 Normalization

1. PBS.
2. Paraformaldehyde (PFA) 4% in PBS.
3. Triton X-100 0.5% in PBS.
4. Hoechst.
5. Image analysis software (e.g., Fiji [20]).

2.5 Data Analysis

1. seahorseanalytics.agilent.com

3 Methods

3.1 Seahorse Plate Coating and Cell Seeding

Quiescent mouse primary muscle stem cells are isolated from the muscles of mice using magnetic-activated cell sorting (MACS), and expanded for 3 days in culture. The following steps are performed in sterile conditions under a tissue culture hood.

1. Dilute Matrigel 1/10 in cold plain DMEM/F-12, GlutaMAX medium.
2. Mix by pipetting up and down using precooled pipet tips.
3. Add 50 μL diluted Matrigel *per* well in a 96-well Agilent Seahorse XFe96/XF Cell Culture Microplate (*see* **Note 1**).
4. Incubate for 30 minutes (min) at 37°C in a humidified incubator.
5. Trypsinize primary muscle stem cell progeny and quantify cell number.
6. Resuspend cells at 10,500 cells in 180 μL growth medium *per* well.
7. Gently add PBS to each well to dilute supernatant and unpolymerized Matrigel.
8. Aspirate supernatant from each well without disturbing the Matrigel coating.
9. Add 180 μL of cell suspension *per* well, and allow cells to attach and recover for at least 6 h in a humidified incubator (*see* **Notes 2–4**).
10. Incubate the cells for 12 h after plating in proliferation conditions before initiating analysis of primary muscle stem cell progeny.
11. Optional: To induce differentiation into myotubes, remove growth medium, and replace with differentiation medium. Incubate the cells for 48 h after addition of differentiation medium before initiating analysis of primary myotubes.

3.2 Seahorse Experiment Preparation

The following steps should be initiated 1 day prior to the Seahorse analysis.

1. Hydrate the cartridge:
   a. Separate the utility plate and the sensor cartridge, and place the sensor cartridge upside down next to the utility plate.
   b. Fill each well of the utility plate with 180 μL of sterile water.
   c. Lower the sensor cartridge onto the utility plate, submerging the sensors in the water.
   d. Place at 37°C in a humidified, non-CO_2 incubator overnight.
2. Place an aliquot of 20 mL XF Calibrant solution at 37°C in a humidified, non-CO_2 incubator overnight.
3. Turn on the Agilent Seahorse XFe-96 Analyzer (US00411612), and let it warm up overnight (at least 5 h) (*see* **Note 5**).
4. Open the Wave Controller software and design your experiment.
   a. Setup the template in Wave program—Select the Seahorse XF Cell Mito Stress test to design your experiment.
   b. Clarify the treatment and control wells (name the groups)—Define your groups according to the potential pretreatments, the assay used, the cell type, etc. You can name your groups according to your experiments and adapt the colors.
   c. Plate Map (the cell densities, treatment, conditions, etc.)—Click on Plate map to indicate where your different cells are plated in the 96-well Cell Culture Microplate. All the wells with no cells should be labeled as "background." In the tab "protocol," you can see and/or modify the injection strategies if you want a specific assay different from the Seahorse XF Cell Mito Stress test.
   d. Click on "Run Assay" to save your template. A message will pop on the screen, to start the calibration of the cartridge after drug loading, when you are ready.

3.3 Seahorse Analysis

3.3.1 Preparing the Calibration Plate

1. Take the cartridge out of the humidified non-CO_2 incubator.
2. Remove the sensor cartridge from the utility plate, and place the sensor cartridge upside down next to the utility plate.
3. Discard the water from the utility plate using multichannel pipet.
4. Fill each well with 180 μL of XF Calibrant solution at 37°C that has incubated in the humidified, non-CO_2 incubator overnight.
5. Lower the sensor cartridge onto the utility plate, submerging the sensors in the XF Calibrant solution.
6. Place at 37°C in a humidified, non-CO_2 incubator for 1 h prior to loading the injection ports of the sensor cartridge.

3.3.2 Preparing Assay Medium

The following steps are performed in sterile conditions under a tissue culture hood. Prepare the working solutions for each compound in assay medium to reach the final analytic concentrations of:

1. For 1 μM Oligomycin injectable in the port, prepare a 10 μM Oligomycin solution by adding 300 μL of Oligomycin 100 μM stock to 2.7 mL of assay medium.
2. For 5 μM FCCP injectable in the port, prepare a 50 μM solution by adding 700 μL of FCCP 100 μM stock to 700 μL of assay medium (*see* **Note 6**).
3. For 0.5 μM Rot/A injectable in the port, prepare a 5 μM Rot/A solution by adding 300 μL of Rot/A 50 μM stock to 2.7 mL of assay medium.

3.3.3 Loading Solutions into the Cartridge

The following steps could be performed in close proximity to the Seahorse device. All the drug solutions and assay medium should be at 37°C.

1. Remove the hydrated sensor cartridge from the non-CO_2 incubator.
2. Orient the Agilent Seahorse Assay Cartridge. Place well labels (lettered A–H) to the left. The triangular notch will be in the bottom left-hand corner (*see* **Note 7**).
3. Gently dispense 20 μL of Oligomycin working solution into port A (or according to your plate/group layout), using a 100 or 200 μL pipette (*see* **Notes 8–10**).
4. Gently dispense 22 μL of FCCP working solution into port B (or according to your plate/group layout), using a 100 or 200 μL pipette.
5. Gently dispense 25 μL of Rot/A working solution into port C (or according to your plate/group layout), using a 100 or 200 μL pipette.
6. Visually inspect the injection ports for even loading.

3.3.4 Running Seahorse Assay

1. Place the calibration plate with the sensor cartridge without the lid into the tray of the analyzer.

   1. Press *Close Tray* when ready to begin the calibration, and press *Continue* to begin the assay.
   2. Take the Agilent Seahorse XF Cell Culture Microplate with cells from the CO_2 incubator.
   3. Check the cells under the microscope to verify that there is no air bubble or cell detachment.
   4. Remove medium carefully, without disturbing the cell layer by pipetting slowly.
   5. Wash cells twice with 180 μL of prewarmed assay medium.

6. Fill all wells with 180 μL of prewarmed assay medium.
7. Place the plate at 37°C in a humidified, non-CO_2 incubator for at least 45 min.
8. Once the calibration is complete, press *Open Tray* to present the calibration utility plate.
9. Remove the calibration utility plate and place the Cell Culture Microplate without the lid into the tray of the analyzer.
10. Run the assay (around 1.5–2 h running time).

3.4 Normalization

1. After the Seahorse program finishes, incubate the analysis plate for 1 h without disturbance to allow reattachment of potentially detached cells due to mixing and probe measurements.
2. Retrieve the analysis plate and discard the cartridge.
3. Carefully remove supernatant.
4. Add 4% paraformaldehyde (PFA) [Room Temperature (RT), 10 min] to fix the cells.
5. Wash cells with phosphate-buffered saline (PBS) (RT, 5 min), three times.
6. Add 0.5% Triton X-100 in PBS (RT, 10 min) to permeabilize cells.
7. Wash cells with PBS (RT, 5 min), three times.
8. Add Hoechst 1/1000 (RT, 15 s).
9. Wash cells with PBS (RT, 5 min), once.
10. Store the plate at 4°C in the dark until imaging.
11. Acquire an image of the stained nuclei by imaging/scanning the entire well at 4–10× (*see* **Note 11**).
12. Quantify the number of nuclei for each well, manually or using an automated quantification of your preference (*see* **Note 12**).

3.5 Data Analysis

1. In the Wave program, open the experiment (.xfd) and verify the plating information. Analyze the OCR of the experiment from selected wells.
2. Load your experiment file into the online software seahorseanalytics.agilent.com, where you can view, modify, normalize, and analyze your data.
3. In the normalization tab, introduce the normalization value obtained in Sect. 3.4, and select the appropriate unit.
4. Results can be exported as GraphPad Prism files and various XF reports can be generated, including OCR (pmol/min), basal respiration (pmol/min), proton leak (pmol/min), maximal respiration (pmol/min), spare respiratory capacity (pmol/

min), non-mitochondrial respiration (pmol/min), ATP production (pmol/min), coupling efficiency (%), and spare respiratory capacity (%). Select the appropriate statistical tests depending on the experiment.

4 Notes

1. The wells on the edge of the 96-well culture microplate should not contain cells due to the risk of evaporation and subsequent variability. There is no need to coat these wells with Matrigel.
2. As the Seahorse device is very sensitive to potential variables such as bubbles in the well or incorrect drug injection, it is recommended that you perform at least six technical replicates *per* condition.
3. To ensure good homogeneity and avoid making bubbles, it is advisable to prepare the correct number of cells for all the replicates in the correct volume in advance (180 μL per well × 6 technical replicates for instance), before plating the cells in each corresponding well.
4. Fill each empty well with PBS to prevent any evaporation.
5. The instrument will not start the experiment if it has not have 5 h of calibration.
6. The Seahorse XF Cell Mito Stress Test does not provide enough FCCP to fill up all the wells of the plate at this concentration.
7. There is a risk of reversing the orientation of the 96-well plate when seeding cells or loading drug solutions. Be vigilant about the orientation of your cell culture plate and cartridge when manipulating it, and do not hesitate to annotate the plate to avoid mistakes. If inversion occurs, you can modify the plate map on the instrument.
8. To help identify which port corresponds to port A, the Wave program provides a port-by-well map in the Injection Strategies tab.
9. To ensure a proper injection, all the ports of the same series (for e.g., port A) should contain the same volume of drug. This concerns also the background wells, so all the wells should have ports loaded with the same volume of solution.
10. Do not force the tips completely into the ports. Avoid air bubbles and never tap any portion of the cartridge to attempt to alleviate air bubbles.

11. Images should be acquired between the three dots that are guides for the sensing probes when they are lowered into the well. These guides delimit the OCR measurement area. If cell detachment is observed, it may be preferable to count the entire well.
12. Here, we advise to correct for the total number of nuclei. Depending on your experimental conditions and scientific question, other normalization parameters may be used, such as protein content, differentiation status, or mitochondrial content measures.

Competing Interests

The authors declare no competing interests.

References

1. Romanello V, Sandri M (2021) The connection between the dynamic remodeling of the mitochondrial network and the regulation of muscle mass. Cell Mol Life Sci CMLS 78:1305–1328
2. Relaix F, Bencze M, Borok MJ, Der Vartanian A, Gattazzo F, Mademtzoglou D, Perez-Diaz S, Prola A, Reyes-Fernandez PC, Rotini A, Taglietti (2021) Perspectives on skeletal muscle stem cells. Nat Commun 12: 692
3. Sartorelli V, Ciuffoli V (2025) Metabolic regulation in adult and aging skeletal muscle stem cells. Genes Dev 39:186–208
4. Zeng W, Yue L, Lam KSW, Zhang W, So W-K, Tse EHY, Cheung TH (2022) CPEB1 directs muscle stem cell activation by reprogramming the translational landscape. Nat Commun 13: 947
5. Machado L, Esteves de Lima J, Fabre O, Proux C, Legendre R, Szegedi A, Varet H, Ingerslev LR, Barrès R, Relaix F, Mourikis P (2017) In situ fixation redefines quiescence and early activation of skeletal muscle stem cells. Cell Rep 21:1982–1993
6. Ryall JG, Dell'Orso S, Derfoul A, Juan A, Zare H, Feng X, Clermont D, Koulnis M, Gutierrez-Cruz G, Fulco M, Sartorelli V (2015) The NAD(+)-dependent SIRT1 deacetylase translates a metabolic switch into regulatory epigenetics in skeletal muscle stem cells. Cell Stem Cell 16:171–183
7. Fukada S, Uezumi A, Ikemoto M, Masuda S, Segawa M, Tanimura N, Yamamoto H, Miyagoe-Suzuki Y, Takeda S (2007) Molecular signature of quiescent satellite cells in adult skeletal muscle. Stem Cells Dayt Ohio 25:2448–2459
8. Pala F, Di Girolamo D, Mella S, Yennek S, Chatre L, Ricchetti M, Tajbakhsh S (2018) Distinct metabolic states govern skeletal muscle stem cell fates during prenatal and postnatal myogenesis. J Cell Sci 131:jcs212977.
9. L'honoré A, Commère P-H, Negroni E, Pallafacchina G, Friguet B, Drouin J, Buckingham M, Montarras D (2018) The role of Pitx2 and Pitx3 in muscle stem cells gives new insights into P38α MAP kinase and redox regulation of muscle regeneration. eLife 7:e32991
10. Rodgers JT, King KY, Brett JO, Cromie MJ, Charville GW, Maguire KK, Brunson C, Mastey N, Liu L, Tsai C-R, Goodell MA, Rando TA (2014) mTORC1 controls the adaptive transition of quiescent stem cells from G0 to G(Alert). Nature 510:393–396
11. Theret M, Gsaier L, Mounier R (2018) AMPK α 1-LDHA, a new metabolic pathway, regulating stem cell fate. Cell Cycle 17:403–404
12. Theret M, Gsaier L, Schaffer B, Juban G, Ben Larbi S, Weiss-Gayet M, Bultot L, Collodet C, Foretz M, Desplanches D, Sanz P, Zang Z, Yang L, Vial G, Viollet B, Sakamoto K, Brunet A, Chazaud B, Mounier R (2017) AMPKα1-LDH pathway regulates muscle stem cell self-renewal by controlling metabolic homeostasis. EMBO J 36:1946–1962
13. Hughes SM, Chi MM, Lowry OH, Gundersen K (1999) Myogenin induces a shift of enzyme activity from glycolytic to oxidative metabolism

in muscles of transgenic mice. J Cell Biol 145:633–642
14. Kraft CS, LeMoine CMR, Lyons CN, Michaud D, Mueller CR, Moyes CD (2006) Control of mitochondrial biogenesis during myogenesis. Am J Physiol Cell Physiol 290:C1119–1127
15. Zheng X, Boyer L, Jin M, Mertens J, Kim Y, Ma L, Ma L, Hamm M, Gage FH, Hunter T (2016) Metabolic reprogramming during neuronal differentiation from aerobic glycolysis to neuronal oxidative phosphorylation. eLife 5:e13374
16. Wüst S, Dröse S, Heidler J, Wittig I, Klockner I, Franko A, Bonke E, Günther S, Gärtner U, Boettger T, Braun T (2018) Metabolic maturation during muscle stem cell differentiation is achieved by miR-1/133a-mediated inhibition of the Dlk1-Dio3 mega gene cluster. Cell Metab 27:1026–1039.e6
17. Shintaku J, Peterson JM, Talbert EE, Gu J-M, Ladner KJ, Williams DR, Mousavi K, Wang R, Sartorelli V, Guttridge DC (2016) MyoD regulates skeletal muscle oxidative metabolism cooperatively with alternative NF-κB. Cell Rep 17:514–526
18. Baker N, Wade S, Triolo M, Girgis J, Chwastek D, Larrigan S, Feige P, Fujita R, Crist C, Rudnicki MA, Burelle Y, Khacho M (2022) The mitochondrial protein OPA1 regulates the quiescent state of adult muscle stem cells. Cell Stem Cell 29:1315–1332.e9
19. Hong X, Isern J, Campanario S, Perdiguero E, Ramírez-Pardo I, Segalés J, Hernansanz-Agustín P, Curtabbi A, Deryagin O, Pollán A, González-Reyes JA, Villalba JM, Sandri M, Serrano AL, Enríquez JA, Muñoz-Cánoves P (2022) Mitochondrial dynamics maintain muscle stem cell regenerative competence throughout adult life by regulating metabolism and mitophagy. Cell Stem Cell 29:1506–1508
20. Schindelin J, Arganda-Carreras I, Frise E, Kaynig V, Longair M, Pietzsch T, Preibisch S, Rueden C, Saalfeld S, Schmid B, Tinevez J-Y, White DJ, Hartenstein V, Eliceiri K, Tomancak P, Cardona A (2012) Fiji: an open-source platform for biological-image analysis. Nat Methods 9:676–682

Part V

Transcriptomic and Proteomic Profiling

Chapter 15

Low Input RNA Sequencing of Freshly Isolated Muscle Stem Cells

Fayez Issa and Joana Esteves de Lima

Abstract

Bulk RNA sequencing allows for the detection and quantification of mRNA transcripts from pooled cell populations, tissue sections, and biopsies. However, working with rare populations with limited amount of input requires specialized handling to ensure high-quality data. Here, we describe a detailed, optimized low-input protocol for bulk RNA sequencing on FACS-isolated satellite cells from mouse hindlimb skeletal muscle tissues. This protocol covers sample preparation, library preparation, sequencing, and data analysis.

Keywords Bulk RNA seq, Low-input RNA sequencing, Satellite cells, FACS, Transcriptomics, Library preparation

1 Introduction

High-throughput RNA sequencing (RNA-seq) involves the deep sequencing of large cDNA libraries to quantify gene expression within a tissue or cell population. Since its buildout in 2008, this technique has revolutionized the field of molecular biology and transcriptomics [1–3]. Isolated RNA is retrotranscribed into cDNA libraries followed by sequencing on next-generation platforms to detect both the presence and abundance of isolated transcripts.

Traditionally, transcriptome profiling is achieved in three ways: (a) hybridization-based microarrays, (b) bulk RNA sequencing, and (c) single-cell RNA sequencing [4, 5]. While single-cell RNA sequencing provides cellular resolution, bulk RNA sequencing is still widely used due to its high sensitivity, cost-efficiency, and ability to detect low-abundance transcripts that often fall below the detection limit of single-cell counterparts [5, 6].

Skeletal muscle is a highly plastic tissue with regenerative capacity that relies on a specialized subset of stem cells, referred to as satellite cells (SC), due to their anatomical position under the basal lamina of the fiber [7–9]. In homeostatic conditions, satellite cells reside in

Nathalie Didier and Caroline E. Brun (eds.), *Muscle Stem Cells*, Methods in Molecular Biology, vol. 3093,
https://doi.org/10.1007/978-1-0716-5643-3_15,

a state of reversible cell cycle arrest known as quiescence (G0) and represent less than 5% of total muscle nuclei [10–21]. Due to this dormant state, satellite cells possess inherently low RNA content compared to activated or proliferating cells. Consequently, performing transcriptomic analysis on these populations requires technical adaptations [22, 23].

Hereafter, we describe a detailed step-by-step protocol for bulk RNA sequencing on FACS-isolated SC from mouse hindlimb skeletal muscle tissue, covering (Sect. 3.1) Sample preparation, (Sect. 3.2) Library preparation, (Sect. 3.3) Sequencing, and (Sect. 3.4) Data analysis.

2 Materials

Prepare all solutions using ultrapure water and analytical-grade reagents. Prepare and store all reagents at room temperature (RT) unless indicated otherwise.

2.1 Muscle Digestion and Cell Sorting

1. Dissection board, pins, 70% ethanol spray, and curved scissors.
2. Digestion buffer: HBSS (Gibco) containing 0.2% BSA, 2 μg/mL Collagenase A, 3.25 μg/mL Dispase II, 0.4 mM $CaCl_2$ and 5 mM $MgCl_2$. Prepare fresh immediately before use.
3. Wash solution: HBSS containing 0.2% BSA. Keep on ice.
4. Sterile cell strainers (100 and 40 μm) and FACS tubes with 35 μm cell strainer caps.
5. Antibody mix: Wash solution with 10 μg/mL DNAse I to which the antibodies are added: Alexa700-anti-ITGA7 (R&D Systems, FAB3518N), BV421-anti-CD34 (BD Pharmingen, 562608), PE-Cy7-anti-TER119 (BD Pharmingen, 557853), PE-Cy7-anti-CD45 (BD Pharmingen, 552848), and PE-anti-Ly-6/E (SCA1) (BD Pharmingen, 553108). Keep protected from light at 4°C.
6. Viability dye 7-AAD (eBioscience).
7. BD FACS-Aria Fusion (or equivalent) calibrated for 85 μm nozzle sorting.

2.2 RNA Extraction and Quality Control

1. TRIzol Reagent (Invitrogen). Perform all work in a fume hood.
2. Chloroform. Perform all work in a fume hood.
3. Glycogen (20 mg/mL). Store at −20°C.
4. 100% Isopropanol.
5. 75% Ethanol.
6. Qubit RNA High Sensitivity Assay Kit (Thermo Fisher Scientific).

7. Agilent RNA 6000 Pico Kit (Invitrogen).
8. RNaseZap decontamination solution (Invitrogen).

2.3 Library Preparation

1. Dynabeads™ mRNA Purification Kit (Invitrogen, 61006).
2. NEBNext Poly(A) mRNA Magnetic Isolation Module (New England Biolabs, E7490) First-strand synthesis: 50 ng/μL Random hexamers, 10 mM dNTP Mix, 5× First-Strand Buffer, 0.1 M DTT, 40 U/μL RNaseOUT™, and 200 U/μL SuperScript™ III Reverse Transcriptase, all from Invitrogen. Store at −20°C.
3. Second-Strand Synthesis: 5× Second-Strand Buffer, *E. coli* DNA Ligase (10 U/μL), *E. coli* DNA Polymerase I (10 U/μ L), and *E. coli* RNase H (2 U/μL), all from Invitrogen. Store at −20°C.
4. Purification Beads: AMPure XP beads (Beckman Coulter). Equilibrate to RT for 30 min before use.
5. End Repair Mix: 10× Phosphorylation Buffer, 3 U/μL T4 DNA Polymerase, 5 U/μL Klenow Fragment, and 10 U/μL T4 Polynucleotide Kinase, all from NEB.
6. A-Tailing Mix: 10× NEBuffer 2, 1 mM dATP, and 5 U/μL Klenow Fragment (3'→5' exo–), all from NEB.
7. Ligation: 2× Rapid/Quick Ligation Buffer, Illumina Adapters (Unique Dual Indices), and Quick T4 DNA Ligase (NEB).
8. Library Amplification: 2× High-Fidelity PCR Master Mix (e.g., KAPA HiFi) and 25 μM Illumina Primers P5 and P7 (Roche).

2.4 Sequencing

1. Library QC: Agilent High Sensitivity DNA Kit (Agilent Technologies) [13].
2. Denaturation: Freshly prepared 0.2 N NaOH and 200 mM Tris-HCl, pH 7.0.
3. Sequencing Reagents: Illumina NextSeq 500/550 High Output Kit v2.5 (75 Cycles) and HT1 Hybridization Buffer (Illumina). Store cartridge at −20°C until use.

3 Methods

3.1 Sample Preparation

3.1.1 Muscle Dissection, Digestion, and Sorting Preparation

1. Euthanize the mice using cervical dislocation, pin the mouse belly up on a dissection board, and sterilize the hindlimbs with 70% ethanol.

2. Lift the skin at the anterior ankle (just above the foot), make a small horizontal incision then insert scissors under the skin but above the muscle fascia. Cut vertically up the leg towards the hip and discard the skin.
3. Isolate the hindlimb muscles and carefully clean fat excess, tendon, and fur.
4. Chop the muscles using curved scissors until you get a homogeneous slurry/paste consistency.
5. Add the chopped muscles to a 15 mL falcon containing 10 mL of ice-cold Digestion buffer [10].
6. Place the falcons in a 37°C water bath with agitation. After 45 min of digestion, pipette the lysate up-and-down (~10–15 times) using a 10 mL sterile pipette to promote mechanical digestion, place the falcons back at 37°C for another 45 min and repeat the up-and-down process.
7. After the second up-and-down, one should obtain a solution with no visible muscle debris left.
8. Fill the 15 mL falcon with ice-cold wash solution to stop the digestion and transfer the lysate to a new 50 mL falcon.
9. Wash the previous 15 mL falcon with 10 mL of wash solution, flick the tube up-and-down then transfer to the new 50 mL falcon.
10. Fill the 50 mL falcon up to 30 mL with wash solution and keep on ice.
11. Filter into a new 50 mL falcon using a sterile 100 μm cell strainer.
12. Wash the previous falcon with 10 mL of wash solution, flick up-and-down to recover the maximum of cells and filter it again through the 100 μm strainer containing 30 mL of filtered digested muscle.
13. Keep on ice and centrifuge for 5 min at 600 × *g*, 4°C.
14. Discard the supernatant, resuspend the pellet with 1 mL of wash solution then add 30 mL of Wash solution.
15. Filter into a new 50 mL falcon using a sterile 40 μm cell strainer.
16. Wash the previous falcon with 10 mL of wash solution, flick up-and-down to recover the maximum of cells and filter it again through the 40 μm strainer containing 30 mL of filtered digested muscle.
17. Keep on ice and centrifuge for 5 min at 600 × *g*, 4°C.
18. Discard the supernatant, resuspend the pellet with 300 μL of wash solution then transfer to a 15 mL falcon.

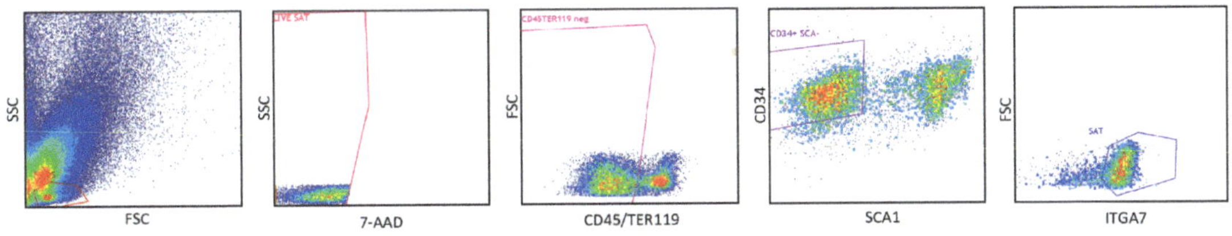

Fig. 1 Gating strategy used to sort satellite cells from C57BL/6N mouse hindlimb muscles from (CD34+ ITGA7+CD45-TER119-SCA1−)

19. Prepare an antibody mix using the following antibodies. For HL muscles, combine 12 μL of Alexa700-anti-ITGA7, 8 μL of BV421-anti-CD34, 5 μL of PE-Cy7-anti-TER119, 5 μL of PE-Cy7-anti-CD45, and 5 μL of PE-anti-Ly-6/E. The volumes below have been adapted to an entire hindlimb muscles of one mouse and are enough for whole musculature, adapt accordingly if using more or fewer muscles.
20. Resuspend the pellet with 250 μL (to be adjusted based on cell number) of filtered wash solution with 10 μg/mL DNAse I.
21. Add antibody mix and incubate for 20 min at room temperature.
22. Transfer the mix to a FACS tube while filtering using the FACS tube cap filter.
23. 5–10 min before sorting, add 5 μL 7-AAD viability dye.

3.1.2 RNA Extraction

1. After calibrating the ARIA Fusion (BD Biosciences), load the tube into the sorter.
2. Sort using CD45-TER119-Sca1-CD34+α7-Integrin+ panel. Cells are sorted by gating CD34+ITGA7+ double-positive lineage as shown in Fig. 1.
3. Pellet the sorted cells by centrifugating 5 min at 600 × *g*, 4°C, and carefully remove the supernatant with a p200.
4. Until RNA pellet resuspension, all steps must be performed under a fume hood. RNA is very sensitive to RNAse, to avoid degradation, use only RNAse free reagents and thoroughly clean all working surfaces and pipettes with RNaseZap in advance.
5. Add 500 μL of TRIzol reagent directly to the cell pellet and pipette up-and-down 10–15 times to thoroughly homogenize the lysate.
6. Incubate the homogenate for 5 min at RT to allow complete dissociation of nucleoprotein complexes.
7. Add 100 μL of Chloroform. Close the tube securely and shake vigorously by hand for 15 s. Do not vortex.
8. Incubate for 2–3 min at RT.

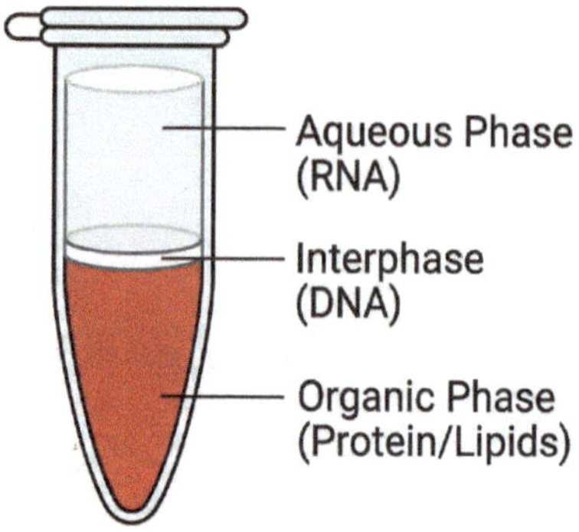

Fig. 2 Phase separation following phenol: chloroform nucleic acid extraction strategy

9. Centrifuge at 12,000 × *g* for 15 min at 4°C.
10. The mixture separates into a lower red phenol-chloroform phase, a white interphase, and a colorless upper aqueous phase containing the RNA [12] (Fig. 2).
11. Carefully transfer the clear upper aqueous phase (approximately 200 μL) into a new, sterile 1.5 mL microcentrifuge tube. Critical: Angle the tube and pipet slowly to avoid drawing up any of the white interphase or red organic layer.
12. Add 1 μL of Glycogen (20 mg/mL) to the aqueous phase and briefly vortex (*see* **Note 1**).
13. Add 250 μL of 100% Isopropanol and incubate at −20°C overnight (recommended for maximum yield with low input) or for 10 min at RT.
14. Centrifuge at 12,000 × *g* for 15–20 min at 4°C. Observation: A small white pellet should be visible at the bottom of the tube.
15. Carefully remove and discard the supernatant without disturbing the pellet.
16. Add 500 μL of freshly prepared 75% Ethanol and gently flick the tube to dislodge the pellet (it should float).
17. Centrifuge at 7500 × *g* for 5 min at 4°C.
18. Carefully remove the supernatant.
19. Optional: Repeat the wash step once if high purity is required; though one wash is usually sufficient for low-input samples to prevent pellet loss.
20. Air-dry the RNA pellet for 5–10 min at RT.
21. Caution: Do not over-dry the pellet (it should not turn clear/translucent). Over-dried pellets are difficult to dissolve.

22. Resuspend the pellet in 10–15 μL of nuclease-free water. Pipette the solution up-and-down gently. If the pellet does not dissolve immediately, incubate at 55°C for 10 min. Immediately place on ice for quality control or store at −80°C.

3.2 RNA Quality Control

3.2.1 RNA Quantification

1. Working solution: dilute the Qubit RNA HS Reagent 1:200 in Qubit RNA HS Buffer.
2. Calculation: (N samples + 2 standards) × 200 μL.
3. Standards: add 190 μL of Working solution to two tubes.
4. Add 10 μL of Standard #1 to the first tube and 10 μL of Standard #2 to the second. Vortex for 2–3 s.
5. Samples: add 198 μL of Working Solution to labeled tubes. Add 2 μL of your RNA sample (*see* **Note 2**).
6. Allow tubes to sit at room temperature for 2 min (*see* **Note 3**).

3.2.2 Integrity Assessment

The Pico chip is extremely sensitive. Ensure all reagents are at RT before starting. It is advised to equilibrate at RT for 30 min.

1. Before use, the electrodes must be cleaned to avoid ribonuclease contamination. Fill a dedicated cleaning chip with 350 μL of RNaseZap and place it in the Bioanalyzer for 1 min. Remove the chip and repeat the process with a second cleaning chip filled with nuclease-free water for 1 min to rinse the pins.
2. Open the Bioanalyzer lid for 30 s to allow the pins to air-dry, or insert a dry empty cleaning chip for 1 min to ensure the electrode pins are completely dry.
3. Prepare the gel-dye mix: pipette 550 μL of RNA gel matrix into a spin filter and centrifuge at 1500 × *g* for 10 min. Aliquot 65 μL of the filtered gel into a new tube, add 1 μL of Blue Dye Concentrate, vortex thoroughly, and centrifuge at 13,000*g* for 10 min. Priming: pipette 9 μL of Gel–Dye mix into the well-marked G (black background) and press the plunger station for 30 s.
4. Pipette 9 μL of Gel–Dye mix into the remaining G wells.
5. Pipette 9 μL of Conditioning Solution into the well-marked CS (Conditioning Solution well).
6. Load Marker and samples: pipette 5 μL of RNA Pico Marker (green cap) into all sample wells and the ladder well.

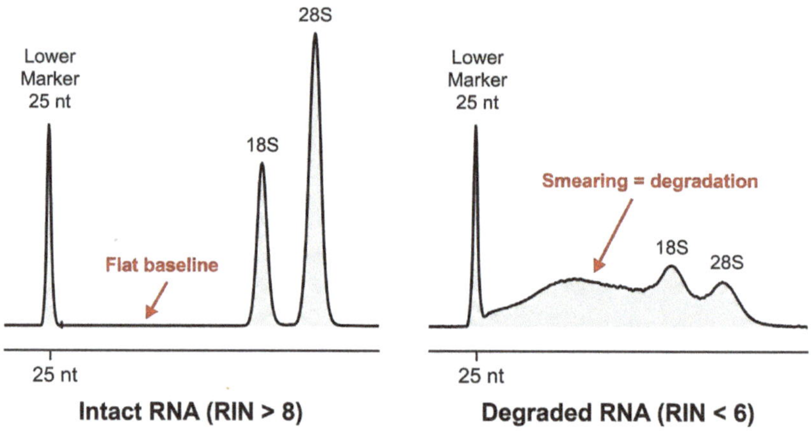

Fig. 3 Assessment of Total RNA Integrity. Representative electropherograms comparing intact RNA (left) and degraded RNA (right) from FACS-isolated satellite cells. Intact samples are characterized by a flat baseline and two distinct peaks corresponding to the 18 S and 28 S ribosomal RNA subunits, yielding a RIN > 8. Degraded samples exhibit significant smearing between the lower marker and the rRNA peaks, along with reduced peak heights, resulting in a RIN < 6

7. Heat Denaturation: heat RNA samples and the RNA Ladder at 70°C for 2 min to remove secondary structures, then snap-cool on ice.
8. Pipette 1 μL of RNA Ladder into the ladder well.
9. Pipette 1 μL of RNA sample into each sample well, and 1 μL of Marker into any unused wells.
10. Preparation: ensure the Bioanalyzer is ON, the computer is ON, and the software shows the "Instrument is Ready" status.
11. Run: vortex the chip for 1 min at 2400 × *g* using the IKA vortex.
12. Run the chip immediately after vortex (within 5 min).

3.2.3 Data Interpretation

1. Electropherogram: lower marker: 25 nt peak and clear 18 S and 28 S peaks (Fig. 3)
2. Flat baseline means RNA is intact while smearing is a sign of RNA degradation.
3. RIN Values: >8.0: high quality. 6–8: usable with caution. <6: degraded.
4. Expected SC RNA: yield: aprox.100 ng from 100 K cells. Concentration: 8–10 ng/μL. RIN: >8.

3.3 Library Preparation

Total RNA extracted from murine tissue is predominantly composed of ribosomal RNA (rRNA), which accounts for approximately 80%–90% of the total cellular RNA, whereas protein-coding messenger

RNA (mRNA) typically represents only 1%–5% of the total pool [14, 15]. Sequencing of the total RNA without an enrichment for mRNA would result in the vast majority of reads mapping to non-informative rRNA species [5, 17]. Therefore, a Poly-A selection strategy using oligo(dT) magnetic beads is performed. This method specifically targets the 3' polyadenylated tail common to eukaryotic mature mRNA selectively eliminating rRNA [17, 18].

3.3.1 mRNA Purification (Poly-A Selection)

1. Dilution: dilute 100 ng of total RNA with Nuclease-Free water to a final volume of 50 µL.
2. Binding: add 50 µL of Oligo(dT) magnetic beads.
3. Denaturation: incubate RNA at 65°C for 5 min to denature secondary structures. Snap-cool on ice.
4. Hybridization: incubate RNA at RT for 5 min to allow poly-A tails to bind to the beads.
5. Washing: place RNA samples on a magnetic stand. Remove supernatant. Wash beads 2× with Bead Washing Buffer to remove rRNA.
6. Elution: resuspend beads in 10–15 µL of 10 mM Tris-HCl (pH 7.0–7.5). Incubate at 80°C for 2 min to release the purified mRNA. Immediately place on the magnetic rack and collect the supernatant.

3.3.2 Fragmentation and Priming

1. Add 10× fragmentation buffer to the eluted mRNA. Incubate at 94°C for 5-6 min to obtain 300 bp pieces.
2. Add 2 µL of RNA Fragmentation Stop Solution and place on ice immediately.

3.3.3 First Strand cDNA Synthesis

1. In a sterile 0.2 mL PCR tube, mix 1 µL random primers (50 ng/µL) (do not add random primers if random primers are already present in the fragmentation buffer from Sect. 3.2.2), 1 µL dNTP mix (10 mM), up to 11 µL mRNA (eluted in Sect. 3.2.1), and nuclease-free water to a final volume of 13 µL.
2. Incubate at 65°C for 5 min.
3. Immediately place on ice for 2 min.
4. Prepare a master mix per sample by mixing 4 µL 5× first-strand buffer, 1 µL DTT (0.1 M), 1 µL RNaseOUT (40 U/µL), and 1 µL SuperScript III RT (200 U/µL).
5. Add 7 µL of master mix to each of the samples and gently mix by pipetting.
6. Incubate samples: 5 min at 25°C, 50 min at 50°C, and 15 min at 70°C.

3.3.4 Second Strand cDNA Synthesis

1. Add the following to the 20 µL first strand reaction tube: 91 µL Nuclease-free water, 30 µL 5× Second-Strand Buffer, 3 µL

dNTP Mix (10 mM), 1 μL DNA Ligase (10 U/μL), 4 μL DNA Polymerase I (10 U/μL), and 1 μL RNase H (2 U/μL). Total volume should be 150 μL.

2. Incubate at 16°C for 2 h.
3. Purify immediately using 1.8× AMPure XP beads (e.g., for 150 μL of sample add 270 μL of beads and elute in 20 μL nuclease-free water).

3.3.5 End Repair

1. To the 20 μL purified cDNA (from Sect. 3.2.5), add 70 μL Nuclease-free water, 10 μL 10× Phosphorylation buffer, 4 μL dNTP mix (10 mM), 5 μL T4 DNA Polymerase (3 U/μL), 1 μL Klenow Fragment (5 U/μL), and 5 μL T4 Polynucleotide kinase (10 U/μL).
2. Incubate at 20°C for 30 min.
3. Purify immediately using 1.6× AMPure XP beads (add 184 μL beads to the 115 μL reaction volume).
4. Elute in 17 μL nuclease-free water.

3.3.6 A-tailing

1. To the 17 μL of cDNA (from Sect. 3.2.5), add 2 μL 10× NEBuffer 2 (or Klenow Buffer), 1 μL dATP (10 mM), 3 μL Klenow Fragment (3'→5' exo–) (5 U/μL).
2. Incubate at 37°C for 30 min.
3. Purify immediately using 1.8× AMPure XP beads.
4. Elute in 22 μL nuclease-free water.

3.3.7 Adapter Ligation

1. Dilute the Illumina adapter stock 1:10 in 10 mM Tris-HCl (pH 7.5) before use [16].
2. Add the following to the 22 μL A-tailed cDNA (from Sect. 3.2.6): 25 μL 2× Quick ligation buffer, 1 μL Diluted Illumina Adapter, 2 μL T4 DNA Ligase.
3. Incubate at 20°C for 15 min.
4. Add 5 μL 0.5 M EDTA to stop the reaction.
5. Purify using 0.8× AMPure XP beads (add 44 μL beads to the 55 μL reaction volume).
6. Elute in 23 μL nuclease-free water.

3.3.8 PCR Library Amplification

1. Add the following to the 23 μL ligated cDNA (from Sect. 3.2.7) [24]: 25 μL NEBNext Ultra II Q5 2× Master Mix, 1 μL Illumina Primer P5 (25 μM), 1 μL Illumina Primer P7 (25 μM).
2. Use the following PCR program (or the program provided by your PCR supplier):
   - 30 s at 98°C
   - 13 cycles of: 10 s at 98°C; 30 s at 60°C; 30 s at 72°C

- 5 min at 72°C
- Hold at 4°C

3. Allow the AMPure XP beads to sit at RT for at least 30 min before use [25].
4. Add 50 µL of resuspended beads to your 50 µL PCR reaction (this is the 1.0× ratio).
5. Pipette mix ten times and incubate at room temperature for 5 min.
6. Place on a magnetic stand for 2–3 min until the liquid is clear. Discard the supernatant.
7. Add 200 µL of freshly prepared 80% Ethanol. Wait 30 s, then remove. Repeat for a total of two washes.
8. Air-dry the beads on the magnet for 2 min. Stop when the pellet looks matte/dull. Do not wait until it cracks.
9. Remove from the magnet. Resuspend the beads in 20 µL of nuclease-free water or 10 mM Tris-HCl (pH 8.0).
10. Incubate for 2 min, place back on the magnet, and transfer the clear supernatant (your final library) to a new tube [19].

3.4 Sequencing (NextSeq 500)

3.4.1 Library Validation and Pooling

1. Quality control: run 1 µL of each library on Agilent Bioanalyzer to verify size (approx. 300 bp).
2. Quantification: measure concentration of each library using Qubit.
3. Normalization: dilute each library to 4 nM using 10 mM Tris-HCl.
4. Pooling: combine equal volumes of all 4 nM libraries into a single tube.

3.4.2 Standard Normalization

1. Mix 5 µL of 4 nM pooled library and 5 µL of 200 mM NaOH.
2. Vortex and incubate 5 min at RT.
3. Add 5 µL of 200 mM Tris-HCl. Vortex briefly.
4. Dilute to 20 pM by adding 985 µL of prechilled HT1 buffer.
5. Dilute to 1.5 pM by mixing 98 µL of the 20 pM denatured library from previous step and 1202 µL of HT1 buffer (total volume = 1.3 mL).

3.4.3 NextSeq 500

For this step, we are using the NextSeq 500/550 High Output Kit v2.5 as previously described [26].

1. Thaw the reagent cartridge in a water bath for approx. 1 h.
2. Load the 1.3 mL library (from Sect. 3.3.2) into reservoir #10. "Load Library Here" is labeled on the cartridge.
3. Open NextSeq Control Software.
4. Set read type to "Single-Read."
5. Set read length to 76 bp (Read 1) + 6 bp or 8 bp (Index 1) + 0 bp (Index 2).
6. Start Run [20].

3.5 Data Analysis

1. The Illumina FASTQ files are uploaded in Galaxy and formatted as Sanger using the FASTQ groomer tool v1.2+galaxy0 [27].
2. Quality control of raw sequencing data is analyzed with FastQC tool v0.72 using default parameters v0.74+galaxy1 [28].
3. FASTQ Sanger files are then aligned to the genome of interest, in our case GRC39/mm39 mouse genome, using the built-in index of BOWTIE2 2.5.4+galaxy0 using default parameters [29].
4. Genes are counted using featureCounts v2.1.1+galaxy0 [30].
5. Differently expressed genes are determined by DESeq2 v2.11.40.8+galaxy1 [31].

4 Notes

1. Glycogen is essential for visualizing the pellet from low cell numbers and doesn't interfere with downstream RNA usage.
2. If you are extremely low on volume, you can use 1 μL sample + 199 μL solution, but 2 μL reduces pipetting error.
3. Select "RNA High Sensitivity" on the Qubit and measure standards, then samples. Calculate: If 2 μL are used, the stock concentration is the Qubit value multiplied by dilution factor (automatic on newer machines).

Funding

F.I. PhD scholarship is funded by ANR (Agence Nationale pour la Recherche) project funding MyoID–ANR-22-CE13-0003-01 granted to J.E.d.L.

Acknowledgments We thank lab members for reading and commenting on the MS. We thank all members of the lab for constructive scientific discussion.

Competing Interests The authors declare no competing interests.

References

1. Mortazavi A, Williams BA, McCue K et al (2008) Mapping and quantifying mammalian transcriptomes by RNA-Seq. Nat Methods 5:621–628
2. Nagalakshmi U, Wang Z, Waern K et al (2008) The transcriptional landscape of the yeast genome defined by RNA sequencing. Science 320:1344–1349
3. Wang Z, Gerstein M, Snyder M (2009) RNA-Seq: a revolutionary tool for transcriptomics. Nat Rev Genet 10:57–63
4. Stark R, Grzelak M, Hadfield J (2019) RNA sequencing: the teenage years. Nat Rev Genet 20:631–656
5. Hrdlickova R, Toloue M, Tian B (2017) RNA-Seq methods for transcriptome analysis. Wiley Interdiscip Rev: RNA 8:e1364
6. Arora R et al (2020) Bulk RNA-seq analysis reveals the molecular mechanism of muscle regeneration. Front Genet 11:568
7. Mauro A (1961) Satellite cell of skeletal muscle fibers. J Biophys Biochem Cytol 9:493–495
8. Relaix F, Zammit PS (2012) Satellite cells are essential for skeletal muscle regeneration: the cell on the edge returns centre stage. Development 139:2845–2856
9. Yin H, Price F, Rudnicki MA (2013) Satellite cells and the muscle stem cell niche. Physiol Rev 93:23–67
10. Liu L, Cheung TH, Charville GW et al (2015) Isolation of viable quiescent skeletal muscle satellite cells. Nat Protoc 10:1612–1624
11. Pasut A, Chang NC, Gurriaran-Rodriguez U et al (2016) Notch signaling rescues loss of satellite cells lacking Pax7 and promotes brown adipogenic differentiation. Cell Rep 16:333–343
12. Rio DC et al (2010) RNA purification by phenol-chloroform extraction. CSH Protoc. 2010:pdb-prot5439
13. Agilent Technologies (2016) Agilent RNA 6000 Pico Kit Guide. Waldbronn, Germany
14. O'Neil D, Glowatz H, Schlumpberger M (2013) Ribosomal RNA depletion for efficient use of RNA-seq capacity. Curr Protoc Mol Biol 103:4.19.1–4.19.8
15. Westermann AJ, Gorski SA, Vogel J (2012) Dual RNA-seq of pathogen and host. Nat Rev Microbiol 10:618–630
16. Illumina (2021) Illumina adapter sequences document (# 1000000002694 v15). San Diego, CA
17. Zhao S et al (2018) Comparison of RNA-Seq by poly (A) capture, ribosomal RNA depletion, and DNA microarray for expression profiling. BMC Genom 19:1–15
18. Cui P et al (2010) A comparison between ribo-minus RNA-sequencing and polyA-selected RNA-sequencing. Genomics 96:259–265
19. Illumina (2017) TruSeq stranded mRNA reference guide. San Diego, CA
20. Illumina (2019) NextSeq 500 and NextSeq 550 sequencing systems: denature and dilute libraries guide. San Diego, CA
21. Cheung TH, Rando TA (2013) Molecular regulation of stem cell quiescence. Nat Rev Mol Cell Biol 14:329–340
22. van Velthoven CT et al (2017) Transcriptional profiling of quiescent muscle stem cells in vivo. Cell Rep 21:1994–2004
23. Machado L et al (2017) Quiescence-associated transcriptional profile of muscle stem cells. Skelet Muscle 7:1–15
24. New England Biolabs (2020) Protocol for PCR library amplification using NEBNext Ultra II Q5 master mix. Ipswich, MA
25. Beckman Coulter (2020) Agencourt AMPure XP PCR purification process guide. Indianapolis, IN
26. Esteves de Lima J et al (2021) HIRA stabilizes skeletal muscle lineage identity. Nat Commun 12:3450
27. Blankenberg D et al (2010) FASTQ Groomer: a flexible and efficient estimator of next-generation seqencing quality. Bioinformatics 26:1253–1261
28. Andrews S (2010) FastQC: a quality control tool for high throughput sequence data. Babraham, UK
29. Langmead B, Salzberg SL (2012) Fast gapped-read alignment with Bowtie 2. Nat Methods 9:357–359
30. Liao Y, Smyth GK, Shi W (2014) featureCounts: an efficient general purpose program for assigning sequence reads to genomic features. Bioinformatics 30:923–930
31. Love MI, Huber W, Anders S (2014) Moderated estimation of fold change and dispersion for RNA-seq data with DESeq2. Genome Biol 15:550

Check for updates

Chapter 16

Low-Input RNA-Sequencing of In Situ Fixed Adult Murine Muscle Stem Cells

Indigo T. C. Chan, Wenxin Zhang, Gary J. He and Tom H. Cheung

Abstract

Muscle stem cells (MuSCs) play a crucial role in skeletal muscle regeneration. Under homeostatic conditions, the majority of MuSCs remain quiescent, a reversible, nonproliferative state with minimal cell activity. Upon muscle injury, MuSCs rapidly activate, proliferate, differentiate, and form nascent muscle fibers. Notably, the tissue dissociation process in conventional cell isolation protocols disrupts the stem cell niche environment, exposes the MuSCs to extracellular stimuli, and swiftly changes the transcriptomic profiles of MuSCs, making it challenging to study the regulation of MuSC quiescence. In vivo fixation allows the preservation of MuSC quiescence for further analysis and profiling. This chapter will focus on the detailed downstream RNA-seq library preparation workflow for fixed MuSCs, including RNA purification and decrosslinking, cDNA synthesis, and tagmentation-based library construction.

Keywords In situ fixation, RNA-seq, Muscle stem cell, Quiescence

1 Introduction

Somatic stem cells possess the ability to proliferate and differentiate into specialized cell types for tissue repair, homeostatic maintenance and self-renew to repopulate the stem cell pool [1]. Most somatic stem cell populations, including muscle stem cells (MuSCs), harbor a population of resting stem cells that lie in a reversible G0 phase, referred to as quiescence [2, 3]. The quiescent population serves as a reserve pool of stem cells maintained throughout the life of the organism. During muscle injury or stress, the landscape of MuSCs rapidly changes, enabling them to promptly activate, proliferate, differentiate, and form nascent myofibers to repair damaged regions or respond to stress [4, 5]. Studies demonstrated that tissue damage induces stress responses in MuSCs, and their transcriptome and chromatin accessibility are altered as early as 30 min after muscle injury [6, 7]. However, protocols for dissociating MuSCs from bulk skeletal muscle typically last 4–6 h [8, 9], during which MuSCs are already

Nathalie Didier and Caroline E. Brun (eds.), *Muscle Stem Cells*, Methods in Molecular Biology, vol. 3093,
https://doi.org/10.1007/978-1-0716-5643-3_16,

considered early-activated. Therefore, there is a need to develop new techniques to preserve the genuine quiescent state of MuSCs for downstream analysis.

In situ fixation is performed by transcardial perfusion of the mouse with PFA (paraformaldehyde) [7, 10–14]. PFA creates covalent cross-links between nucleic acids and proteins, thereby preserving the quiescent signature of MuSCs. Notably, a mouse line with an internal fluorescent marker is required for this procedure, as antibodies are unavailable to detect the surface markers for fluorescent-activated cell sorting. Hindlimb muscles are dissected and dissociated according to the standard muscle stem cell isolation protocols [8], with a modification in which 20,000 U of collagenase II is used in the primary digestion. Subsequently, purified quiescent MuSCs are snap-frozen for library preparation.

The miRNeasy FFPE kit from Qiagen is used with minor modifications [15]. The kit provides essential steps for reversing formalin-cross-linked RNA, enabling high-quality RNA extraction. Purified RNA is then subjected to the SMART-seq2 (Switching Mechanism At the end of the 5′-end of the RNA Transcript sequencing) protocol for cDNA synthesis [16]. cDNA quality can be verified by fragment analysis [17]. cDNAs are tagmented by the Tn5 transposome, which randomly fragments the cDNAs and attaches adapter sequences for barcode assignments. The size of the tagmented cDNA is selected with beads to obtain 200–500 bp fragments, and the quality is verified by fragment analysis. Eventually, the DNA samples are pooled together to prepare the library for sequencing on Illumina platforms or MGISEQ platforms. In this book chapter, ssDNA circularization is performed for DNBSEQ (DNA nanoball sequencing) when using the MGISEQ platforms [18]. An additional adapter conversion step is performed to introduce a phosphate group at the 5′ end of the DNA fragments, facilitating the generation of the circular ssDNA library.

2 Materials

Ultrapure water is used for all steps in this protocol. Reagents are stored at −20°C unless otherwise specified. The solution is prepared for a single sample unless otherwise indicated. Solutions added to one sample tube are in a small volume. Prepare all solutions freshly, as master mixes, on ice.

2.1 RNA Extraction and Purification

1. miRNeasy FFPE kit (Qiagen): The kit is used to extract and purify PFA-fixed RNAs. Store the reagents based on the manufacturer's instructions.
2. Qubit RNA HS assay kit: Prepare the master mix of RNA HS reagent and RNA HS buffer in the ratio of 1:200. For the sample tube, combine 1 μL of sample with 199 μL of the master

mix. For standards 1 and 2, combine 10 μL of the standard and 190 μL of the master mix. Vortex briefly and incubate for 2 min at room temperature. Measure the RNA concentration using the Qubit fluorometer. Store the reagents according to the manufacturer's instructions.

2.2 cDNA Conversion

1. Oligo-dT primer: /5biotin/AAGCAGTGGTATCAACGCAGAGTACT$_{30}$VN-3′ (*see* **Note 1**).
2. TSO (template-switching oligo): /5biotin/AAGCAGTGGTATCAACGCAGAGTACATrGrG+G-3′ (*see* **Note 1**). After reconstitution, store the TSO at −80°C. Avoid freeze–thaw cycles.
3. IS PCR primer: /5biotin/-AAGCAGTGGTATCAACGCAGAGT-3′ (*see* **Note 1**).
4. Lysis mix: Mix 1 μL of 10 μM Oligo-dT primer, 1 μL of 10 μM dNTP, 0.1 μL of 40 U/μL RNase inhibitor, and 0.04 μL of 10% Triton X-100. Add the lysis mix to 1 ng of purified RNA. Top up the sample to 4 μL with ultrapure water. Triton X-100 is stored at room temperature.
5. RT (reverse transcription) mix: Mix 2 μL of 5× Superscript II first strand buffer, 2 μL of 5 M betaine, 0.9 μL of 100 mM magnesium chloride, 0.25 μL of 100 mM DTT, 0.1 μL of 100 μM TSO, 0.25 μL of 40 U/μL RNase inhibitor, and 0.5 μL of 200 U/μL Superscript II reverse transcriptase. Betaine is stored at 4°C.
6. PCR preamplification mix: Mix 12.5 μL of 2× KAPA HiFi Hot Start ReadyMix, 0.25 μL of 10 μM IS PCR primer, and 2.25 μL of ultrapure water.
7. AMPure XP beads/SPRIselect DNA size selection reagent: They are used for DNA cleanup and size selection at a later stage. Keep at 4°C and room temperature, respectively.
8. Qubit dsDNA HS assay kit: Prepare the master mix of dsDNA HS reagent and dsDNA HS buffer in the ratio of 1:200. For the sample tube, combine 1 μL of sample with 199 μL of the master mix. For standards 1 and 2, combine 10 μL of the standard and 190 μL of the master mix. Vortex briefly and incubate for 2 min at room temperature. Measure the dsDNA concentration using the Qubit fluorometer. Store the reagents according to the manufacturer's instructions.

2.3 DNA Quality Verification and Size Determination

In this section, the fragment analyzer (Agilent) is used with the DNF-474 high-sensitivity NGS fragment analysis kit (1 bp–6000 bp). A compatible 96-well plate is required. Each run detects one row of the 96-well plate (i.e., 12 wells). Store the reagents according to the manufacturer's instructions.

1. Fragment analysis gel: Add 1 μL of intercalating dye to 10 mL of gel for one row. For each extra row, add an additional 0.5 μL of intercalating dye and 5 mL of gel. Mix well by inverting the tube. Wrap the tube with aluminum foil to avoid light exposure. Equilibrate the gel to room temperature for at least 30 min. Only keep the mixture at 4°C for up to 2 weeks.

2.4 cDNA Tagmentation

1. 5× Tagmentation buffer: Mix 50 mM Tris-HCl, pH 8.0, 25 mM $MgCl_2$, and 50% (v/v) dimethylformamide (DMF). The buffer can be stored at −20°C for at least 6 months.
2. Transposome mix: The transposome is a complex constituted by the Tn5 dimers and oligonucleotides. The oligonucleotides assembled in the transposome consist of the 19-bp mosaic end (ME) sequence recognized by the Tn5 transposase with partial sequences of the Illumina adaptors. The Tn5 transposome can be assembled in-house [19] or ordered from commercial sources, such as the Tagment DNA Enzyme (TDE1), previously included in the discontinued Nextera DNA library preparation kits.
   Tn5MErev:
   5′-[phos]CTGTCTCTTATACACATCT-3′;
   Tn5ME-A (Illumina FC-121-1030):
   5′-TCGTCGGCAGCGTCAGATGTGTATAAGAGA-CAG-3′;
   Tn5ME-B (Illumina FC-121-1031):
   5′-GTCTCGTGGGCTCGGAGATGTGTATAAGAGA-CAG-3′
3. Tagmentation reaction mix: Mix 0.8 μL of 5× Tagmentation buffer and 0.1 μL of Transposome mix on ice. Add the tagmentation reaction mix to 1 ng of purified cDNA. Top up the sample to a total volume of 4 μL with ultrapure water.
4. PCR amplification mix: Mix 5 μL of 5× KAPA HiFi fidelity buffer, 0.75 μL of 10 mM KAPA dNTP mix, 0.5 μL of 1 U/μL KAPA HiFi DNA polymerase, 1 μL of 10 μM Tn5-forward primer, 1 μL of 10 μM Tn5-reverse primer, and 11.75 μL of ultrapure water.

2.5 Adapter Conversion and ssDNA Circularization

In this section, the MGIEasy universal library conversion kit (app-A) is used. Other circularization kits might also be applicable. All reagents are prepared on ice.

1. Adapter conversion PCR mix: Mix 25 μL of AC-PCR Amplification master mix and 3 μL of AC-PCR primer. Add the adapter conversion PCR mix to 50 ng of pooled DNA. Top up the sample to a total volume of 50 μL with ultrapure water.

2. ssDNA (single-stranded DNA) circularization mix: Mix 11.6 μL of App-A splint buffer and 0.5 μL of ligation enzyme.
3. Exo digestion reaction mix: Mix 1.4 μL of digestion buffer and 2.6 μL of digestion enzyme.
4. Qubit ssDNA assay kit: Prepare the master mix of ssDNA reagent and ssDNA buffer in the ratio of 1:200. For the sample tube, combine 1 μL of sample with 199 μL of the master mix. For standards 1 and 2, combine 10 μL of the standard and 190 μL of the master mix. Vortex briefly and incubate for 2 min at room temperature. Measure the ssDNA concentration using the Qubit fluorometer. Store the reagents according to the manufacturer's instructions.

3 Methods

Quiescent MuSCs (QSCs) are obtained by in situ fixation followed by fluorescent-activated cell sorting (FACS). The sorted cells are then transferred to a 1.5 mL tube and centrifuged at 2000 g, 4°C for 10 min. Supernatant is carefully removed (*see* **Note 2**). The cell pellet is snap-frozen in liquid nitrogen and transferred to −80°C for long-term storage. A minimum of 100 K cells is recommended.

In this protocol, carry out all procedures at room temperature unless otherwise specified. For steps involving RNA samples, wear a mask, and sterilize the workbench and pipettes with 70% ethanol, then with RNase AWAY.

3.1 RNA Extraction and Purification

1. Add 100 μL of Buffer PKD to the cell pellet, then vortex to mix.
2. Centrifuge at 11,000 *g* for 1 min at room temperature.
3. Add 10 μL of Proteinase K. Mix gently by pipetting up and down.
4. Incubate the sample at 56°C and 1000 rpm for 1 h (*see* **Note 3**).
5. Transfer the lower, colorless phase into a new 2 mL microcentrifuge tube.
6. Incubate on ice for 3 min.
7. Centrifuge at 20,000 *g* for 20 min at room temperature.
8. Transfer the supernatant to a new tube without disturbing the pellet.
9. Add 16 μL of DNase booster buffer and 10 μL of DNase I stock solution. Mix by inverting the tube. Centrifuge briefly.
10. Incubate at room temperature for 15 min.

11. Add 320 μL of Buffer RBC and mix the lysate by pipetting up and down thoroughly.
12. Add 1120 μL of 100% ethanol to the sample and mix by pipetting up and down. Do not centrifuge.
13. Transfer 700 μL of the sample to the RNeasy MiniElute spin column. Centrifuge at 8000 *g* for 30 s at room temperature. Discard the flow through.
14. Repeat Step 13 until the entire sample has passed through the column.
15. Add 500 μL Buffer RPE to the column. Centrifuge at 8000 *g* for 30 s at room temperature. Discard the flow through.
16. Repeat Step 15 once.
17. Transfer the spin column to a new 2 mL collection tube (*see* **Note 4**). Centrifuge at full speed for 5 min.
18. Transfer the spin column to a new 1.5 mL microcentrifuge tube. Add 15 μL of ultrapure water directly to the spin column membrane (*see* **Notes 5 and 6**).
19. Incubate for 2 min. Centrifuge at full speed for 1 min at room temperature to elute the RNA.
20. Measure the RNA concentration using the Qubit RNA HS assay kit with 1 μL of RNA.
21. Store the RNA samples at −80°C for long-term storage.

3.2 cDNA Conversion

1. Add the lysis mix to 1 ng of total RNA (*see* **Notes 7 and 8**). Top up to a total volume of 4 μL with ultrapure water.
2. Incubate at 72°C for 3 min, then immediately place on ice.
3. Centrifuge briefly, then place on ice.
4. Add 6 μL of RT mix into the sample. Mix well, centrifuge briefly, and place it on ice.
5. Perform reverse transcription using the following parameters (Table 1).

Table 1
PCR program setup for reverse transcription. Hot lid temperature: 105°C

Temperature	Duration	Number of cycles
42°C	90 min	1 cycle
50°C	2 min	5 cycles (*see* **Note 9**)
42°C	2 min	
70°C	15 min	1 cycle
4°C	Infinite	

Table 2
PCR program setup for preamplification. Hot lid temperature: 105°C

Temperature	Duration	Number of cycles
98°C	3 min	1 cycle
98°C	15 s	14 cycles (*see* **Notes 7 and 8**)
67°C	20 s	
72°C	6 min	
72°C	5 min	1 cycle
4°C	Infinite	

6. Add the PCR preamplification mix to the sample. Mix well, centrifuge briefly, and put it on ice.
7. Perform preamplification using the following parameters (Table 2).
8. Add 20 μL (0.8×) of AMPure XP Beads to the sample (*see* **Note 10**). Mix well and incubate for 10 min.
9. Place the tube on a magnetic stand and incubate for 2 min to separate the beads from the supernatant. Discard the supernatant.
10. Add 200 μL of freshly prepared 80% ethanol to the sample while the tube remains on the magnetic stand. Incubate for 30 s and discard the supernatant.
11. Repeat Step 10 once more.
12. Completely remove the residual ethanol and air-dry the beads for 2 min while the tube remains on the magnetic stand with the tube lid open (*see* **Note 11**).
13. Remove the tube from the magnetic stand. Add 20 μL of ultrapure water and mix thoroughly with the beads by gentle vortexing. Incubate for 10 min.
14. Place the tube back in the magnetic stand for 2 min to separate the beads from the supernatant. Transfer the supernatant into a new tube. Discard the beads.
15. Measure the cDNA concentration using the Qubit dsDNA HS Assay Kit with 1 μL of DNA.
16. Store the cDNA samples at −20°C for long-term storage.

3.3 cDNA Quality Verification using the Fragment Analyzer

1. Add 2 μL of 1–5 ng purified cDNA to each well of a row of the 96-well plate (*see* **Notes 12–14**).
2. Add 2 μL of ladder to column 12 of the same row.

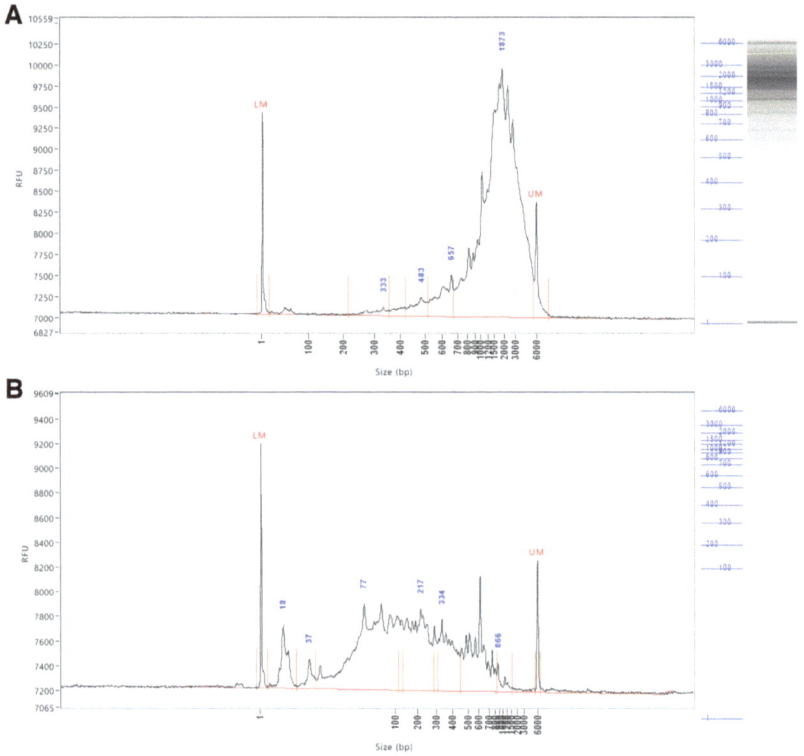

Fig. 1 (**A**) Representative image of a good-quality distribution. The cDNA fragments have an average length of 1000–3000 bp. (**B**) Representative image of a poor-quality size distribution. Substantial cDNA signal at shorter fragment sizes indicates underlying RNA degradation.

3. Add 22 μL of diluent marker to each well. Mix well by pipetting up and down.
4. Centrifuge briefly to ensure no air bubbles.
5. Place the gel solution and the 96-well plate in the designated positions in the fragment analyzer.
6. Select the "DNF-474-33 HS NGS Fragment 1-6,000 bp" protocol from the fragment analyzer and start the run.
7. Conduct the smear analysis to determine the DNA abundance at 1000 bp or higher (*see* **Note 15** and Fig. 1).

3.4 Tagmentation

1. Add the tagmentation reaction mix to 1 ng cDNA. Top up to a total volume of 4 μL with ultrapure water. Mix well with brief centrifugation.
2. Incubate at 55°C for 10 min, then immediately place on ice.
3. Add 1 μL of 0.1% freshly prepared sodium dodecyl sulfate (SDS) to the sample. Mix well with brief centrifugation.
4. Incubate at 55°C for 5 min to terminate tagmentation.

Table 3
PCR program setup for tagmented cDNA amplification. Hot lid temperature: 105°C

Temperature	Duration	Number of cycles
72°C	5 min	1 cycle
98°C	3 min	1 cycle
98°C	20 s	
60°C	30 s	12 cycles
72°C	1 min	
72°C	10 min	1 cycle
4°C	Infinite	

5. Add the PCR amplification mix to the sample. Mix well, then centrifuge briefly in a microcentrifuge.
6. Perform PCR amplification using the following parameters (Table 3).
7. Store the tagmented cDNA at 4°C for short-term storage or −20°C for long-term storage.

3.5 Size Selection

1. Top up the sample to 50 μL with ultrapure water.
2. Add 27.5 μL (0.55×) of AMPure XP beads to the sample (*see* **Note 10**). Mix well, centrifuge briefly, and incubate for 10 min.
3. Place the tube onto a magnetic stand for 2 min. Transfer the supernatant to a new tube without bead carryover. Discard the beads.
4. Add 7.5 μL (0.7×) of AMPure XP beads to the supernatant. Mix well, centrifuge briefly, and incubate for 10 min.
5. Place the tube onto a magnetic stand for 2 min. Remove the supernatant.
6. Add 200 μL of freshly prepared 80% ethanol to the tube while it remains on the magnetic stand, without disturbing it. Incubate for 30 s and discard the supernatant.
7. Repeat Step 6 once more.
8. Completely remove the residual ethanol and air-dry the beads for 2 min while the tube remains on the magnetic stand with the tube lid open (*see* **Note 11**).
9. Remove the tube from the magnetic stand. Add 20 μL of ultrapure water and mix thoroughly with the beads by gentle vortexing. Incubate for 10 min.

10. Place the tube onto the magnetic stand for 2 min to separate the beads from the supernatant. Transfer the supernatant into a new tube. Discard the beads.
11. Measure the purified DNA concentration using the Qubit dsDNA HS assay kit with 1 μL of DNA.
12. Store the purified DNA at 4°C for short-term storage or −20°C for long-term storage.

3.6 Size Determination

1. Add 2 μL of 1–5 ng purified DNA to each well of a row of the 96-well plate (*see* **Notes 12–14**).
2. Add 2 μL of ladder to column 12 of the same row.
3. Add 22 μL of diluent marker to each well. Mix well by pipetting up and down.
4. Centrifuge briefly to ensure no air bubbles.
5. Place the gel solution and the 96-well plate in the designated positions in the fragment analyzer.
6. Select the "DNF-474-33 HS NGS Fragment 1-6,000 bp" protocol from the fragment analyzer and start the run.
7. Conduct the smear analysis to determine the DNA abundance within 100 bp and 1000 bp. Determine the average size of each sample (*see* **Notes 16 and 17**, and Fig. 2).
8. Libraries are ready for Illumina sequencing after this step. For MGISEQ sequencing, please follow the Steps outlined in 3.7 and 3.8.

3.7 Adapter Conversion

1. Determine the molar quantity of each of the DNA samples using the following equation:

$$1\ \text{pmol of DNA} = \frac{\text{Average size of DNAfragment (bp)}}{1000\ \text{bp}} \times 660\ \text{ng}$$

2. Pool all the DNA samples together with equal molar quantities, or according to the desired read distribution across samples.
3. Measure the pooled DNA concentration using the Qubit dsDNA HS assay kit with 1 μL of DNA.
4. Add the Adapter conversion PCR mix to 50 ng of the pooled DNA. Top up to a total volume of 50 μL with ultrapure water (*see* **Note 18**). Mix well with brief centrifugation.
5. Perform the adapter conversion PCR using the following parameters (Table 4)
6. Add 60 μL of (1.2×) AMPure XP beads to the sample (*see* **Note 10**). Mix well, centrifuge briefly, and incubate for 10 min.

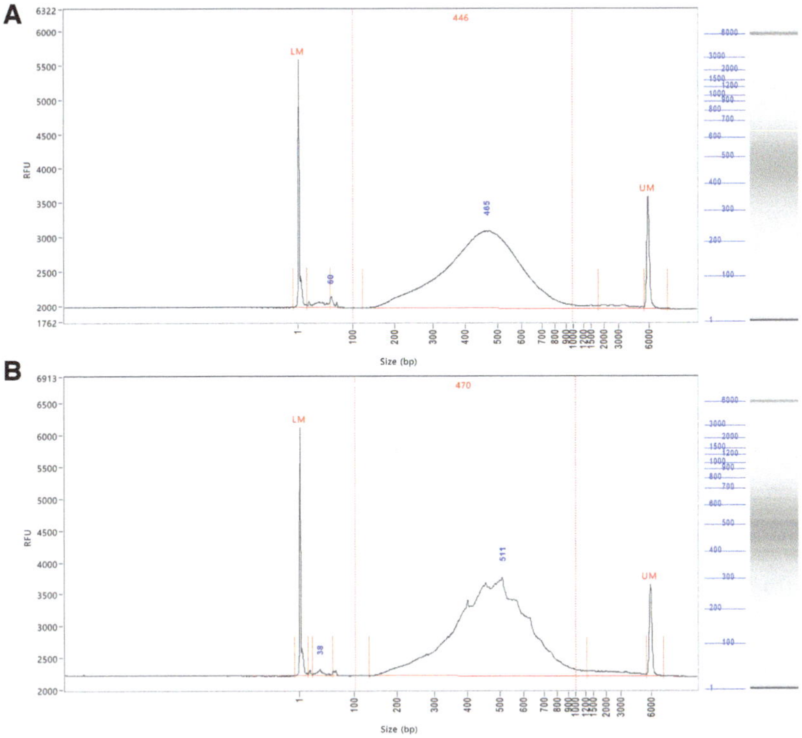

Fig. 2 (**A**) Representative image of a good-quality distribution. The average DNA size ranges from 100 bp to 1000 bp. (**B**) Representative image of a bad-quality size distribution. Spiky curves are observed

Table 4
PCR program setup for adapter conversion. Hot lid: 105°C

Temperature	Duration	Number of cycles
98°C	3 min	1 cycle
98°C	30 s	
62°C	15 s	5 cycles
72°C	1 min	
72°C	5 min	1 cycle
4°C	Infinite	

7. Place the tube onto a magnetic stand for 2 min. Remove the supernatant.
8. Add 200 μL of freshly prepared 80% ethanol to the tube while the tube remains on the magnetic stand. Incubate for 30 s and discard the supernatant.
9. Repeat Step 8 once more.

10. Completely remove the residual ethanol and air-dry the beads for 2 min while the tube remains on the magnetic stand with the tube lid open (*see* **Note 11**).
11. Remove the tube from the magnetic stand. Add 30 μL of ultrapure water and mix thoroughly with the beads by gentle vortexing. Incubate for 10 min.
12. Place the tube back onto the magnetic stand for 2 min to separate the beads from the supernatant. Transfer the supernatant into a new tube. Discard the beads.
13. Measure the purified DNA concentration using the Qubit dsDNA HS assay kit with 1 μL of DNA.
14. Store the purified DNA at 4°C for short-term storage or −20°C for long-term storage.

3.8 ssDNA Circularization (Only Applicable to DNBSEQ Platform)

1. Transfer 1.5 pmol of the pooled DNA to a new tube. Top up to a total volume of 48 μL with ultrapure water.
2. Incubate the tube at 95°C for 3 min and then immediately place it on ice for 2 min.
3. Add the circularization reaction mix to the denatured DNA. Mix well with brief centrifugation.
4. Incubate at 37°C for 30 min, then immediately place on ice.
5. Add 4 μL of the Exo digestion mix. Mix well with brief centrifugation.
6. Incubate at 37°C for 30 min.
7. Add 7.5 μL of digestion stop buffer. Mix well with brief centrifugation.
8. Add 170 μL of (2.4×) AMPure XP beads to the sample (*see* **Note 10**). Mix well, centrifuge briefly, and incubate for 10 min.
9. Place the tube onto a magnetic stand for 2 min. Remove the supernatant.
10. Add 200 μL of freshly prepared 80% ethanol to the tube while the tube remains on the magnetic stand. Incubate for 30 s and discard the supernatant.
11. Repeat Step 10 once more.
12. Completely remove the residual ethanol and air-dry the beads for 2 min while the tube remains on the magnetic stand with the tube lid open (*see* **Note 11**).
13. Remove the tube from the magnetic stand. Add 25 μL of ultrapure water and mix thoroughly with the beads. Incubate for 10 min.

14. Place the tube back onto the magnetic stand for 2 min to separate the beads from the supernatant. Transfer the supernatant into a new tube. Discard the beads.
15. Measure the purified DNA concentration with the Qubit ssDNA assay kit with 1 μL of ssDNA.
16. The final yield of the ssDNA library should be more than 400 fmol. Determine the molar quantity of the ssDNA library with the following equation:

$$400 \text{ fmol of circulars sDNA} = \frac{\text{Average size of ssDNA (nt)}}{1000 \text{ bp}} \times 330 \text{ ng}$$

17. Store the library at 20°C.

4 Notes

1. The 5′ biotin serves as a blocking moiety on the oligo-dT primer, TSO, and IS-PCR primer, preventing the formation of concatemers and primer-dimers that would otherwise interfere with DNA amplification [16, 20].
2. The cells form a small pellet along the lower side wall of the 1.5 mL microcentrifuge tube. Tilt the tube so that the pellet remains at the lowest point. Use a pipette with double tips to aspirate the supernatant gently, avoiding disturbance of the cell pellet. Do not use an aspirator.
3. If the agitation function is not available, it is recommended to pipette up and down to mix the sample every 3–5 min to ensure thorough lysis.
4. This step removes all ethanol residue from the column. Avoid contacting the flow-through when transferring the column to the new tube to prevent carryover.
5. The recommended elution volume is 15–30 μL. If the input cell number is low, use the lowest elution volume. If the input cell number is high, it is preferable to use a higher elution volume to improve RNA yield.
6. The pipette tip should not touch the spin column membrane.
7. If the RNA concentration is low, the maximum amount of RNA input is 1.86 μL without adding ultrapure water. If the input is still less than 1 ng, increase the number of PCR cycles in Sect. 3.2, Step 7 accordingly.
8. It is optional to increase the RNA input to 2–5 ng. The higher RNA input should be accompanied by a reduction in the number of PCR cycles in Sect. 3.2, Step 7 to 12 cycles.

9. Typically, five cycles are sufficient to complete reverse transcription. To increase the product yield, increase the number of cycles to 10.
10. AmPure XP beads are kept at 4°C. Equilibrate the beads to room temperature for at least 30 min before use. Alternatively, use SPRIselect DNA size selection reagent, which is stored at room temperature.
11. The air-drying time should not be too long or too short. Over-drying the beads reduces elution efficiency. Any residual ethanol from insufficient drying will also affect elution efficiency.
12. The detection limit of the DNF-474 high-sensitivity NGS fragment analysis kit is 1–5 ng. Further dilute the cDNA samples if necessary.
13. The fragment analysis is sensitive to the presence of beads. Ensure there is no bead carryover in the DNA samples.
14. For each row, one well should be filled with the ladder for size calibration. Therefore, a single row can accommodate up to 11 samples. If there are fewer than 11 samples, add 24 μL of Blank solution to the empty well.
15. A good cDNA library profile has a major peak at ~1.5–2 kb with a small amount of fragments below 500 bp and primer dimers (Fig. 1A). A high amount of short fragment smear (Fig. 1B) indicates severe RNA degradation caused by over-fixation or RNase contamination. Repeat the preparation of the fluorescence-activated cell sorting-isolated fixed muscle stem cells.
16. The DNA abundance within 100 bp and 1000 bp should be higher than 90%, and the average size should stand within 400 bp and 600 bp (Fig. 2A). If the sample has a significantly long cDNA portion, rerun Sect. 3.5. A spike around 100 bp corresponds to the primer-dimer, which is difficult to eliminate by further size selection. Proceed to Sect. 3.6.
17. A spiky curve (Fig. 2B) indicates overamplification of certain fragments during library construction. The library complexity is relatively low. Rerun the protocol starting from Sect. 3.3.
18. If the concentration of the pooled DNA is low such that more than 22 μL is needed for 50 ng of DNA, conduct an extra round of DNA selection with 1.2× beads and elute the DNA with a lower elution volume to concentrate the pooled DNA.

Acknowledgments

We acknowledge the Laboratory Animal Facility and Biosciences Central Research Facility at HKUST (Clear Water Bay) for their support. This work was funded by the Hong Kong Research Grant Council (GRF16100322, GRF16102021, GRF16101524, RFS2425-6S01, C6018-19G, AoE/M-604/16, T13-605/18W, T13-602/21N), the CAS-Croucher Funding Scheme for Joint Laboratories (CAS24SC02), and the Hong Kong Epigenomics Project (Lo Ka Chung Charitable Foundation). This study was supported by the InnoHK initiative of the Innovation and Technology Commission of the Hong Kong Special Administrative Region Government. This study was also supported in part by the Innovation and Technology Commission (ITCPD/17-9). T.H.C. is the S H Ho Professor of Life Science at HKUST.

Competing Interests The authors have no competing interests to declare.

References

1. de Morree A, Rando TA (2023) Regulation of adult stem cell quiescence and its functions in the maintenance of tissue integrity. Nat Rev Mol Cell Biol 24(5):334–354
2. Cheung TH, Rando TA (2013) Molecular regulation of stem cell quiescence. Nat Rev Mol Cell Biol 14(6):329–340
3. Urbán N, Cheung TH (2021) Stem cell quiescence: the challenging path to activation. Development 148(3):dev165084
4. Le Grand F, Rudnicki MA (2007) Skeletal muscle satellite cells and adult myogenesis. Curr Opin Cell Biol 19(6):628–633
5. Yin H, Price F, Rudnicki MA (2013) Satellite cells and the muscle stem cell niche. Physiol Rev 93(1):23–67
6. Machado L, Geara P, Camps J et al. (2021) Tissue damage induces a conserved stress response that initiates quiescent muscle stem cell activation. Cell Stem Cell 28(6):1125–1135.e7
7. Dong A, Liu J, Lin K et al. (2022) Global chromatin accessibility profiling analysis reveals a chronic activation state in aged muscle stem cells. iScience 25(9):104954
8. Liu L, Cheung TH, Charville GW, Rando TA (2015) Isolation of skeletal muscle stem cells by fluorescence-activated cell sorting. Nat Protoc 10(10):1612–1624
9. Elizalde G, Munoz AZ, Almada AE (2023) Protocol for the isolation of mouse muscle stem cells using fluorescence-activated cell sorting. STAR Protoc 4(4):102656
10. Yue L, Wan R, Luan S, Zeng W, Cheung TH (2020) Dek modulates global intron retention during muscle stem cells quiescence exit. Dev Cell 53(6):661–676.e6
11. Yue L, Cheung TH (2020) Protocol for isolation and characterization of *In Situ* fixed quiescent muscle stem cells. STAR Protoc 1(3):100128
12. Dong A, Chan ITC, Cheung TH (2023) ATAC-seq protocol for the profiling of chromatin accessibility of *in situ* fixed quiescent and activated muscle stem cells. STAR Protoc 4(3):102376
13. Zeng W, Zhang W, Tse EHY et al. (2023) Restoration of CPEB4 prevents muscle stem cell senescence during aging. Dev Cell 58(15):1383–1398.e6
14. Zhang W, Zeng W, Wong PS, Cheung TH (2023) Protocol for low-input proteomic analysis of *in situ* fixed adult murine muscle stem cells. STAR Protoc 4(4):102750
15. miRNeasy FFPE Kit | FFPE Tissue microRNA Extraction | QIAGEN. https://www.qiagen.com/us/products/discovery-and-translational-research/dna-rna-purification/rna-purification/mirna/mirneasy-ffpe-kit
16. Picelli S, Faridani OR, Björklund ÅK, Winberg G, Sagasser S, Sandberg R (2014) Full-length RNA-seq from single cells using Smart-seq2. Nat Protoc 9(1):171–181

17. Agilent Technologies Inc. (2022). Quality assessment of NGS libraries using Agilent automated electrophoresis systems
18. MGIEasy Universal Library Conversion Kit (APP-A) user manual 3.0.
19. Picelli S, Björklund ÅK, Reinius B, Sagasser S, Winberg G, Sandberg R (2014) Tn5 transposase and tagmentation procedures for massively scaled sequencing projects. Genome Res 24(12): 2033–2040
20. Picelli S (2016) Single-cell RNA-sequencing: the future of genome biology is now. RNA Biol 14(5):637–650

Check for updates

Chapter 17

Low-Input Proteomic Analysis of In Situ Fixed Adult Murine Muscle Stem Cells

Wenxin Zhang, Indigo T. C. Chan, Ning-Sum To and Tom H. Cheung

Abstract

Preserving the quiescent state of adult stem cells is challenging because they rapidly activate in response to external stimuli. Conventional isolation methods for muscle stem cells (MuSCs), such as enzymatic digestion followed by fluorescence-activated cell sorting (FACS), are time-consuming, resulting in early activation and loss of quiescent features. Here, we introduce an in situ fixation protocol for the dissociation and proteomic characterization of quiescent MuSCs using trapped ion-mobility time-of-flight (timsTOF) mass spectrometry. This protocol encompasses mouse perfusion, FACS preparation and sorting, protein extraction, decrosslinking, digestion, and subsequent liquid chromatography–mass spectrometry (LC–MS) analysis. Our approach effectively preserves the quiescent features of MuSCs, enabling precise proteomic analysis with limited sample amounts and providing novel insights into stem cell biology.

Keywords Mass spectrometry, Stem cell, Quiescence, Proteomics, In situ fixation

1 Introduction

Adult stem cell quiescence is a critical cellular state that maintains tissue homeostasis and regeneration capacity while preserving the stem cell pool in low-turnover tissue throughout the lifetime [1, 2]. Contrary to a passive, inactive state, quiescence is actively maintained as a homeostatic condition. In skeletal muscle, adult muscle stem cells (MuSCs), commonly known as satellite cells, respond rapidly to injury [3, 4]. Early activation of MuSCs involves a transition from a quiescent state with low global protein synthesis to one with increased translational machinery and protein synthesis [5]. For instance, quiescent MuSCs retain MyoD and Dek transcripts while post-transcriptionally inhibiting their translation [6, 7]. Upon injury, MuSCs promptly enhance the translation activity, facilitating a rapid activation response essential for tissue regeneration.

Nathalie Didier and Caroline E. Brun (eds.), *Muscle Stem Cells*, Methods in Molecular Biology, vol. 3093, https://doi.org/10.1007/978-1-0716-5643-3_17,

While the rapid activation response of MuSCs is crucial for tissue regeneration, it poses challenges for studying proteomic features in quiescent MuSCs. Studies have found that MuSCs rapidly lose their quiescent features after isolation from muscle tissue, and freshly isolated MuSCs already exhibit an early activation signature [3, 4]. To overcome this challenge, paraformaldehyde (PFA) is used as a fixative in many protocols to preserve the quiescent features. PFA functions as a cross-linking agent that covalently binds proteins and nucleic acids, preserving cells and tissues at the molecular level and effectively freezing the cell state at the moment of fixation [8]. However, fixation increases cell stiffness and leads to cell loss during the isolation process [9]. The limited number of MuSCs constrains the downstream analysis of the purified stem cell population. In this protocol, we present a low-input liquid chromatography–mass spectrometry-based proteomic approach to elucidate the protein profile of a small number of MuSCs while preserving their quiescent signature [10–12].

2 Materials

Animals used in this protocol were 3-month-old Pax7-nGFP C57BL/6 J male mice.

2.1 Materials for Sample Preparation

1. Phosphate-buffered saline (PBS) 10× stock solution (2 L): 26.8 mM KCl (4 g), 10 mM $Na_2HPO_4{\cdot}2H_2O$ (35.6 g), 17.6 mM KH_2PO_4 (4.8 g), 1.37 M NaCl (160 g) in Milli-Q H_2O.
2. PBS 1× buffer (pH 7.4): Diluted in Milli-Q H_2O and filtered using a 0.22 μm sterile filter. Precool at 4°C before use.
3. 0.5% Paraformaldehyde (PFA) in PBS 1×. Prepare fresh before use. Prepare 30 mL for one mouse.
4. 2 M Glycine: Diluted in PBS 1× and filtered using a 0.22 μm sterile filter. Precool at 4°C before use.
5. Pentobarbitone: 20 mg/mL pentobarbital sodium diluted in sterile 0.9% NaCl. Store at 4°C for up to 1 month or follow the manufacturer's instructions.
6. Collagenase II stock solution: 3000 U/mL Collagenase II diluted in PBS 1×. Store at −20°C for up to 1 year (1 mL is needed for 1 mouse).
7. Dispase stock solution: 33 U/mL dispase diluted in PBS 1×. Store at −20°C for up to 1 year (1 mL is needed for 1 mouse).
8. Wash medium: 10% horse serum (HS), 1% penicillin/streptomycin (P/S) diluted in Ham's F-10 Nutrient Mix. Store at 4°C for up to a week (500 mL is needed for two mice).

9. Muscle dissociation buffer: 2000 U/mL Collagenase II diluted in wash medium. Prepare fresh (10 mL is needed for one mouse).
10. A perfusion pump with a rotational speed of 70 rpm is required to stimulate blood circulation during fluid delivery.

2.2 Protein/Peptide Digestion and Purification

1. **RIPA lysis buffer (2% SDS, 1% EDTA):** 50 mM Tris-HCl (1 M, pH 7.4), 150 mM NaCl, 1.5% sodium deoxycholate, 1% Triton X-100, 1 mM EDTA, 2% SDS. Store at 4°C (*see* **Note 1**).
2. **80% ethanol:** Freshly prepared just before use.
3. **2% DMSO:** Mix 20 μL DMSO (LC–MS grade) with 980 μL deionized water.
4. **Preomics iST 96× kit** (Supplier, catalogue number P.O.00027)
5. **SP3 beads:** Vortex 200 μL E3 and E7 beads until no aggregation is observed. Wash twice with 500 μL UltraPure water. Top up to a total volume of 1 mL with water and mix.
6. **20% formic acid:** Prepare 20% formic acid solution by diluting the commercial stock solution in water.

3 Methods

3.1 Mouse Perfusion

1. Aliquot 30 mL of PBS 1×, 30 mL of 0.5% PFA, and 30 mL of 2 M Glycine into 50 mL Falcon tubes and place on ice.
2. Set up the perfusion pump at 70 rpm (5 mL/min) and the dissection stage.
3. Anesthetize a mouse with 20 mg/mL pentobarbitone via intraperitoneal injection (*see* **Notes 2 and 3**).
4. Lay the anesthetized mouse on the dissection stage in the supine position and immobilize it with surgical tape/needles. Open the chest and expose the heart for perfusion (Fig. 1A).
5. Lower the pipe to PBS 1×. Turn on the pump to drain the air from the butterfly needle and pause. Create a small incision on the right atrium of the heart for fluid removal (Fig. 1B). Pierce the heart in the left ventricle with the butterfly needle (Fig. 1C).
6. Perfuse the mouse with cold PBS 1× for 5 min, then 0.5% PFA for 5 min, and 2 M Glycine for 5 min to quench the formaldehyde (*see* **Notes 4 and 5**).

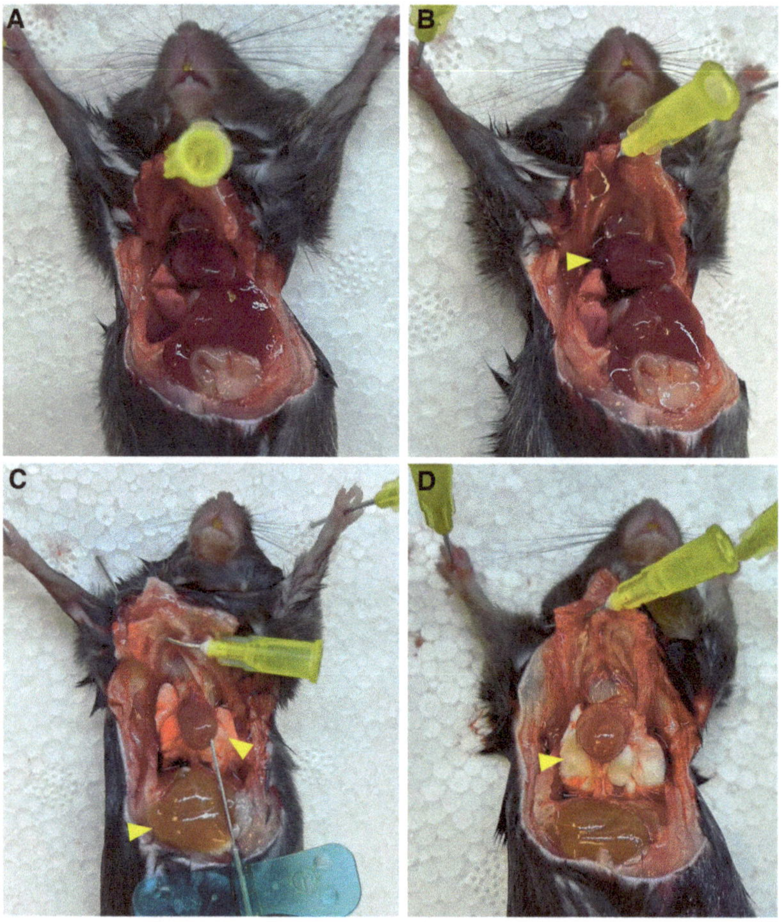

Fig. 1 Graphical illustration of perfusion. (**A**) and (**B**) Before perfusion. (**C**) After perfusion. (**D**) Perfusion in lungs

3.2 Muscle Slurry Preparation

1. Prepare a 10 cm glass Petri dish with 10 mL of cold wash medium for each mouse.
2. Dissect the hindlimb muscles and place them into the Petri dish.
3. Transfer the muscle to the lid of glass dishes and mince the muscle into fine pieces (~1 mm in diameter) in 2 mL of muscle dissociation buffer.
4. Transfer the minced muscle pieces together with the dissociation buffer to the 50 mL Falcon tube containing the remaining 8 mL of dissociation buffer and seal the tube with Parafilm.
5. Incubate the tube in a 37°C shaking water bath for 90 min at 65 rpm.
6. Top up the tube volume to 50 mL with cold wash medium.
7. Centrifuge the tissue suspension at 500 g for 10 min at 4°C. Aspirate the supernatant until only 20 mL of the buffer remains in the tube.

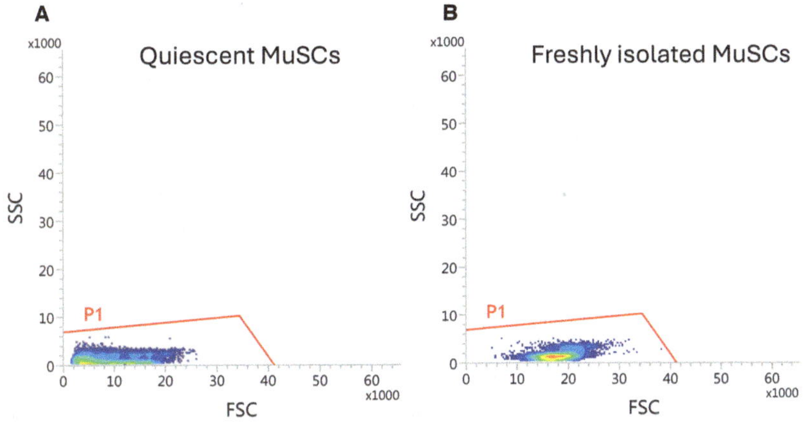

Fig. 2 Sort plot of quiescent MuSCs. (**A**) and freshly isolated MuSCs (**B**). X-axis FSC forward scatter, Y-axis SSC side scatter

8. Add 1 mL of stock collagenase II solution and 1 mL of stock dispase solution to the tube and resuspend the pellet by pipetting up and down.
9. Seal the tube with Parafilm and incubate in a 37°C shaking water bath for 30 min at 65 rpm.
10. After digestion, use a 20 mL syringe and a 20-gauge needle to aspirate and eject the muscle suspension 10 times (*see* **Note 6**).
11. Place a 40 μm nylon cell strainer onto a new 50 mL Falcon tube, and filter the cell suspension into the new tube.
12. Rinse the original tube with cold wash medium and transfer to the new tube with the strainer. Top up the tube volume to 50 mL with cold wash medium.
13. Centrifuge the tube at 500 g for 10 min at 4°C. Remove all the supernatant without disturbing the cell pellet.
14. Resuspend the cell pellet in 1 mL of wash medium, and transfer it to the FACS loading tube through a 35 μm nylon mesh filter cap.
15. Set up the cell sorter following the manufacturer's specifications with a 70 μm nozzle and sort out MuSCs (*see* Fig. 2 and **Note 7**).
16. After sorting, cells can be directly processed or stored at −80°C. To store the cells, centrifuge at 4°C, 2000*g* for 10 min, remove the supernatant, add ice-cold PBS, and centrifuge again at 4°C, 2000*g* for 10 min. Carefully remove the supernatant, snap-freeze the tube in liquid nitrogen, and store at −80°C (*see* **Note 8**).

3.3 Protein and Peptide Preparation

In this protocol, protein and peptide single-pot solid-phase-enhanced sample preparation (SP3) are applied to achieve reproducible peptide cleanup for mass spectrometry.

1. Add cold RIPA lysis buffer to the cell pellet. Ensure the final protein concentration is higher than 0.1 μg/μL. Mix thoroughly and incubate on ice for 10 min.
2. Incubate the protein lysate at 70°C for 2 h to de-crosslink the protein.
3. Measure the protein concentration and normalize across samples using the Protein BCA assay.
4. Aliquot 5 μg of protein into a 1.5 mL LoBind tube, add 20 μL of Preomic LYSE buffer, and denature the sample at 83°C for 10 min, 1100 rpm with the Eppendorf Sealing Mat (*see* **Note 9**).
5. Cool down samples to room temperature, and spin down briefly at 100 g.
6. Add 10 μL of prewashed SP3 magnetic beads to samples and vortex to mix thoroughly (*see* **Note 10**).
7. Add 90 μL of acetonitrile to the cooled denatured samples immediately. Incubate at 25°C, 1100 rpm for 10 min with the Eppendorf Sealing Mat.
8. Incubate the tube on a magnetic stand until the solution becomes clear. Remove the solution.
9. Add 200 μL of freshly prepared 80% ethanol. Wash at 25°C, 1100 rpm for 10 min. Remove all liquid from the magnetic stand. Repeat this step once more.
10. Add 200 μL of acetonitrile. Wash at 25°C, 1100 rpm for 10 min. Remove all of the liquid from the magnetic stand.
11. Add 10 μL of Preomic LYSE buffer and 10 μL of Preomics DIGEST to SP3-beads-aggregated samples and incubate at 37°C for 1 h with the Eppendorf Sealing Mat.
12. Add 450 μL of acetonitrile immediately for peptide precipitation. Shake at 25°C, 1100 rpm for 10 min without the sealing mat (*see* **Note 11**).
13. Place the sample on a magnetic stand for 2 min and remove all of the liquid.
14. Add 200 μL of acetonitrile. Wash at 25°C, 1100 rpm for 10 min. Remove all of the liquid and air-dry the samples for 2 min on the magnetic stand.
15. Elute the peptides with 15–20 μL 2% DMSO. Incubate at 25°C, 1100 rpm for 10 min.
16. Put the samples on a magnetic stand for 2 min.
17. Filtered the eluted peptide solution with a Corning Costar 0.22 μm CA filter.

18. Acidify the peptide solution by adding sufficient 20% formic acid to a final concentration of 0.1% formic acid and transfer to the LC vial (*see* **Note 12**).

3.4 Liquid Chromatography-Mass Spectrometry Analysis

Samples are analyzed using the TimsTOF Pro mass spectrometer (Bruker Daltonics) coupled with a NanoElute system (Bruker Daltonics) using solvent A (0.1% formic acid in water) and solvent B (0.1% formic acid in acetonitrile).

1. Tryptic peptides are loaded on the IonOpticks Aurora Ultimate CSI 25 × 75 C18 UHPLC column. The elution gradient is set as shown in Fig. 3 with a flow rate of 0.3 μL/min and a column oven set to 50°C.
2. Eluted peptides are measured in diaPASEF mode with the base method m/z range 100–1700 and $1/k_0$ range of 0.85–1.30 V × s × cm^{-2}. PASEF m/z window and mobility range are listed in Table 1 with 100 ms cycle time at 100% duty cycle. The source parameters are 1400 V for capillary voltage, 3.0 L/min for dry gas, and 180°C for dry temperature using Captive Spray (Bruker Daltonics). 27 eV and 45 eV of collision energies are allocated for $1/K_0 = 0.85$ Vs/cm^2 and $1/K_0 = 1.30$ Vs/cm^2, respectively.

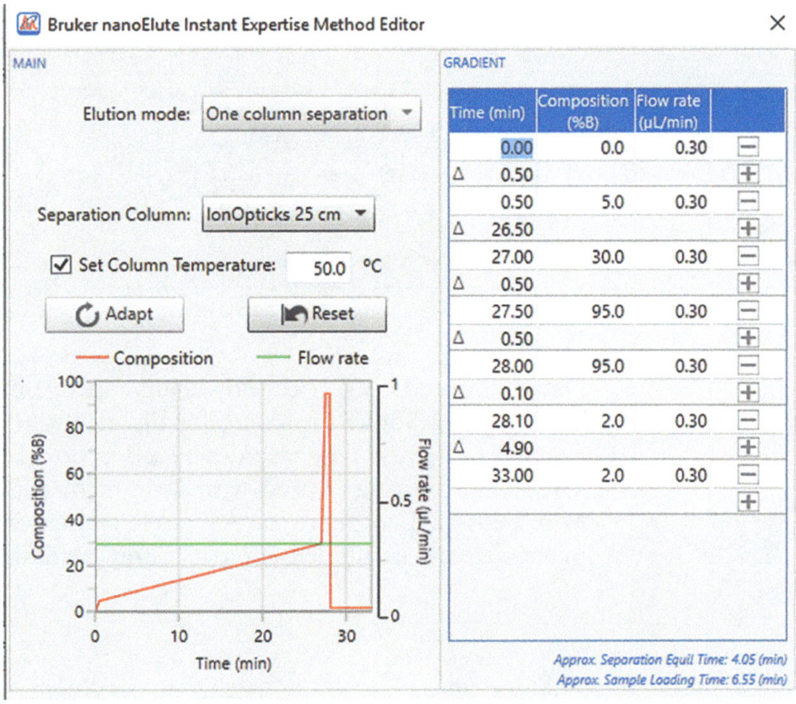

Fig. 3 Set up of IonOpticks Aurora Ultimate

Table 1
PASEF setup

#MS type	Cycle Id	Start IM [1/K0]	End IM [1/K0]	Start mass [m/z]	End mass [m/z]	CE [eV]
MS1	0	-	-	-	-	-
PASEF	1	1.0465	1.1981	875	900	-
PASEF	1	0.893	1.0446	675	700	-
PASEF	1	0.85	0.8911	475	500	-
PASEF	2	1.0657	1.2173	900	925	-
PASEF	2	0.9122	1.0638	700	725	-
PASEF	2	0.85	0.9103	500	525	-
PASEF	3	1.0849	1.2365	925	950	-
PASEF	3	0.9314	1.083	725	750	-
PASEF	3	0.85	0.9295	525	550	-
PASEF	4	1.1041	1.2557	950	975	-
PASEF	4	0.9506	1.1022	750	775	-
PASEF	4	0.85	0.9487	550	575	-
PASEF	5	1.1233	1.2749	975	1000	-
PASEF	5	0.9698	1.1214	775	800	-
PASEF	5	0.85	0.9679	575	600	-
PASEF	6	0.989	1.1406	800	825	-
PASEF	6	0.85	0.9871	600	625	-
PASEF	7	1.0082	1.1598	825	850	-
PASEF	7	0.8547	1.0063	625	650	-
PASEF	8	1.0273	1.179	850	875	-
PASEF	8	0.8738	1.0255	650	675	-

3. Data captured is processed using Spectronaut (Version 18.6.231227.55695) software for directDIA search with the Swiss-Prot Mouse database downloaded on March 3, 2023. The default setting is used with a slight modification of the Min Peptide Length to 6. A search archive with in-house microglia data is supplemented as library extension runs.

4 Notes

1. Add EDTA and an MS-safe proteinase inhibitor cocktail just before use.
2. The suggested dosage for mice anesthetization: 0.05–0.1 mL for a 20 g mouse, 0.07–0.13 mL for a 30 g mouse, and 0.1–0.18 mL for a 40 g mouse.
3. Pinch the mouse by the toe to check whether it is fully anesthetized. Provide extra pentobarbitone if it elicits a pain reaction.
4. Observe the lung and the liver during the perfusion. The liver will turn white as an indication of successful blood removal (Fig. 1C). The enlargement of the lungs and/or fluid overflowing from the nose indicates that the needle is pinched too deeply into the right ventricle, and the liquid has entered the pulmonary circulation (Fig. 1D), which can lead to insufficient fixation. To avoid insufficient fixation, the needle should be inserted to a depth of about 3 mm in the parallel direction to the mouse, and the PBS perfusion time can be prolonged until the liver turns white.
5. Pause the pump before changing the solution to prevent air bubbles from entering the circulatory system.
6. The syringing process aims to break the DNA and remove undigested tissue that clogs the needle.
7. The size of fixed MuSCs is slightly smaller than the size of freshly isolated MuSCs, which can be reflected from the FSC of the sort plot (Fig. 2).
8. The serum in the wash medium will interfere with the BCA measurement of protein concentration. PBS washing is necessary for effective sample normalization.
9. The volume of the protein lysate should not exceed 20% of the LYSE buffer. The minimum amount of input protein is 1 μg.
10. The captioned SP3 beads amount only caters to 1 μg–10 μg input proteins. The recovery rates of 1 μg and 10 μg protein input are around 25% and 50%, respectively.
11. Remove the sealing mat. Magnetic beads will stick to the mat.
12. In case the LC system has never injected 2% DMSO solution, disconnect the LC column and make a 2% DMSO injection for cleaning the sample loop before any sample injection.

Acknowledgments

We acknowledge the Laboratory Animal Facility and Biosciences Central Research Facility at HKUST (Clear Water Bay) for their support. This work was funded by the Hong Kong Research Grant Council (GRF16100322, GRF16102021, GRF16101524, RFS2425-6S01, C6018-19G, AoE/M-604/16, T13-605/18W, T13-602/21N), the CAS-Croucher Funding Scheme for Joint Laboratories (CAS24SC02), and the Hong Kong Epigenomics Project (Lo Ka Chung Charitable Foundation). This study was supported by the InnoHK initiative of the Innovation and Technology Commission of the Hong Kong Special Administrative Region Government. This study was also supported in part by the Innovation and Technology Commission (ITCPD/17-9). T.H.C. is the S H Ho Professor of Life Science at HKUST.

Competing Interests The authors declare no competing interests.

References

1. De Morree A, Rando TA (2023) Regulation of adult stem cell quiescence and its functions in the maintenance of tissue integrity. Nat Rev Mol Cell Biol 24:334–354
2. Urbán N, Cheung TH (2021) Stem cell quiescence: the challenging path to activation. Development 148:dev165084.
3. Machado L, Esteves de Lima J, Fabre O et al. (2017) In situ fixation redefines quiescence and early activation of skeletal muscle stem cells. Cell Rep 21:1982–1993
4. Zeng W, Yue L, Lam KSW et al. (2022) CPEB1 directs muscle stem cell activation by reprogramming the translational landscape. Nat Commun 13: 947
5. Cheung TH, Quach NL, Charville GW et al. (2012) Maintenance of muscle stem-cell quiescence by microRNA-489. Nature 482:524–528
6. De Morrée A, Van Velthoven CTJ, Gan Q et al. (2017) Staufen1 inhibits MyoD translation to actively maintain muscle stem cell quiescence. Proc Natl Acad Sci USA 114(43):E8996–E9005
7. Yue L, Wan R, Luan S et al. (2020) Dek modulates global intron retention during muscle stem cells quiescence exit. Dev Cell 53(661-676):e6
8. Kim S-O, Kim J, Okajima T, Cho N-J (2017) Mechanical properties of paraformaldehyde-treated individual cells investigated by atomic force microscopy and scanning ion conductance microscopy. Nano Converg 4: 5
9. Kang W, Robitaille MC, Merrill M et al. (2020) Mechanisms of cell damage due to mechanical impact: an in vitro investigation. Sci Rep 10: 12009
10. Dong A, Chan ITC, Cheung TH (2023) ATAC-seq protocol for the profiling of chromatin accessibility of in situ fixed quiescent and activated muscle stem cells. STAR Prot 4:102376
11. Yue L, Cheung TH (2020) Protocol for isolation and characterization of in situ fixed quiescent muscle stem cells. STAR Protoc 1:100128
12. Zhang W, Zeng W, Wong PS, Cheung TH (2023) Protocol for low-input proteomic analysis of in situ fixed adult murine muscle stem cells. STAR Protoc 4:102750

Check for updates

Chapter 18

Bioinformatic Workflow for the Analysis of Single-Nucleus RNA-Sequencing Data from Skeletal Muscle

Chiara D'Ercole , Laura Virtanen and Lorenzo Giordani

Abstract

Understanding cellular diversity within a tissue is essential to elucidate how distinct cell populations cooperate to maintain homeostasis, regenerate, and respond to aging or disease. Unlike bulk methods that measure averaged gene expression, single-cell RNA sequencing (scRNA-seq) enables high-resolution profiling of the transcriptomes of individual cells, providing a powerful tool for analyzing cellular heterogeneity. However, its application in skeletal muscle remains limited to the mononucleated cellular fraction, as standard single-cell approaches cannot efficiently capture muscle fibers due to their large and multinucleated nature. Single-nucleus RNA sequencing (snRNA-seq), which profiles individual nuclei rather than intact cells, offers a more suitable and technically robust approach, enabling comprehensive characterization of the transcriptional landscape across the entire tissue. In this chapter, we present a workflow for the bioinformatic analysis of snRNA-seq data generated from mouse skeletal muscle using the 10x Genomics Chromium platform. The chapter focuses on computational processing using R software and the *Seurat* package, including quality control, normalization, dimensionality reduction, clustering, and cell-type annotation. By applying this pipeline, researchers can identify distinct muscle-resident cell populations and explore the complexity of skeletal muscle. The described framework also illustrates how to perform downstream analyses such as differential gene expression and cell–cell communication analysis.

Keywords Skeletal muscle, Single-nucleus RNA-seq, Bioinformatics

1 Introduction

Skeletal muscle is a complex tissue primarily composed of multinucleated myofibers, which are organized to form an efficient contractile apparatus. Tissue homeostasis is maintained through the dynamic interplay among stem cells, muscle fibers, and interstitial cells. The interstitial compartment contains a diverse array of cell types, including immune cells, mesenchyme-derived cells, and nerve

Chiara D'Ercole and Laura Virtanen contributed equally to this work.

Supplementary Information The online version contains supplementary material available at https://doi.org/10.1007/978-1-0716-5643-3_18

Nathalie Didier and Caroline E. Brun (eds.), *Muscle Stem Cells*, Methods in Molecular Biology, vol. 3093,
https://doi.org/10.1007/978-1-0716-5643-3_18,

cells. Under pathological conditions, genetic and epigenetic alterations can profoundly influence the behavior of resident cells. These changes not only affect the cells directly involved but also impact the surrounding microenvironment, ultimately altering the overall muscle homeostasis [1].

Single-nuclear RNA sequencing (snRNA-seq) is a powerful technique that enables the analysis of gene expression at single-cell resolution. Unlike bulk RNA sequencing, which only measures the average gene expression of entire cell populations, snRNA-seq enables the exploration of cellular heterogeneity, the identification of rare or previously unrecognized cell types, and the study of cell dynamics in development or disease [2]. In recent years, these technologies have been widely adopted in the field of skeletal muscle, with a large and rapidly growing body of studies employing scRNA-seq, snRNA-seq, and scATAC-seq, across both human and mouse models under a wide range of physiological and pathological conditions.

The analysis of scRNA-seq or snRNA-seq data typically involves multiple steps, including normalization, dimensionality reduction, and cell clustering using various computational methods. Techniques such as PCA and UMAP simplify high-dimensional datasets and allow visualization of complex cellular relationships. Clustering reveals subpopulations and functional states, supporting the discovery of disease biomarkers, drug-resistant clones, and immune cell diversity. Moreover, advanced tools such as CellChat [3] enable the mapping of cell-to-cell communication, uncovering key ligand–receptor interactions and signaling pathways. Overall, snRNA-seq provides an essential framework for understanding complex biological systems, refining cellular classifications, and guiding the development of targeted therapeutic strategies [4].

Here, to illustrate the different steps of the analysis, we use a publicly available dataset from a murine model of Duchenne muscular dystrophy, originally published in the seminal study by Chemello et al. [5]. Along with the protocol, we provide a script in the notes to reproduce the different steps and the figures presented, which can be adapted by users to customize the workflow for the analysis of their own datasets.

2 Materials

2.1 Cell Preparation

1. Microdissection scissors (F.S.T).
2. Forceps (F.S.T., Dumont #7 and #5).
3. Dry ice.
4. Liquid nitrogen.
5. Isopentane (2-Methylbutane).
6. Nuclei Isolation Reagents (10X, PN-1000447).

2.2 Nuclei Encapsulation and Library Preparation

1. Chromium Next GEM Single Cell 3′ Reagent Kits v3 (10X, PN-1000269).
2. Chromium Controller machine.

2.3 Main Software and Packages for Data Analysis

1. Cell Ranger v10.0 or greater.
2. R version 4.0 or greater.
3. Seurat v5.
4. CellChat v2.

Most of the data analysis steps can be performed on a standard laptop. However, Cell Ranger pipeline must be run on a Linux system (64-bit CentOS/RedHat 8.0 or Ubuntu 20.04) and requires substantial computational resources, including at least 64 GB of RAM (128 GB recommended), an 8-core Intel or AMD processor (64 cores recommended), and approximately 1.5 TB of free disk space.

3 Methods

A full description of the experimental procedures, including nuclei extraction, barcoding, and sequencing, goes beyond the scope of this protocol; however, we will briefly outline the key steps, including FASTQ alignment preceding data analysis.

Several commercial kits and protocols are available for cell isolation and barcoding. In the present protocol, we will refer to the standard 10x Genomics workflow (10x Chromium Nuclei Isolation Protocol, CG000505). It should be noted that *ad hoc* procedures specifically optimized for skeletal muscle have also been developed to perform snRNA-seq of myonuclei, enabling the inclusion of multinucleated cells while minimizing transcriptional artifacts induced by cell dissociation [6].

After isolation, nuclei are processed using the Chromium Next GEM Single Cell 3′ Reagent Kit, following the manufacturer's protocol (CG000315). Next, nuclei are encapsulated into nanoliter-scale Gel Beads-in-Emulsion (GEMs) using a Chromium Controller. Following encapsulation, cDNA is generated, amplified, and used for library preparation prior to sequencing.

3.1 Cell Preparation

1. Dissect mouse hindlimb muscles and immediately flash-freeze them in isopentane cooled with liquid nitrogen. Cut frozen tissues into 3–50 mg fragments.
2. Process samples according to protocol CG000505 using a dedicated lysis buffer combined with mechanical dissociation.

3. Centrifuge the resulting lysate and collect nuclei from the pellet after removal of the supernatant.

3.2 Nuclei Encapsulation and Library Preparation

1. Encapsulate nuclei into the GEMs, where, in the presence of primers containing a unique barcode sequence and a master mix, full-length barcoded cDNA is generated from the mRNA.
2. After incubation, remove the GEMs. Process and amplify the cDNA to generate the 3′ gene expression library.

3.3 Sequencing

The generated libraries are paired-end, dual-indexed, and compatible with Illumina sequencing platforms. A minimum sequencing depth of 20,000 read pairs per cell is recommended, with the following number of sequencing cycles:

Sequencing read	Number of cycles
Read 1	28 cycles
i7 Index	10 cycles
i5 Index	10 cycles
Read 2	90 cycles

3.4 BCL—FASTQ Processing Using Cell Ranger

Cell Ranger is a suite of analysis pipelines that processes Chromium Single Gene Expression data to perform read alignment, generate feature-barcode matrices, and carry out downstream analyses such as clustering, dimensionality reduction, and gene expression profiling. Both Binary Base Call (BCL) files and FASTQ files can be used as input.

As of Cell Ranger v10.0, the *cellranger mkfastq* pipeline, which converts BCL files to FASTQ format, has been deprecated and will be removed in future releases. It is therefore recommended to use Illumina's BCL Convert software to generate Cell Ranger–compatible FASTQ files.

Cell Ranger outputs also include a ".cloupe" (Loupe) file, which enables interactive visualization and exploration of the dataset using the Loupe Browser.

Raw sequencing data are aligned against a reference genome using the STAR aligner. For each read, STAR calculates a mapping quality score (MAPQ), which represents the confidence of the alignment between the read and the reference genome. Exonic reads are aligned to annotated transcripts; if a read is compatible with a single gene annotation, it is considered uniquely mapped. Only uniquely mapped reads are used for UMI counting. Output files will appear in the outs/subdirectory within pipeline output directory. Cell Ranger pipeline outputs both unfiltered (raw) and filtered feature-barcode matrices in two file formats: the Market Exchange Format (MEX) and the Hierarchical Data Format (HDF5 or H5). Feature-barcode matrices in MEX format are zipped and contain a barcode, a feature and a matrix file. Feature and barcode sequences correspond to row and column indices, respectively. In contrast, the H5 format is provided as a single file that stores the matrix entries as well as the features and cell-barcodes associated with the matrix rows and columns. The unfiltered feature-barcode matrix contains every barcode from the fixed list of known-good barcode sequences that has at least one read. This includes background and cell-associated barcodes, while the filtered feature-barcode matrix contains only detected cell-associated barcodes.

These files constitute the primary output used for downstream single-nucleus RNA-seq data analysis.

3.5 Data Analysis

Using the R-based tools Seurat [7, 8] and CellChat [3], we present the main steps of snRNA-seq data analysis, from preprocessing and quality control to dataset integration, cell type annotation, and differential gene expression analysis. To do so, we will use publicly available snRNA-seq data from WT (C57/BL6N) and ΔEx51 mouse skeletal muscle [5].

Integrative analysis enables the alignment of shared cell types across different samples, increasing statistical power and supporting accurate comparative studies. Furthermore, we apply CellChat to predict intercellular communication networks. More detailed information about individual Seurat and CellChat functions is available on the Seurat website (https://satijalab.org/seurat/) or CellChat GitHub repository (https://github.com/jinworks/CellChat), respectively. Together, these approaches provide a comprehensive framework for exploring cellular heterogeneity and communication in skeletal muscle.

The FASTQ files and processed feature-barcode matrices are available in the GEO repository under accession number GSE156498. In addition to the original Cell Ranger outputs, the authors have also provided SoupX-corrected feature-barcode matrices as processed data.

3.5.1 Preprocessing with SoupX (Optional)

A potential source of background signal in both snRNA-seq and scRNA-seq is the sequencing of cell-free RNA present in the input solution, which can become admixed with a cell or nucleus within its enclosing droplet. In skeletal muscle snRNA-seq datasets, this background contamination is often observed as the unspecific expression of highly abundant transcripts, such as myosin genes, across multiple mononucleated resident cell populations. To mitigate this effect, specific tools such as SoupX [9] or CellBender [10] have been developed. In this protocol, we briefly describe the application of SoupX, as it was the method used by Chemello and colleagues in their original publication. The workflow consists of three main steps. First, SoupX estimates the ambient mRNA expression profile from empty droplets. Next, it calculates (or allows the user to manually set) the contamination fraction, defined as the proportion of UMIs in each cell that originate from background RNA. Finally, it corrects the expression profile of each cell using the estimated ambient mRNA profile and contamination fraction. As output, SoupX produces a modified count matrix that can be used in place of the original matrix for downstream analysis. Chemello et al. have already made the SoupX-processed matrices available on GEO as supplementary files. Therefore, this step in the script is marked as optional. SoupX requires clustering information, which can be loaded directly from Cell Ranger output, in order to automatically estimate the contamination.

1. Load Cell Ranger output into a SoupChannel object using *sc=load10X()*. This command will also load clustering information and dimensional reduction coordinates from the 10× output if present.
2. Estimate contamination fraction running *sc=autoEstCont(sc)*.
3. Generate the adjusted matrix using *out = adjustCounts(sc)*. The output produced by SoupX is a sparse matrix with non-integer values. If integer counts are required for downstream applications, such as differential gene expression analysis using methods like DESeq, the option *roundToInt = TRUE* can be used. In this case, SoupX rounds each value to the nearest integer using a probabilistic approach, where the probability of rounding up corresponds to the fractional part of the original value.
4. The new matrix can be used directly to create a Seurat object or exported using *write10xCounts()* from *DropletUtils* package

3.5.2 Processing with Seurat

1. Import the filtered feature-barcode matrix generated by Cell Ranger (or the SoupX-corrected matrix) using the *Read10X()* function, and create a Seurat object by setting the imported matrix as counts with the *CreateSeuratObject()* function (using the *counts* argument). Seurat objects store both raw data (count matrix) and later added information such as dimensionality reduction (PCA), clustering, and cell-type annotations.
2. Calculate the proportion of mitochondrial gene expression using the *PercentageFeatureSet()* function with the pattern "^mt-" (for mouse data, "^MT-" for human). Low-quality or dying cells often exhibit elevated levels of mitochondrial transcripts. Moreover, since this is a single nuclei dataset, mitochondrial contamination is expected to be minimal. For this reason, cells containing more than 5% mitochondrial transcripts are typically filtered out, although appropriate thresholds should be determined empirically for each dataset.
3. Assess data quality using the *VlnPlot()* function to visualize the number of detected features, total RNA counts, and

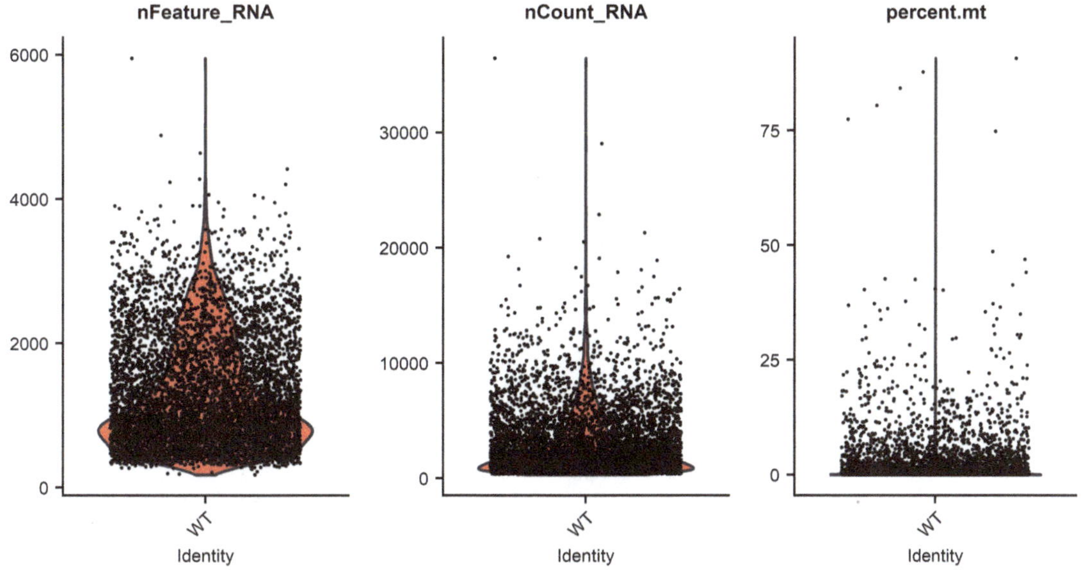

Fig. 1 Quality control of WT sample showing the number of detected features, RNA, and mitochondrial counts per cell

mitochondrial gene expression per cell (Fig. 1). These metrics are used to identify and remove low-quality cells and potential doublets. In the original study, cells expressing fewer than 200 genes were considered low-quality or empty droplets and were excluded, whereas cells expressing more than 4000 genes or more than 15,000 total mRNA transcript counts were classified as doublets and removed.

4. Perform data normalization using *NormalizeData()*. For each cell, feature expression is normalized by the total expression, multiplied by a scale factor (default: 10,000), and subsequently log-transformed.
5. Identify highly variable features using the *FindVariable Features()* function to highlight genes with substantial expression differences across the data (default: 2000 features).
6. Center and scale expression values using *ScaleData()* to ensure each gene has a mean of zero and variance of one. This step prevents highly expressed genes from disproportionately influencing subsequent analyses. By default, only variable features are scaled, but additional features can be scaled by changing the features argument.
7. Perform Principal Component Analysis (PCA) using the *RunPCA()* function. PCA reduces the dimensionality of the dataset while retaining significant variation, focusing on the previously identified highly variable genes. By default,

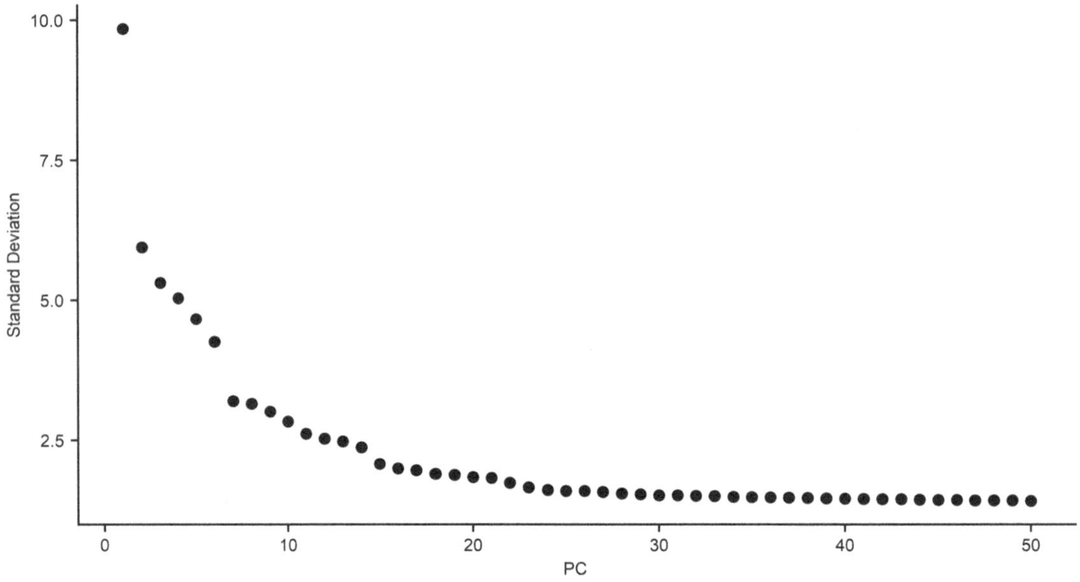

Fig. 2 Elbow plot of WT sample showing standard deviation of the principal components (PC)

only these variable features are used as input. The number of principal components to include in downstream analyses can be evaluated using *ElbowPlot()*, which ranks components based on the percentage of variance explained. Here, we selected the first 25 principal components based on the ElbowPlot (Fig. 2), as variance explained by additional components plateaued beyond PC25, indicating minimal contribution of higher-order components.

8. Use the *FindNeighbors()* function to construct a k-nearest neighbor (KNN) graph based on Euclidean distances in PCA space, capturing the similarity between cells. Specify the number of principal components to use for graph construction with the *dims* argument.

9. Cluster the cells using *FindClusters()* function, which implements a community detection algorithm (default: Louvain method). The number of clusters can be adjusted by modifying the resolution parameter (default: 0.8), where lower values yield fewer clusters and higher values produce more. Clusters will be stored in the *seurat_clusters* column in the metadata and automatically set as object's *Idents*.

10. To visualize the dataset in two-dimensional space (Fig. 3), perform a nonlinear dimensionality reduction using the UMAP algorithm (Uniform Manifold Approximation and Projection) with the *RunUMAP()* function. Cells belonging to the same graph-based clusters are expected to localize in close proximity on the UMAP projection. Visualize cluster

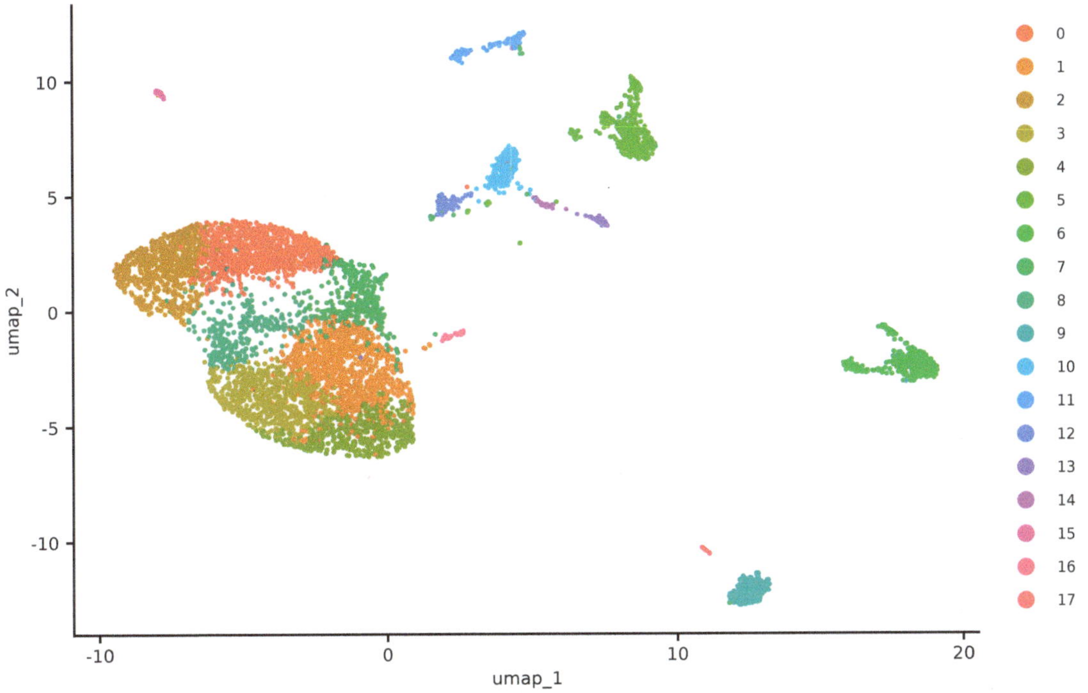

Fig. 3 Uniform Manifold Approximation and Projection (UMAP) of the pre-processed WT sample

assignments using the *DimPlot()* function. By default, the active identities (*Idents*) are displayed on the UMAP; however, this can be changed to any metadata column by modifying the *group.by* parameter. Similarly, the dimensional reduction used for visualization can be specified using the *reduction* parameter.

11. Save the processed Seurat objects using the saveRDS() function.

3.5.3 Samples Integration

1. Load the preprocessed Seurat objects using the *readRDS()* function.
2. Prior to integration, create a list containing the individual Seurat objects. Merge the objects into a single Seurat object using the *merge()* function setting the object.list parameter. After merging, reprocess the combined dataset by running *NormalizeData()*, *FindVariableFeatures()*, *ScaleData()*, and *RunPCA()*.
3. Integration is performed using the *IntegrateLayers()* function, specify PCA as the original reduction, and select an integration method. It is recommended to test multiple methods to evaluate their impact on a given dataset. Here, we explore two methods: CCAIntegration and Harmony. CCAIntegration

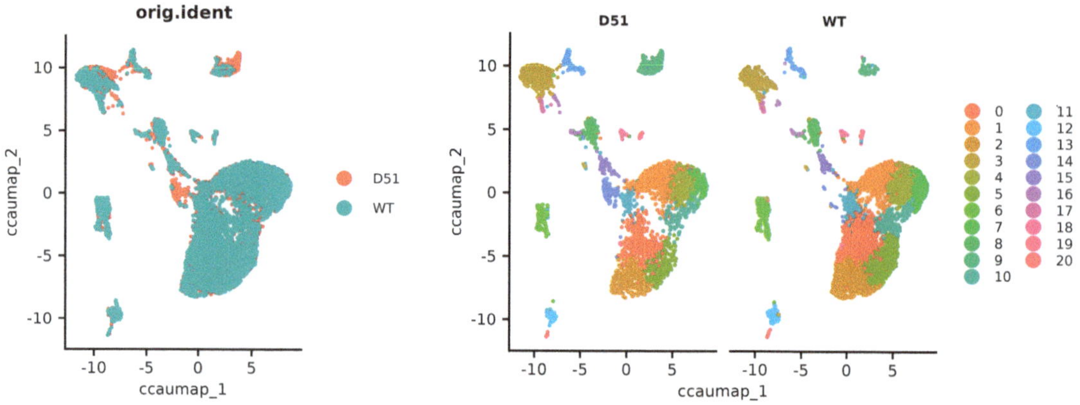

Fig. 4 UMAP visualization of the integrated dataset showing batch-corrected embeddings using cca.integration (left) with *group.by* = *"orig.ident"* and corresponding cell clusters with *split.by* = *"orig.ident"* (right)

aligns datasets by finding shared linear correlations (anchors) between cells in gene expression space, correcting batch effects while preserving biological variation. Starting from PCA embeddings, Harmony applies an iterative algorithm to correct batch effects, accounting for both linear and nonlinear sources of variation and generating batch-corrected embeddings for clustering and visualization. Additional available methods include RPCAIntegration, FastMNNIntegration, and scVIIntegration.

4. Following integration, run *FindNeighbors()*, *FindClusters()*, and *RunUMAP()* for each integration method separately. Use the *DimPlot()* function to visualize the resulting UMAP embeddings, allowing comparison of integration outputs and selection of the most suitable method. To assess the overlap of clusters (batch effect) between WT and D51 samples, use *group.by* = *"orig.ident"* or *split.by* = *"orig.ident"* when plotting the UMAP embeddings (Fig. 4).
5. Finally, join the different layers in the integrated Seurat object using *JoinLayers()* function.
6. Save the integrated Seurat objects using the saveRDS() function.

3.5.4 Annotation and Differential Gene Expression Analysis

1. A list of annotation features can be defined based on the literature; for consistency, we used the same markers reported in the original paper. Use the *DotPlot()* function to visualize their expression across clusters, grouping the dataset according to the selected clustering method (e.g., cca_clusters; Fig. 5A).
2. Set the clusters that need to be annotated (in our case cca_clusters) as Idents using *Idents()* function.

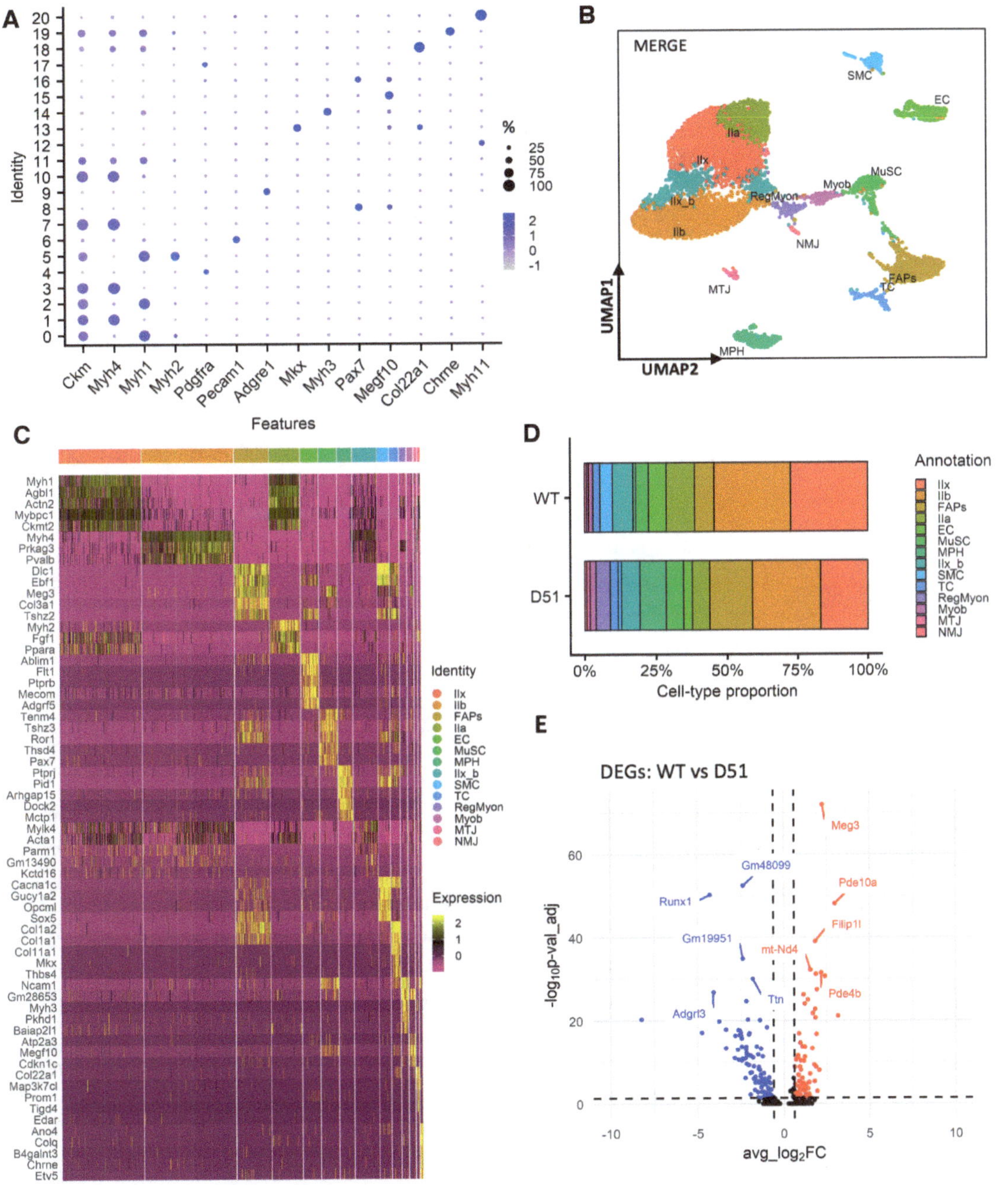

Fig. 5 Analysis of snRNA-seq of TA muscle from WT and ΔEx51 mice. (**A**) Dotplot showing average expression of canonical markers across CCA clusters. (**B**) UMAP visualization of all the nuclei from WT and D51 TA muscle colored by cluster annotation. (**C**) Heatmap showing differentially expressed genes between cell populations. (**D**) A bar plot illustrating the percentage of each cell population in WT and D51 samples. (**E**) A volcano plot showing differentially expressed genes between WT and D51 samples in the myogenesis subset

3. Idents can be annotated using *RenameIdents()* based on their gene expression profiles. Create a new column in the object metadata to store the annotations. Annotated cell groups can be visualized using *DimPlot()* and *FeaturePlot()*, adjusting feature settings as needed (Fig. 5B).
4. Use the *FindAllMarkers()* function to identify differentially expressed genes (DEGs) for each cluster compared with the rest of the dataset, which is particularly useful for annotating clusters with unclear identities. To identify DEGs between selected clusters, use the *FindMarkers()* function. The *min.pct* and *logfc.threshold* parameters can be adjusted to speed up DEG testing and refine the results. The *min.pct* parameter restricts the analysis to genes expressed in at least a specified fraction of cells in either group, while *logfc.threshold* limits the analysis to genes showing at least a specified log-fold change between the compared populations.
5. Differentially expressed genes can be visualized using heat maps generated with the *DoHeatmap()* function (Fig. 5C). Using the dplyr package, we selected the top five genes per cluster.
6. Save the annotated Seurat object using the *saveRDS()* function.

3.5.5 Additional Visualization and Dataset Subset

1. Use the *ggplot2* and *dplyr* packages to visualize the percentage of each cell population in WT and D51 samples using a bar plot (Fig. 5D).
2. To focus specifically on cells belonging to the myogenesis trajectory, extract the MuSC, Myob, and RegMyon clusters from the full dataset using the *subset()* function (with the idents parameter), thereby generating a smaller Seurat object. After subsetting, rejoin the data layers using *JoinLayers()* and rescale the data using the *ScaleData()* function.
3. Identify differentially expressed genes between WT and D51 samples using the *FindMarkers*() function. In addition to standard visualization approaches for marker genes, such as *FeaturePlot()*, *VlnPlot()*, *DotPlot()*, and *DoHeatmap()*, the results can also be displayed as a volcano plot using ggplot2 to highlight genes differentially expressed (Fig. 5E).

3.5.6 Cell–Cell Communication Analysis Using CellChat

1. Load the annotated Seurat object, and subset the cells based on their sample origin (orig.ident). Initial CellChat analysis of WT and D51 samples is conducted separately. Merge data layers using *JoinLayers()*, and scale expression values with *ScaleData()*. The cellular landscape can be visualized using *DimPlot()* with CCA-based UMAP coordinates, grouping cells by the annotation.cca labels.

2. Create a CellChat object from the processed WT Seurat object using the *createCellChat()* function, grouping cells according to their annotated identities.
3. Load the CellChat mouse ligand–receptor database. To restrict the analysis to secreted signaling pathways (optional), set "Secreted Signaling" category via *subsetDB()*.
4. Assign the subset database to the CellChat object for downstream analysis.
5. Next, subset the data matrix to keep only genes relevant to the selected L–R pairs using *subsetData()*. Parallelization can be enabled with the future package to accelerate computation by increasing the multisession workers.
6. Identify overexpressed genes using *identifyOverExpressedGenes()*.
7. Based on overexpressed genes, identify overexpressed ligand–receptor interactions using *identifyOverExpressedInteractions()*.
8. Compute cell–cell communication probabilities with *computeCommunProb()* and filter low-abundance cell groups using *filterCommunication(min.cells = 10)* to ensure robust estimation of interactions.
9. Derive pathway-level communication probabilities with *computeCommunProbPathway()*.
10. Aggregate global cell–cell communication networks using *aggregateNet()*. Cell group sizes were extracted for downstream visualization and interpretation.
11. Save the WT CellChat object using *saveRDS()* for subsequent analysis and comparison with D51 samples.
12. Perform the same analysis on D51 sample.

3.5.7 Comparison of Cell–Cell Communication Between WT and D51 Samples

1. Load previously saved CellChat objects for WT and D51 using *readRDS()*.
2. Match cell identities across datasets using the *liftCellChat()* function to ensure consistent labeling between WT and D51. This step aligns clusters that may be present in one dataset but not the other, facilitating a direct comparison.
3. Compute network centrality metrics for each dataset independently using *netAnalysis_computeCentrality*() on the aggregated communication network slot (netP).
4. Merge the WT and D51 CellChat objects using *mergeCellChat()* to generate a combined object and save it for downstream comparative analysis using *saveRDS()*.

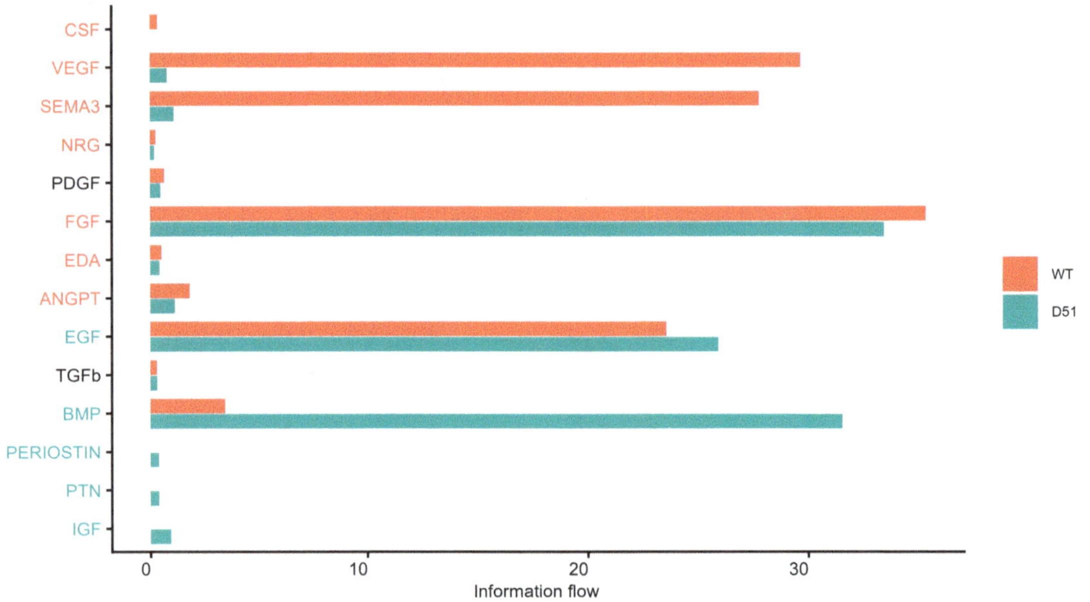

Fig. 6 Rank plot showing the relative activity of signaling pathways in WT and D51 samples. Pathways are ranked by their overall communication probability, highlighting differences in pathway activity between conditions

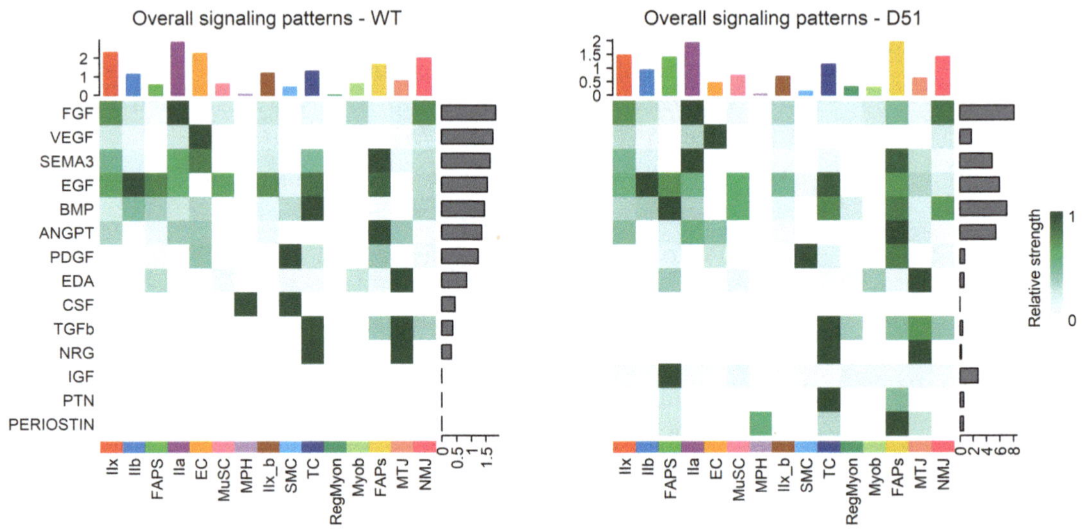

Fig. 7 Signaling role heatmaps showing predicted overall signaling patterns for each cell group across all pathways in WT and D51 samples

5. Compare communication differences using *rankNet()*, which provides a ranked overview of signaling pathways and enables statistical assessment of changes between WT and D51 (Fig. 6).
6. To compare pathway-specific signaling roles between WT and D51, extract pathway-level networks (netP) from each CellChat object. The union of all detected pathways across both datasets ensures that downstream analyses are performed on an identical pathway set.
7. Generate signaling role heatmaps using *netAnalysis_signalingRole_heatmap()*, which summarizes the relative strength of outgoing, incoming, or overall signaling for each cell group (Fig. 7). These heatmaps allow direct comparison of how individual clusters contribute to or receive signaling in WT versus D51, highlighting shifts in pathway dominance and cellular communication patterns between conditions.
8. Visualize selected signaling pathways (e.g., *Vegf*) using *netVisual_aggregate()*. In these plots, nodes represent cell clusters and edges represent interactions, with edge thickness proportional to communication strength. Edge weights are normalized across datasets using *getMaxWeight()*, which calculates the maximum communication probability for the selected pathway across all datasets in object.list (Fig. 8).
9. Save the merged CellChat object using *saveRDS()*.

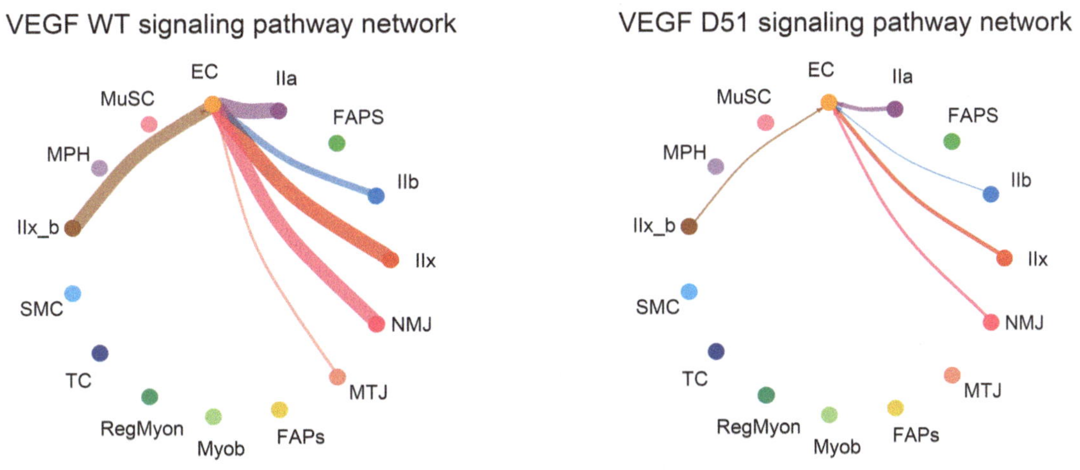

Fig. 8 Visualization of *Vegf* signaling interaction probabilities between different cell populations in WT and D51 TA muscle samples. In these plots, nodes represent cell clusters and edges represent interactions, with edge thickness proportional to communication strength

4 Notes

Below, we provide a script to perform all the proposed analysis steps. The script is also available for download as an R file so that users can customize the workflow for their own datasets (see Electronic Supplementary Material).

To run the script, specify a base directory where the output folders will be created. Within this base directory, store the Cell Ranger outputs in a folder named *Cellranger_output*.

5 Main Packages Used in the Script

Seurat v5.4.0 [11].
CellChat v2.1.2 [12].

6 Additional Packages Used in the Script

Future v1.33.2 [13].
future.apply v1.11.2 [13].
dplyr v1.1.4 [14].
ggplot2 v3.5.1 [15].
SoupX v1.6.2 [9].
DropletUtils v1.24.0 [16].
ggrepel v0.9.5 [17].
patchwork v1.2.0 [18].
ComplexHeatmap v2.20.0 [19].

```
### Single-nucleus RNA sequencing workflow ###
#Tested with Seurat_5.4.0


##The analysis includes:
#1. Library loading
#2. Setting Global path and creating directories
#3. Single Dataset preprocess
#4. Dataset Integration
#5. Annotation and Visualization
#6. Cellchat Analysis


# Set seed for reproducibility
set.seed(1234)

###1. LOADING LIBRARIES ###
library(Seurat)
library(future)
library(dplyr)
```

```
library(ggplot2)
library(patchwork)


##Required for SoupX
#library(DropletUtils)
#library(SoupX)


###2. SETTING GLOBAL PATHS AND CREATING DIRECTORIES ###

base_dir <- "~/Documents/Applications/2025/Methods for
SingleNuclei Analysis/test/"
cellranger_output.folder <- paste0(base_dir,
"Cellranger_output/")


processed_dir <- paste0(base_dir, "Processed/")
figures_dir <- paste0(base_dir, "Figures/")
QC_dir <- paste0(base_dir, "QC/")


# Creating folders - If the folder is already present
it will not overwrite it
dir.create(processed_dir, showWarnings = FALSE)
dir.create(figures_dir, showWarnings = FALSE)
dir.create(QC_dir, showWarnings = FALSE)


# To run the script save the cellranger output folders
you want to process
# inside Cellranger_output folder in your base folder


# This will list all the files and or folder in
Cellranger_output folder.
Samples.path=list.dirs(cellranger_output.folder, full.
names = FALSE,recursive = FALSE)


###3. SINGLE DATASET PREPROCESS ###
# This will load each sample, perform SoupX
(optional), create process and save each Seurat object
Samples.list <- lapply(X = Samples.path, FUN =
function(X) {


 message("Processing ", X)
### SoupX Optional
#loading
#sc = load10X(paste0(cellranger_output.folder, X,"/
outs/"),verbose = T)
#Estimate Rho and Clean data
#sc = autoEstCont(sc)
```

```
#out = adjustCounts(sc, roundToInt=TRUE)


#Exporting the adjusted matrix
#DropletUtils::write10xCounts(path =
paste0(cellranger_output.folder, X, "/outs/strained_
feature_bc_matrix/"), x=out)


#inputdata.10x=out
### END of Soup X module
## Since Chemello and colleagues have provided the
SoupX adjusted version of their matrix we will use
those in the protocol.
## We created an additional folder in /outs/ called
strained_feature_bc_matrix to store the SoupX-adjusted
matrices.
# Load the 10x Genomics count matrix - Remove this
step if the SoupX module has been run.
# Use /outs/filtered_feature_bc_matrix/ to proceed
without SoupX correction.
inputdata.10x <- Read10X(paste0(cellranger_output.
folder,X,"/outs/strained_feature_bc_matrix/"))


# Create Seurat object
DATA <- CreateSeuratObject(
    counts = inputdata.10x,
    project = X,
    assay = "RNA")
  # Calculate percent mitochondrial genes - use "^MT-"
for human datasets
  DATA[["percent.mt"]] <- PercentageFeatureSet(DATA,
pattern = "^mt-")
  # Quality Control Visualization
  p1=VlnPlot(DATA, features = c("nFeature_RNA",
"nCount_RNA", "percent.mt"), ncol = 3) &
    theme(
      text = element_text(size = 8),
      axis.text = element_text(size = 8),
      plot.title = element_text(size = 9))


  # This will export the violin plot for QC
  ggsave(
    plot = p1,
    filename = paste0(QC_dir, X, "_QC.tiff"),
    device = "tiff", dpi = 600, width = 18, height =
10, units = "cm")


  # Quality Control Filters
  DATA <- subset(DATA,
                 subset = nCount_RNA < 15000 &
```

```
                    nFeature_RNA > 200 &
                    nFeature_RNA < 4000 &
                    percent.mt < 5)


  # Normalization and Pre-processing
  DATA <- NormalizeData(object = DATA)
  DATA <- FindVariableFeatures(object = DATA)
  DATA <- ScaleData(object = DATA)
  DATA <- RunPCA(object = DATA)


  # ElbowPlot for PCA selection
  p1=ElbowPlot(DATA, ndims = 50) &
    theme(
      text = element_text(size = 8),
      axis.text = element_text(size = 8),
      plot.title = element_text(size = 9))


  ggsave(
    plot = p1,
    filename = paste0(QC_dir, X, "_ElbowPlot.tiff"),
    device = "tiff", dpi = 600, width = 18, height =
10, units = "cm")


  DATA <- FindNeighbors(object = DATA, dims = 1:25)
  DATA <- FindClusters(object = DATA)
  DATA <- RunUMAP(object = DATA, reduction = "pca",
dims = 1:25)


  p1=DimPlot(DATA, reduction = "umap") &
    theme(
      text = element_text(size = 8),
      axis.text = element_text(size = 8),
      plot.title = element_text(size = 9))


  ggsave(
    plot = p1,
    filename = paste0(QC_dir, X, "_Dimplot.tiff"),
    device = "tiff", dpi = 600, width = 18, height =
12, units = "cm")


  # Save processed object
  saveRDS(DATA,paste0(processed_dir,X,"_processed.
rds"))

  return(DATA)
})
```

```
###4. DATASET INTEGRATION ###
# This will load the files in the Processed folder
# Use in case you have run the PREPROCESSING BLOCK
INDEPENDENTLY


#Samples.list <- lapply(X = Samples.path, FUN =
function(X){
#
# DATA <- readRDS(paste0(processed_dir,X,"_processed.
rds"))
# return(DATA)
#})


# Merge WT and D51 Seurat objects
integrated <- merge(Samples.list[[1]], y = Samples.
list[-1])


# Reprocess after merging
integrated <- NormalizeData(integrated)
integrated <- FindVariableFeatures(integrated)
integrated <- ScaleData(integrated)
integrated <- RunPCA(object = integrated)


 # Integrate samples using different batch correction
methods
integrated <- IntegrateLayers(object = integrated,
method = CCAIntegration, orig.reduction = "pca", new.
reduction = "integrated.cca",verbose = FALSE)
integrated <- IntegrateLayers(object = integrated,
method = HarmonyIntegration,orig.reduction = "pca",
new.reduction = "harmony",verbose = FALSE)
# Clustering of CCA integration
integrated <- FindNeighbors(integrated, reduction =
"integrated.cca", dims = 1:25)
integrated <- FindClusters(integrated, cluster.name =
"cca_clusters")
integrated <- RunUMAP(integrated, reduction =
"integrated.cca", dims = 1:25, reduction.name = "cca.
umap", return.model = TRUE)
#Exporting Dimplots to assess cca integration
p1=DimPlot(integrated, reduction = "cca.umap", group.
by = "orig.ident") &
  theme(
    text = element_text(size = 8),
    axis.text = element_text(size = 8),
    plot.title = element_text(size = 9))


ggsave(plot = p1,
       filename = paste0(figures_dir, "merged_cca_umap_
samples_Dimplot.tiff"),
```

```
      device = "tiff", dpi = 600, width = 8, height =
8, units = "cm")


p2=DimPlot(integrated, reduction = "cca.umap", split.
by = "orig.ent") &
  theme(
    text = element_text(size = 8),
    axis.text = element_text(size = 8),
    legend.key.size = unit(0.2, "cm"),
    plot.title = element_text(size = 9))


ggsave(
  plot = p2,
  filename = paste0(figures_dir, "merged_cca_umap_split_
Dimplot.tiff"),
  device = "tiff", dpi = 600, width = 12, height = 8,
units = "cm")


p1+p2
# Clustering of Harmony integration
integrated <- FindNeighbors(integrated, reduction =
"harmony", dims = 1:25)
integrated <- FindClusters(integrated, cluster.name =
"harmony_clusters")
integrated <- RunUMAP(integrated, reduction =
"harmony", dims = 1:25, reduction.name = "h.umap",
return.model = TRUE)


p3=DimPlot(integrated, reduction = "h.umap", group.by
= "orig.ident") &
  theme(
    text = element_text(size = 8),
    axis.text = element_text(size = 8),
    plot.title = element_text(size = 9))


ggsave(plot = p3,
      filename = paste0(figures_dir, "merged_harmony_
umap_samples_Dimplot.tiff"),
      device = "tiff", dpi = 600, width = 8, height =
8, units = "cm")
p4=DimPlot(integrated, reduction = "h.umap", split.by
= "orig.ident") &
  theme(
    text = element_text(size = 8),
    axis.text = element_text(size = 8),
    legend.key.size = unit(0.2, "cm"),
    plot.title = element_text(size = 9))
```

```
ggsave(
  plot = p4,
  filename = paste0(figures_dir, "merged_harmony_umap_
split_Dimplot.tiff"),
  device = "tiff", dpi = 600, width = 12, height = 8,
units = "cm")


p3+p4
# Join different data layers together
integrated <- JoinLayers(integrated)


# Save integrated and processed seurat object
saveRDS(integrated, file = paste0(processed_dir,"RNA_
integrated_processed.rds"))


###5. ANNOTATION AND VISUALIZATION ###
#This will load the integrated file - use in case you
have run the INTEGRATION BLOCK INDEPENDENTLY

#integrated <- readRDS("/shared/ifbstor1/projects/
multiome/D51_Multiome/RNA_integrated_processed.rds")
# Annotation features selected based on Chemello
et al., 2020
annotation.features=c("Ckm","Myh4","Myh1","Myh2","Pdg-
fra","Pecam1","Adgre1","Mkx","Myh3","Pax7","Megf10","C
ol22a1","Chrne","Myh11")


p1=DotPlot(integrated,features = annotation.features,
group.by = "cca_clusters", scale.min = 10, dot.scale =
2) &
  theme(
    text = element_text(size = 8),
    axis.text = element_text(size = 8),
    axis.text.x = element_text(angle = 45, hjust = 1),
    legend.key.size = unit(0.2, "cm"))


ggsave(plot = p1,
       filename = paste0(figures_dir, "merged_cca_umap_
Dotplot.tiff"),
       device = "tiff", dpi = 600, width = 12, height =
8, units = "cm"
)

#Annotation based on canonical markers on cca
clustering
#check annotation as cluster number and relative size
can fluctuate between runs
Idents(integrated)="cca_clusters"
integrated <- Renameeents(integrated,
```

```
                              "0"="IIx",
                              "1"="IIb",
                              "2"="IIx",
                              "3"="IIb",
                              "4"="FAPs",
                              "5"="IIa",
                              "6"="EC",
                              "7"="IIb",
                              "8"="MuSC",
                              "9"="MPH",
                              "10"="IIx_b",
                              "11"="IIx_b",
                              "12"="SMC",
                              "13"="TC",
                              "14"="RegMyon",
                              "15"="Myob",
                              "16"="MuSC",
                              "17" = "FAPs",
                              "18" = "MTJ",
                              "19" = "NMJ",
                              "20" = "SMC")


# Add annotation column to the Seurat object
integrated$annotation.cca=Idents(integrated)


# Visualize annotation in umap
p2=DimPlot(integrated, reduction = "cca.umap", label
= TRUE, repel = TRUE, label.size = 2, group.by =
"annotation.cca") &
theme(
  text = element_text(size = 8),
  axis.text = element_text(size = 8),
  legend.key.size = unit(0.2, "cm"),
  plot.title = element_text(size = 9))


ggsave(plot = p2,
      filename = paste0(figures_dir, "merged_cca_umap_
annotated_Dimplot.tiff"),
      device = "tiff", dpi = 600, width = 10, height =
8, units = "cm")


p3=FeaturePlot(integrated,
               reduction = "cca.umap",
               features = "Pax7",
               split.by = "orig.ident",
               order = TRUE) &
  theme(
    text = element_text(size = 8),
    axis.text = element_text(size = 8),
    plot.title = element_text(size = 9))
```

```
ggsave(plot = p3,
       filename = paste0(figures_dir, "merged_cca_umap_
Pax7_Featureplot.tiff"),
       device = "tiff", dpi = 600, width = 12, height =
6, units = "cm")


# Differentially expressed genes (DEG)
# Find markers for every cluster compared to all
remaining cells
Idents(integrated) = "annotation.cca"
All.markers <- FindAllMarkers(integrated, only.pos =
TRUE)


# Heatmap
All.markers %>%
  group_by(cluster) %>%
  dplyr::filter(avg_log2FC > 1) %>%
  slice_head(n = 5) %>%
  ungroup() -> top5
p4=DoHeatmap(integrated, features = top5$gene,label =
FALSE) &
  theme(
    legend.position = "right",
    text = element_text(size = 7),
    legend.key.size = unit(0.2, "cm"),
    plot.title = element_text(size = 9))


ggsave(plot = p4,
       filename = paste0(figures_dir, "annotated_
Heatmap.tiff"),
       device = "tiff", dpi = 600, width = 10, height =
14, units = "cm")


# Save annotated seurat object
saveRDS(integrated, file = paste0(processed_dir,"RNA_
annotated.rds"))


# Additional visualization and dataset subset
# Barplot
df <- integrated@meta.data %>% dplyr::count(orig.
ident, annotation.cca)
df_prop <- df %>%
  group_by(orig.ident) %>%
  mutate(prop = n / sum(n))


p1=ggplot(df_prop, aes(x = orig.ident, y = prop, fill =
annotation.cca)) +
```

```
  geom_bar(stat = "identity", width = 0.7,
position = position_stack(reverse = FALSE), color=
"black",linewidth = 0.3) +
  scale_y_continuous(labels = scales::percent) +
  theme_classic() +
  theme(
    text = element_text(size = 8),
    legend.text = element_text(size = 6),
    legend.key.size = unit(0.2, "cm")) +
  labs(
    x = "",
    y = "Cell-type proportion",
    fill = "Annotation") +
  coord_flip() +
  theme(legend.position = "right",
        text = element_text(size = 8),
        axis.text = element_text(size = 8),
        legend.key.size = unit(0.2, "cm"),
        plot.title = element_text(size = 9))


ggsave(plot = p1,
       filename = paste0(figures_dir, "annotated_
barplot.tiff"),
       device = "tiff", dpi = 600, width = 9, height =
5, units = "cm")


# Subsetting
Idents(integrated) = "annotation.cca"
Regeneration <- subset(integrated, idents =
c("MuSC","Myob","RegMyon"))
Regeneration <- JoinLayers(Regeneration)
Regeneration <- ScaleData(object = Regeneration)


# Comparison between WT and D51
Idents(Regeneration) = "orig.ident"
DEG <- FindMarkers(Regeneration, ident.1 = "WT",
ident.2 = "D51", verbose = FALSE, min.pct = 0.25,
logfc.threshold = 0.25, only.pos = F)
DEG$gene <- rownames(DEG)
#Volcano Plot
library(ggrepel)


# add a column of NAs
DEG$Expression <- "Unchanged"
# if log2Foldchange > 0.6 and pvalue < 0.05, set as
"UP"
DEG$Expression[DEG$avg_log2FC > 0.6 & DEG$p_val_adj <
0.05] <- "Up-regulated"
# if log2Foldchange < -0.6 and pvalue < 0.05, set as
"DOWN"
```

```
DEG$Expression[DEG$avg_log2FC < -0.6 & DEG$p_val_adj <
0.05] <- "Down-regulated"
DEG$delabel <- NA DEG$delabel[DEG$Expression!=
"Unchanged"] <- DEG$gene[DEG$Expression!= "Unchanged"]


top <- 5
top_genes <- bind_rows(
  DEG %>%
    dplyr::filter(Expression == 'Up-regulated') %>%
    arrange(p_val_adj, desc(abs(avg_log2FC))) %>%
    head(top),
  DEG %>%
    dplyr::filter(Expression == 'Down-regulated') %>%
    arrange(p_val_adj, desc(abs(avg_log2FC))) %>%
    head(top))


p1=ggplot(data=DEG, aes(x=avg_log2FC, y=-log10(p_val_
adj), col=Expression, label=gene)) +
  geom_point(size = 0.5) +
  theme_minimal() +
  geom_vline(xintercept=c(-0.6, 0.6), col=
"black",linetype = "dashed") +
  geom_hline(yintercept=-log10(0.05), col=
"black",linetype = "dashed") +
  geom_label_repel(data = top_genes, mapping =
aes(avg_log2FC, -log10(p_val_adj), label = gene), size
= 2,label.size = NA, max.overlaps = Inf, min.segment.
length = 0, show.legend = FALSE)+
  theme(text = element_text(size = 8))+
  scale_color_manual(values=c("Down-regulated"="blue",
"Unchanged"="black", "Up-regulated"="red")) +
  xlab(expression("avg_log"[2]*"FC")) +
  ylab(expression("-log"[10]*"p-val_adj")) +
  xlim(-10, 10)
p1
ggsave(
  plot = p1,
  filename = paste0(figures_dir, "Myogenic_trajectory_
Volcano.tiff"),
  device = "tiff", dpi = 600, width = 11, height = 7,
units = "cm")


###6. CELLCHAT ANALYSIS ###
library(CellChat)
library(Seurat)
library(patchwork)
library(ggplot2)
library(ComplexHeatmap)
library(future)
library(future.apply)
```

```
Cellchat_dir <- paste0(base_dir, "CellChat/")
dir.create(Cellchat_dir, showWarnings = FALSE)


#this will load the annotated files - use in case you
have already run Separately all the other blocks
integrated <- readRDS(paste0(processed_dir,"RNA_
annotated.rds"))


# Divide WT and D51 samples
Idents(integrated) = "orig.ident"
WT <- subset(integrated, idents = "WT")
WT <- JoinLayers(WT)
WT <- ScaleData(object = WT)
DimPlot(WT, reduction = "cca.umap", label = TRUE,
group.by = "annotation.cca")


# Create CellChat object
cellchat_WT <- createCellChat(object = WT, group.by =
"annotation.cca")


# Creating CellChat Database for mouse
CellChatDB <- CellChatDB.mouse
showDatabaseCategory(CellChatDB)


# Use a subset of CellChatDB for cell-cell
communication analysis
CellChatDB.use <- subsetDB(CellChatDB, search =
"Secreted Signaling")
# Set the used database in the object
cellchat_WT@DB <- CellChatDB.use


# Subset the expression data of signaling genes for
saving computation cost
cellchat_WT <- subsetData(cellchat_WT) # This step is
necessary even if using the whole database
future::plan("multisession", workers = 4) # Increase
workers if needed
cellchat_WT <- identifyOverExpressedGenes(cellchat_WT)
cellchat_WT <
- identifyOverExpressedInteractions(cellchat_WT)


# Compute the cell-cell communication probabilities
cellchat_WT <- computeCommunProb(cellchat_WT, raw.use
= TRUE)
```

```
# Filter out the cell-cell communication if there are
only few number of cells in certain cell groups
cellchat_WT <- filterCommunication(cellchat_WT, min.
cells = 10)


# Compute Pathway-level communication probabilities
and cell-cell communication networks
cellchat_WT <- computeCommunProbPathway(cellchat_WT)
cellchat_WT <- aggregateNet(cellchat_WT)
groupSize <- as.numeric(table(cellchat_WT@idents))
cellchat_WT


saveRDS(cellchat_WT, file=
paste0(Cellchat_dir,"CellChat_WT.rds"))
# Same for D51
# Divide WT and D51 samples
Idents(integrated) = "orig.ident"
D51 <- subset(integrated, idents = "D51")
D51 <- JoinLayers(D51)
D51 <- ScaleData(object = D51)
DimPlot(D51, reduction = "cca.umap", label = TRUE,
group.by = "annotation.cca")
cellchat_D51 <- createCellChat(object = D51, group.by
= "annotation.cca")


# set the used database in the object
cellchat_D51@DB <- CellChatDB.use


# subset the expression data of signaling genes for
saving computation cost
cellchat_D51 <- subsetData(cellchat_D51) # This step
is necessary even if using the whole database
future::plan("multisession", workers = 4)
cellchat_D51 <- identifyOverExpressedGenes(cellchat_
D51)
cellchat_D51 <
- identifyOverExpressedInteractions(cellchat_D51)


# Compute communication probabilities
cellchat_D51 <- computeCommunProb(cellchat_D51, raw.
use = TRUE)
cellchat_D51 <- filterCommunication(cellchat_D51, min.
cells = 10)
cellchat_D51 <- computeCommunProbPathway(cellchat_D51)
cellchat_D51 <- aggregateNet(cellchat_D51)
groupSize <- as.numeric(table(cellchat_D51@idents))
cellchat_D51
```

```
saveRDS(cellchat_D51, file=paste0(Cellchat_
dir,"CellChat_D51.rds"))
# CellChat WT vs D51
#This loads the different cellchat objects
cellchat_WT=readRDS(paste0(Cellchat_dir,"CellChat_
WT.rds"))
cellchat_D51=readRDS(paste0(Cellchat_dir,"CellChat_
D51.rds"))


# Define the cell labels to lift up
group.WT = levels(cellchat_WT@idents)
group.D51 = levels(cellchat_D51@idents)
cellchat_WT <- liftCellChat(cellchat_WT, group.D51)
cellchat_D51 <- liftCellChat(cellchat_D51, group.WT)


# Compute network centrality metrics
cellchat_WT <- netAnalysis_computeCentrality(cellchat_
WT, slot.name = "netP")
cellchat_D51 <- netAnalysis_
computeCentrality(cellchat_D51, slot.name = "netP")


object.list <- list(WT = cellchat_WT, D51 =
cellchat_D51)
cellchat <- mergeCellChat(object.list, add.names =
names(object.list))


saveRDS(cellchat, file=paste0(Cellchat_dir,"cellchat_
merge.rds"))
# Visualization of cell-cell communication
# RankNet plot
p1=rankNet(cellchat, mode = "comparison", stacked = F,
do.stat = TRUE) +
  theme(text = element_text(size = 8))


ggsave(plot = p1,
       filename = paste0(figures_dir, "CellChat_merged_
RankNetplot.tiff"),
       device = "tiff", dpi = 600, width = 18, height =
10, units = "cm")


# Heatmap
i = 1
pathway.union <- union(object.list[[i]]@netP$pathways,
object.list[[i+1]]@netP$pathways)
ht1 = netAnalysis_signalingRole_heatmap(object.
list[[i]],
                                        pattern =
"all",
```

```
                                          signaling =
pathway.union, width = 6,height = 6,
                                          title =
names(object.list)[i])
ht2 = netAnalysis_signalingRole_heatmap(object.
list[[i+1]],
                                      pattern
=                                     "all", width =
6,height = 6,
                                      ignaling = pathway.
union,
                                      title = names(object.
list)[i+1]
)


tiff(file=paste0(figures_dir, "CellChat_merged_Heatmap.
tiff"),res = 600, width = 21, height = 10, units =
"cm")
draw(ht1 + ht2, ht_gap = unit(2, "cm"))
dev.off()


# Chose a pathway for further analysis
pathways.show <- c("VEGF")


# Compute the maximum communication probability
weight.max <- getMaxWeight(object.list, slot.name =
"netP", attribute = pathways.show)


# Hierarchy plot
tiff(file=paste0(figures_dir, "CellChat_merged_Hierarchy_
plot.tiff"),res = 600, width = 21, height = 12, units =
"cm")
par(mfrow = c(1,2), xpd=TRUE)
for (i in 1:length(object.list)) {
  netVisual_aggregate(object.list[[i]],
                      signaling = pathways.show,
                      edge.weight.max = weight.max[1],
                      edge.width.max = 10,
                      signaling.name = paste(pathways.
show,
names(object.list)[i]),
                                            nam-
es(object.list)[i]),
                      pt.title = 8)
}
dev.off()
sessionInfo()
```

Competing Interests

The authors declare no competing interests.

References

1. Mukund K, Subramaniam S (2020) Skeletal muscle: a review of molecular structure and function, in health and disease. WIREs Mech Dis 12:e1462
2. Petrany MJ, Swoboda CO, Sun C et al (2020) Single-nucleus RNA-seq identifies transcriptional heterogeneity in multinucleated skeletal myofibers. Nat Commun 11:6374
3. Jin S, Guerrero-Juarez CF, Zhang L et al (2021) Inference and analysis of cell-cell communication using CellChat. Nat Commun 12:1088. https://doi.org/10.1038/s41467-021-21246-9
4. Arya A, Tripathi P, Dubey N et al (2025) Navigating single-cell RNA-sequencing: protocols, tools, databases, and applications. Genom Inform 23:13
5. Chemello F, Wang Z, Li H et al (2020) Degenerative and regenerative pathways underlying Duchenne muscular dystrophy revealed by single-nucleus RNA sequencing. Proc Natl Acad Sci USA 117:29691–29701
6. Dos Santos M, Gioftsidi S, Backer S et al (2021) Extraction and sequencing of single nuclei from murine skeletal muscles. STAR Protocols 2:100694
7. Stuart T, Butler A, Hoffman P et al (2019) Comprehensive integration of single-cell data. Cell 177(1888-1902):e21
8. Hao Y, Hao S, Andersen-Nissen E et al (2021) Integrated analysis of multimodal single-cell data. Cell 184(3573-3587):e29
9. Young MD, Behjati S (2020) SoupX removes ambient RNA contamination from droplet-based single-cell RNA sequencing data. GigaScience 9:giaa151
10. Fleming SJ, Chaffin MD, Arduini A et al (2023) Unsupervised removal of systematic background noise from droplet-based single-cell experiments using CellBender. Nat Methods 20:1323–1335
11. Hao Y, Stuart T, Kowalski MH et al (2024) Dictionary learning for integrative, multimodal and scalable single-cell analysis. Nat Biotechnol 42:293–304
12. Jin S, Plikus MV, Nie Q (2025) CellChat for systematic analysis of cell–cell communication from single-cell transcriptomics. Nat Protoc 20:180–219
13. Bengtsson H (2021) A unifying framework for parallel and distributed processing in R using futures. R J 13:208
14. Wickham H, François R, Henry L et al (2025) dplyr: a grammar of data manipulation - R package version 1.2.1, https://dplyr.tidyverse.org.
15. Wickham, H, 2016. ggplot2: elegant graphics for data analysis. Springer-Verlag, New York
16. Lun ATL, Riesenfeld S, Andrews T, participants in the 1st Human Cell Atlas Jamboree et al (2019) EmptyDrops: distinguishing cells from empty droplets in droplet-based single-cell RNA sequencing data. Genome Biol 20:63
17. Slowikowski K (2024) ggrepel: automatically position non-overlapping text labels with "ggplot2" - https://github.com/slowkow/ggrepel.
18. Pedersen TL (2025) patchwork: the composer of plots - https://patchwork.data-imaginist.com
19. Gu Z (2022) Complex heatmap visualization. Imeta 1(3), e43.

Part VI

Chromatin and Genome Organization

Chromatin Immunoprecipitation Followed by PCR (ChIP-PCR) from Prospectively Isolated Murine Muscle Stem Cells

Marie-Claude Sincennes

Abstract

Muscle stem cells are essential for skeletal muscle regeneration, but their scarcity makes epigenetic studies technically and financially challenging. Here, we describe a robust chromatin immunoprecipitation (ChIP) protocol compatible with freshly isolated muscle stem cells using as few as 25,000 cells. This method avoids sample pooling, reduces animal use, and enables reliable analysis of chromatin modifications and transcription factor binding.

Keywords Muscle stem cells, Epigenetics, Histone marks, Chromatin immunoprecipitation, Transcription, Gene expression, Myogenesis, Satellite cell, Skeletal muscle

1 Introduction

Muscle stem cells, or satellite cells, govern skeletal muscle tissue regeneration in response to disease or injury. A better understanding of their biology is crucial for the development of regenerative medicine therapies for neuromuscular diseases. Studying muscle stem cells represents a technical challenge, mostly due to their scarcity within the muscle tissue, which makes them difficult to study with molecular biology techniques that usually require a lot of biological material. One such technique is chromatin immunoprecipitation or ChIP. Some epigenetic features of muscle stem cells have been characterized by ChIP and led us to partly decipher the molecular events that take place during aging [1]. These ChIP experiments required 10^6 satellite cells per condition, which involves pooling samples from independent biological replicates to achieve $n = 1$. Therefore, a comprehensive analysis of muscle stem cell epigenetics requires huge financial and animal resources. To circumvent these technical limitations, many elegant studies have been performed in cell types that model muscle stem cells, such as embryonic stem cell-derived muscle progenitors [2, 3] and myoblasts [4, 5] (reviewed in [6]). However, it is difficult

Nathalie Didier and Caroline E. Brun (eds.), *Muscle Stem Cells*, Methods in Molecular Biology, vol. 3093,
https://doi.org/10.1007/978-1-0716-5643-3_19,

to determine to what extent these cell types fairly represent the biology of muscle stem cells.

Recent chromatin profiling methods have been developed, such as Cleavage Under Targets and Tagmentation (CUT&TAG) [7], which are faster and more efficient, in addition to requiring fewer cells and less sequencing. However, CUT&TAG reagents are expensive, and the method is not recommended for weakly-bound chromatin-binding proteins, due to the native and stringent binding conditions in CUT&TAG reactions. In addition, the method relies on using a transposase enzyme that has a higher affinity for accessible chromatin, leading to a bias that could prevent the study of proteins and histone marks associated with closed chromatin. For example, ChIP could be a preferred method to study pioneer transcription factors, such as PAX7 [3, 8], which have the capacity to bind their target sites even in compacted chromatin, when these sites are wrapped inside nucleosomes.

We developed a protocol to perform chromatin immunoprecipitation in freshly isolated muscle stem cells, which would avoid the use of massive animal resources. Our robust protocol allows the detection of chromatin modifications in as few as 25,000 cells, meaning that sample pooling is no longer required. In addition, biological material from a single mouse can be split in order to have a paired IgG control for each single ChIP experiment. Our protocol led us to investigate changes in histone methylation mediated by *Mll1* in quiescent (freshly isolated) and activated muscle stem cells [9]. Our protocol can be used to determine the chromatin status of specific, predetermined sets of genomic loci in freshly isolated satellite cells. The protocol is also compatible with low-input library preparation for sequencing the ChIP DNA [10, 11].

2 Materials

2.1 Muscle Cell Dissociation and Preparation for FACS Sorting

1. Dissection tools and scissors.
2. Plasticware: Sterile 60 × 15 mm tissue culture dishes, conical 15 mL tubes, 5 mL polypropylene tubes.
3. GentleMACS C Tubes (Miltenyi Biotec).
4. 100 μm cell strainers that fit to 15 mL conical tubes (Miltenyi Biotec).
5. 50 μm cell strainers that fit into 5 mL polypropylene tubes (Sysmex).
6. FACS buffer: PBS, 10% FBS, 5 mM EDTA. Store at 4°C.
7. Collagenase/dispase solution: 10 g/L collagenase B (Millipore-Sigma) and 4 g/L dispase II (Millipore-Sigma) in Ham'S F-10 media (Wisent), filter-sterilized. Aliquot and store at −20°C.

   Red blood cell lysis buffer (Millipore-Sigma).

8. FACS antibodies: Alexa647 mouse anti-Integrin alpha7 (clone R2F2) (UBC AbLab); BV421 rat anti-CD31 (BD Biosciences); BV421 rat anti-CD45 (BD Biosciences); BV421 rat anti-CD11b (BD Biosciences); BV421 rat anti-Ly6A/E (Sca-1) (BD Biosciences); Biotin rat anti-CD34 (Miltenyi Biotech); Streptavidin Pe-Cy7 (BD Biosciences).
9. PI (propidium iodide). Make a stock at 50 μg/mL in sterile water. Aliquot and store at 4°C (protect from light).

2.2 Cross-Linking Samples for Chromatin Immunoprecipitation

1. Low retention 1.5 mL tubes
2. Formaldehyde 37% (ThermoFisher). Prior to use, prepare a 1% working solution by diluting 1 mL of formaldehyde stock solution into 36 mL PBS.
3. 1.25 M glycine: Weight 4.7 g of Glycine and complete to 50 mL with deionized water. Filter-sterilize. Store at 4°C.
4. Fetal bovine serum (FBS), filtered.
5. ChIP lysis buffer: 50 mM Tris-HCl pH 8.0, 10 mM EDTA, 0.5% SDS. Store at room temperature.

2.3 Chromatin Immunoprecipitation

1. ChIP-grade antibodies and isotype IgG controls.
2. Protein A/G mix magnetic beads (Millipore Sigma).
3. ChIP dilution buffer: 20 mM Tris-HCl pH 8.0, 1.2 mM EDTA, 1.1% Triton X-100, 200 mM NaCl. Store at room temperature.
4. microTUBE AFA Fiber Pre-Slit Snap Cap 6 × 16 mm (Covaris).
5. ChIP Low Salt Buffer: 20 mM Tris-HCl pH 8.0, 150 mM NaCl, 2 mM EDTA, 0,1% SDS, 1% Triton X-100. Store at room temperature.
6. Proteinase K (D-Mark): Prepare a 10 mg/mL stock solution in glycerol. Aliquot and freeze at −20°C for long-term storage.
7. Elution buffer: 20 mM Tris-HCl pH 7.5, 5 mM EDTA, 50 mM NaCl. Store at room temperature. Prior to use, add: 1% SDS (10 μL of a 10% stock to 1 mL elution buffer) and 50 μg/mL proteinase K (5 μL of a 10 mg/mL stock to 1 mL elution buffer).
8. Magnetic separation rack.
9. NucleoSpin Gel and PCR Clean-Up Kit (Macherey-Nagel).
10. Binding buffer NTB (Macherey-Nagel).
11. Real-time PCR master mix.
12. Validated real-time PCR primers.

2.4 Equipment

1. Sterile tissue culture hood
2. Automated cell dissociator (gentleMACS Octo Dissociator with Heaters, Miltenyi Biotec)
3. Tabletop centrifuge (swing bucket is preferred)
4. Cell sorter (MoFlo XDP, Beckman Coulter, or equivalent)
5. Covaris S220 Focused-ultrasonicator
6. Real-time PCR system

3 Methods

3.1 Muscle Cell Dissociation and Preparation for FACS Sorting

1. Sacrifice mice and dissect the hindlimb muscles using dissection tools. For each mouse, collect the dissected muscle tissue in an individual 60 × 15 mm culture dish filled with sterile, cold PBS and keep on ice. Clean the sample by removing obvious fat paws, tendons, and blood vessels (this step increases the sort efficiency and reduces sort time).
2. Using dissection forceps, transfer muscle tissue into a new, empty 60 × 15 mm culture dish to let the excess PBS flow out.
3. Fill a gentleMACS CTube with 5 mL of the collagenase/dispase solution (*see* **Note 1**), and transfer muscle tissue into the CTube. Using dissection scissors, mince muscles directly into the CTube, to generate pieces of approximately 2 mm in size.
4. Close the CTube securely and turn it upside down. Ensure that all muscle pieces are immersed in the collagenase/dispase solution.
5. Insert the CTube into the gentleMACS Octo Dissociator with Heaters and insert the heater according to the manufacturer's instructions.
6. Use the following program to perform mechanical and enzymatic dissociation of the muscle tissue:
   a. Temp ON
   b. Loop 10×
   c. Spin 360 rpm 5 s
   d. Spin −360 rpm 5 s
   e. End Loop
   f. Spin 60 rpm 3 min
   g. Spin −30 rpm 9 min
   h. Loop 10×
   i. Spin 360 rpm 5 s
   j. Spin −360 rpm 5 s
   k. End loop
   l. Spin −30 rpm 12 min
   m. End

7. Gently remove the heater and the CTube according to the manufacturer's instruction. The cell suspension should be devoid of aggregates or clumps.
8. In a sterile tissue culture hood, open the CTube, and add 2 mL of FBS. Mix with a pipette. It is preferable to perform the subsequent steps in a sterile tissue culture hood.
9. Insert a 100 μm cell strainer over a conical 15 mL tube, and gradually add cell suspension to filter. Rinse the empty CTube with 4–5 mL of PBS and add to filter. Collect total effluent.
10. Centrifuge at 350 × *g* for 10 min. Aspirate carefully the supernatant.
11. Resuspend the cell pellet in 400 μL of red blood cell lysis buffer and resuspend quickly but evenly. Add immediately 10 mL FACS buffer. Mix by inversion and spin at 350 × *g* for 5 min. That procedure lyses around 60%–80% of the red blood cells (this red blood cell lysis step is optional, but it improves the sort efficiency and reduces sort time).
12. Resuspend the cell pellet with 800 μL FACS buffer. Add the different antibodies for cell sorting, for which the combination and quantity should be predetermined by the investigator. Typically, this combination leads to good results: 2 μL CD31, 2 μL CD11b, 2 μL CD45, 2 μL Sca1, 10 μL alpha7-integrin, and 10 μL CD34.
13. Incubate 10 min on ice, and spin at 350 × *g* for 5 min. Remove supernatant.
14. Resuspend the pellet in 1 mL FACS buffer and add 2 μL of streptavidin. Incubate 10 min on ice and spin 350 × *g* for 5 min.
15. Resuspend the pellet in 1 mL FACS buffer, and apply the cell suspension to a 50 μm cell strainer inserted into a 5 mL tube. Rinse the cell strainer with 3 × 1 mL FACS buffer, so the total volume is 4 mL. Mix the tube by inversion.
16. Add 20 μL of the propidium iodide (PI) solution (cell death marker) immediately prior to sort. Mix the tube by inversion.
17. Using the cell sorter, collect the muscle stem cell fraction (PI−, CD31−, CD11b−, CD45−, Sca1−, CD34+, Itga7+).

3.2 Cross-Linking Samples for Chromatin Immunoprecipitation

1. Spin the sorted muscle stem cells at 350 × *g* for 5 min (*see* **Note 2**). The pellet should not be visible. Resuspend in 1 mL PBS supplemented with 1% formaldehyde. Transfer the sample into a low-retention 1.5 mL tube (*see* **Note 3**), and split to have the proper cell number in each tube (*see* **Note 4**).
2. Let the cross-linking reaction take place for 10 min at room temperature.

3. Add 100 µL of 1.25 M glycine, mix by gentle inversion and incubate for 5 min.

4. Add 100 µL of FBS, mix by gentle inversion, and spin in a swing-bucket tabletop centrifuge for 5 min at 350 × *g*. Remove the supernatant carefully using the pipette.

5. Wash the cell pellet (invisible) with 100 µL PBS supplemented with 10% FBS. Spin in a swing-bucket tabletop centrifuge for 5 min at 350 × *g*. Remove the supernatant carefully using the pipette.

6. Resuspend the pellet in 100 µL ChIP Lysis buffer.

7. You can freeze the cross-linked samples at −80°C until all the samples to be compared together are collected. Therefore, the chromatin immunoprecipitation steps can be performed simultaneously with all samples to minimize technical variation.

3.3 Chromatin Immunoprecipitation Protocol

1. Determine how many ChIPs and IgG control reactions will be performed. For each reaction, precouple the antibody or IgG control (2 µg per reaction) (*see* **Note 5**) to protein A/G mix magnetic beads (15 µL per reaction) for 3 h at 4°C in a mix of 50 µL ChIP Lysis buffer and 50 µL ChIP Dilution buffer. The precoupled beads do not need to be washed after coupling, as the quantity of antibody is lower to the binding capacity of the beads (no excess antibody).
2. During the precoupling step, thaw the cross-linked samples on ice and transfer in Covaris microTUBEs. Perform chromatin shearing using the Covaris sonicator according to manufacturer's instructions, using the following specifications: Peak Power 105.0; Duty Factor 5.0; Cycles/Burst 200; Run time 600 s.
3. Transfer the sonicated sample back to a low retention 1.5 mL tube. Add 100 µL ChIP dilution buffer. Mix by inversion.
4. Transfer 10 µL of each sample into a clean, new tube, and keep it at 4°C. That will represent the input control (5% of sample).
5. In the sample to be immunoprecipitated, add the precoupled beads to your sample. The final volume should be of approximately 300 µL. Incubate for 2 h at 4°C (with rotation).
6. Apply the sample to the magnetic separation rack and let sit for 30 s. Remove the supernatant with a pipette. You may keep the unbound fraction (supernatant) as a positive control for real-time PCR (optional).

7. Wash the beads with 500 μL ChIP low salt buffer. Apply to the magnet. Remove supernatant.
8. Repeat Step 7 two more times (total of three washes).
9. Prepare the proper amount of ChIP elution buffer, keeping in mind that 150 μL per sample and 150 μL per input control are needed.
10. Add 100 μL elution buffer to each tube (ChIP sample, IgG control, input control) and incubate for 2 h at 65°C. Mix the tube by inversion every 30 min.
11. Apply the samples to the magnetic separation rack and let sit for 30 s. Keep the supernatant and transfer to a new 1.5 mL tube. The input control sample does not need to be applied on the magnet, as it does not contain any bead.
12. Add another 50 μL elution buffer to the beads and incubate for another for 10 min at 65°C. Apply the samples to the magnetic separation rack and let sit for 30 s. Keep the supernatant and combine with the first eluate (total volume of 150 μL).
13. Recover DNA using the NucleoSpin Gel and PCR Clean-Up Kit according to the manufacturer's instructions. Follow the clean-up protocol that is suitable for samples containing SDS (protocol using NTB binding buffer). For the last elution step, recover DNA in 40 μL H20.
14. Perform real-time PCR reactions using 2 μL of sample with prevalidated ChIP-PCR primers (*see* **Note 6**).
15. Analyze PCR data by normalizing to input signal or to Total H3 binding (*see* **Note 7**).

4 Notes

1. The volumes of collagenase/dispase solution, red blood cell lysis buffer, and FACS buffer are optimized for the efficient dissociation of muscle tissue from the hindlimbs of one 6–8-week-old, uninjured mouse. The investigator should adapt the volumes and incubation times if using mice of different ages and/or genotypes.
2. The optimal centrifugation speed and time may vary depending on the centrifuge model.
3. When working with very small, invisible cell pellets, many considerations have to be taken to avoid losing material due to centrifugation steps. We highly recommend using low-retention tubes. We also recommend, when possible, mixing reagents by

gentle tube inversion instead of pipetting, and adding FBS to the solutions. Finally, swing bucket centrifuges are preferable as it ensures that the cell pellet is formed at the bottom of the tube.

4. The ChIP protocol is suitable for cell numbers as low as 25,000 cells. However, as demonstrated in Fig. 1A, B, while the proportional binding (% input) increases with lower cell numbers, the signal-to-noise ratio decreases. It is recommended to determine the optimal cell numbers to maximize signal-to-noise ratio while increasing binding efficiency.

5. The efficiency of one given antibody could vary between manufacturers and even from one lot to the other. Each new vial of antibody should be tested and the volume needed per reaction should be optimized accordingly. The amount of antibody needed for each chromatin immunoprecipitation (suggested 2 μg) has to be optimized for every condition. The quantity needed depends of the antibody used, of the presence of the

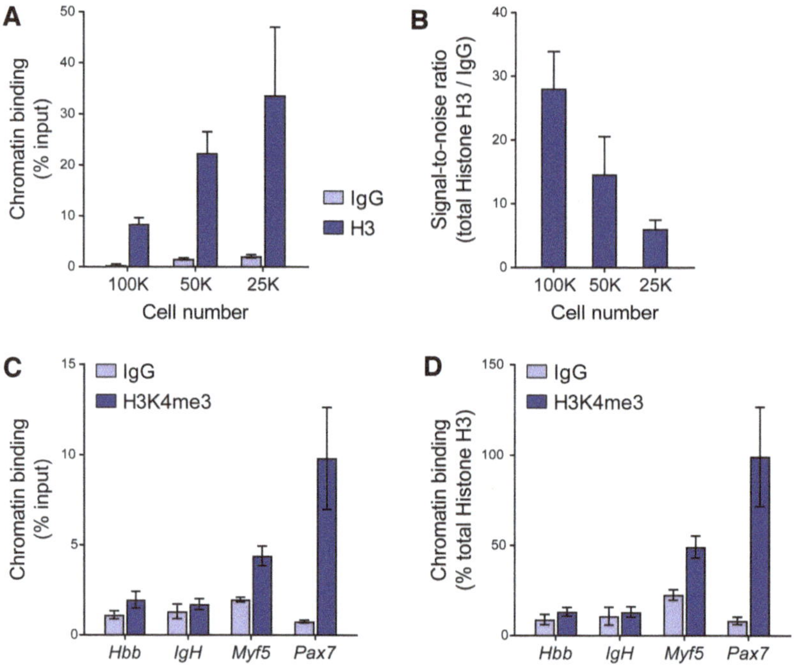

Fig. 1 Histone H3 tri-methylation (H3K4me3) and total histone H3 chromatin immunoprecipitation efficiency in freshly isolated muscle stem cells. **A,** Comparison of ChIP efficiency relative to the number of cells employed as starting material. Data are shown for Total H3 binding to the *Pax7* locus ($n = 4$). **B,** Signal-to-noise ratio (Total H3 binding relative to IgG binding) depending on the amount of starting material ($n = 4$). **C and D,** H3K4me3 ChIP in freshly isolated muscle stem cells (50k cells). The histone modification is present on *Myf5* and *Pax7* loci, while it is absent on non-myogenic genes such as *Hbb* and *IgH*. Data is presented as % input (C) or % total H3 (D) ($n = 4$)

epigenetic mark in the studied cell type, and of the number of cells.

6. ChIP primers should include DNA loci that are expected to be bound by the bait (i.e., *Myf5* and *Pax7* in the example presented in Fig. 1) and loci that are not bound (i.e., *Hbb* and *IgH* in Fig. 1). The IgG isotype control should display minimal binding to all loci (less than 2% input).
7. Chromatin immunoprecipitation data can be normalized over input (% input) or over total H3 binding (% total H3) (Fig. 1C, D). Normalizing over the input DNA, corresponding to the starting material, accounts for the variations in starting material and for technical differences in sample processing. Input DNA is also commonly used as a positive control for real-time PCR. While normalizing to total H3 implies an additional ChIP reaction using an anti-H3 antibody, this control accounts for variations in ChIP efficiency between experiments, and allows correction for differences in chromatin abundance and ensures that the observed differences for a given histone modification are specific to the modification, and do not reflect a difference in the total amount of histones in the samples.

Acknowledgments

The author acknowledges Institut National de la Recherche Scientifique (INRS) and Universite du Quebec a Chicoutimi (UQAC) for their support in the context of the INRS-UQAC Joint Research Unit. These studies were carried out with support of grants to M.C.S. from the Natural Sciences and Engineering Research Council of Canada (RGPIN-2025-05174, DGECR-2025-00421) and from the Canadian Institutes of Health Research (PDI 185642).

Competing Interests The authors declare no competing interests.

References

1. Liu L, Cheung TH, Charville GW et al (2013) Chromatin modifications as determinants of muscle stem cell quiescence and chronological aging. Cell Rep 4:189–204
2. Zhang N, Mendieta-Esteban J, Magli A et al (2020) Muscle progenitor specification and myogenic differentiation are associated with changes in chromatin topology. Nat Commun 11: 6222
3. Lilja KC, Zhang N, Magli A et al (2017) Pax7 remodels the chromatin landscape in skeletal muscle stem cells. PLoS One 12: e0176190
4. Asp P, Blum R, Vethantham V et al (2011) Genome-wide remodeling of the epigenetic landscape during myogenic differentiation. Proc Natl Acad Sci U S A 108:E149–158
5. Soleimani VD, Yin H, Jahani-Asl A et al (2012) Snail regulates MyoD binding-site occupancy to direct enhancer switching and differentiation-specific transcription in myogenesis. Mol Cell 47:457–468
6. Sincennes MC, Brun CE, Rudnicki MA (2016) Concise review: epigenetic regulation of myogenesis in health and disease. Stem Cells Transl Med 5:282–290

7. Kaya-Okur HS, Wu SJ, Codomo CA et al (2019) CUT&Tag for efficient epigenomic profiling of small samples and single cells. Nat Commun 10: 1930
8. Budry L, Balsalobre A, Gauthier Y et al (2012) The selector gene Pax7 dictates alternate pituitary cell fates through its pioneer action on chromatin remodeling. Genes Dev 26:2299–2310
9. Addicks GC, Brun CE, Sincennes MC et al (2019) MLL1 is required for PAX7 expression and satellite cell self-renewal in mice. Nat Commun 10: 4256
10. Adli M, Zhu J, Bernstein BE (2010) Genome-wide chromatin maps derived from limited numbers of hematopoietic progenitors. Nat Methods 7:615–618
11. Shankaranarayanan P, Mendoza-Parra MA, Walia M et al (2011) Single-tube linear DNA amplification (LinDA) for robust ChIP-seq. Nat Methods 8:565–567

Chapter 20

Optimized Low-Input ATAC-Seq Protocol for Freshly Sorted Muscle Stem Cells

Darren M. Blackburn , Korin Sahinyan and Vahab D. Soleimani

Abstract

Skeletal muscle stem cells (MuSCs) are essential for tissue regeneration and repair, yet their scarcity has historically limited routine application of chromatin-level assays. Assay for Transposase-Accessible Chromatin using sequencing (ATAC-seq) provides a powerful means to define the chromatin state and study cellular regulatory landscapes, but standard protocols require cell numbers far exceeding those that can be freshly sorted from muscle, particularly under conditions such as aging, disease, and various genetic models. Here, we present an optimized, low-input ATAC-seq workflow specifically tailored for analysis of MuSCs freshly isolated from skeletal muscle. This method integrates high-purity isolation of MuSCs by fluorescence-activated cell sorting (FACS), refined nuclei purification, and a high-efficiency transposition assay that can be applied to the analysis of chromatin using as few as 5000 cells. We describe critical steps for minimizing technical noise, including optimized mitochondrial depletion, micro-volume tagmentation, low-cycle ATAC-seq library amplification, and quantitative quality-control metrics. The workflow is compatible with FACS-purified MuSCs from young and aged mice, injury models, and genetic knockout models, enabling direct comparison of chromatin accessibility across biological conditions where cell numbers are limiting.

Keywords ATAC-seq, Skeletal muscle stem cells, MuSC, Chromatin accessibility, FACS, Epigenomics

1 Introduction

Skeletal muscle regeneration relies on a population of tissue-resident stem cells known as muscle satellite cells (MuSCs) [1]. These cells remain quiescent under homeostatic conditions but rapidly activate and proliferate in response to injury or disease [2, 3]. Understanding the chromatin landscape of MuSCs is critical for elucidating the epigenetic mechanisms underlying their quiescence, activation, lineage specification, and age-related decline.

ATAC-seq is a widely used, sensitive, and efficient method for mapping genome-wide open chromatin in numerous cell types and tissues [4]. ATAC-seq utilizes the hyperactive Tn5 transposase to

Nathalie Didier and Caroline E. Brun (eds.), *Muscle Stem Cells*, Methods in Molecular Biology, vol. 3093,
https://doi.org/10.1007/978-1-0716-5643-3_20,

simultaneously fragment DNA and insert indexed sequencing adapters into the ends of fragmented DNA [4, 5]. Such features make this technique a powerful tool for studying low-input samples and rare populations such as freshly isolated MuSCs.

However, isolating pure populations of MuSCs poses unique challenges. Muscle tissue is heterogeneous, requiring enzymatic digestion and careful handling to maintain cell viability [6, 7]. Furthermore, the chromatin state can change rapidly during tissue processing. In this chapter, we present a step-by-step protocol optimized for minimizing technical variability and preserving the native chromatin landscape of freshly sorted MuSCs. The workflow integrates standard FACS-based purification strategies with rapid downstream processing for ATAC-seq library preparation and sequencing.

Combining FACS with low-input ATAC-seq is important for accurately assessing the chromatin state of freshly sorted MuSCs because in their in vivo niche environment, they are rare [8]. FACS enables precise isolation of the stem cell population based on surface markers or use of reporter mouse lines [9]. Application of low-input ATAC-seq on MuSCs immediately after isolation captures a high-resolution chromatin accessibility landscape without the need for in vitro expansion, which would otherwise alter their native epigenetic state [10].

2 Materials

1. Mice: Tg(Pax7-EGFP)#Tajb (Pax7-nGFP) [11] was used in this protocol, but any mouse strain can be used.
2. FACS digestion buffer: 0.5 mM $CaCl_2$, 9 U/mL Dispase II, 2.4 U/mL Collagenase D. Dissolve the reagents in unsupplemented Ham's F10 media. Prepare 10 mL of buffer per sample and use immediately.
3. FACS Buffer: 2% FBS, 2 mM EDTA. Reagents are diluted in PBS.
4. Antibody master mix: 4 μg/mL Hoechst 33342 (Thermofisher), 1:200 Anti-mouse ITGA7-Alexa647 (AbLab), 1:150 Anti-mouse VCAM-1 PE-Cyanine7 (Invitrogen), 1:5000 Anti-mouse PE-CD45 (Invitrogen), 1:5000 Anti-mouse PE-CD11b (Invitrogen), 1:5000 Anti-mouse PE-CD31 (BD Pharmigen), 1:5000 Anti-mouse PE-Sca1/Ly6A-E (BD Pharmigen).
5. ATAC-Seq lysis and permeabilization buffer: 10 mM Tris-HCl pH 7.5, 10 mM NaCl, 3 mM $MgCl_2$, 0.1% Tween-20, 0.1% NP-40, 0.01% Digitonin.

6. ATAC-Seq wash buffer: 10 mM Tris-HCl pH 7.5, 10 mM NaCl, 3 mM $MgCl_2$, 0.1% Tween-20.
7. Transposition mix: 5 μL Tagment DNA (TD) buffer (Illumina), 3.2 μL PBS, 0.89 μL Tn5 (Illumina), 0.1% Tween-20, 0.01% Digitonin, 0.75 μL nuclease-free water. Prepare 10 μL per sample.
8. Library prep PCR master mix: 10 μL Q5 buffer, 10 μL Q5 enhancer, 1 μL dNTPs, 2.5 μL i7 index, 2.5 μL i5 index, 3.5 μL nuclease-free water, 0.5 μL Q5 high-fidelity DNA polymerase.
9. Qiagen QIAquick PCR Purification Kit.
10. Nextera XT Index Kit.
11. AMPure XP beads.
12. Fetal Bovine Serum.
13. PBS.
14. 15 mL conical tubes.
15. 50 mL conical tubes.
16. 40 μm cell strainers.
17. 5 mL polypropylene round-bottom tubes.
18. Ethanol.
19. Nuclease-free water.

3 Methods

3.1 Enzymatically Digest Skeletal Muscle

1. Prior to euthanizing the mice, prepare your workstation by cleaning your dissection tools, preparing two 6-cm plates with 2–4 mL of PBS per mouse. Thaw FBS and prepare the digestion buffer (10 mL per mouse).
2. Euthanize the mice following your facility's guidelines.
3. Dissect the muscle from the hindlimbs and place the muscle in a 6-cm plate with PBS. The dish may be left at room temperature until the next step.
4. Clean up the dissected muscle by cutting out the fat and fibrosis from the muscle and transfer the muscle to a fresh 6-cm plate with PBS.
5. Aspirate the PBS from the new plate and replace it with 2 mL of fresh PBS.
6. Mince the muscle with scissors into a slurry (*see* **Note 1**).

7. Transfer the muscle slurry into a 15 mL Falcon tube. Rinse the 6-cm plates with 5 mL of digestion buffer and add the buffer to the 15 mL Falcon tube.
8. Incubate the muscle in a cell incubator at 37°C and 5% CO_2. Place the 15 mL Falcon tube on a miniature rocking platform at 20 RPM for 20–30 min.
9. Centrifuge the muscle digest at 500 × *g* for 10 s to pellet the undigested muscle.
10. Transfer the supernatant to a 50 mL Falcon tube containing 8 mL of FBS (*see* **Note 2**).
11. Add the remaining 5 mL of digestion buffer to the undigested muscle pellet.
12. Triturate the muscle pellet with a P1000 pipette (*see* **Note 3**).
13. Incubate the remaining muscle in a cell incubator at 37°C and 5% CO_2. Place the 15 mL Falcon tube on a miniature rocking platform at 20 RPM for 20–30 min.
14. Add the digested muscle to the same 50 mL Falcon tube containing the FBS and the first round of digested muscle.
15. Filter the muscle digest through a 40 μm cell strainer (*see* **Note 4**).
16. Centrifuge the filtered muscle digest for 10 min at 500 × g at 4°C.

3.2 Antibody Labeling of Muscle Stem Cells

1. If you are using a mouse model without an endogenous fluorescent reporter for muscle stem cells (in this protocol, we used a Pax7 eGFP mouse model), antibody labeling will need to be done to identify the MuSC population. Prepare the antibody master mix during the centrifugation step (Step 3.1.16). If you are using an endogenous fluorescent reporter, move to Step 3.2.6.
2. Prepare 500 μL of master mix per mouse.
3. Aspirate the supernatant and resuspend the pellet in 500 μL of antibody master mix.
4. Agitate the sample protected from light for 15–20 min at room temperature.
5. Wash the sample by adding 8–10 mL of FACS buffer. Centrifuge the samples for 10 min at 500 × *g* at 4°C.
6. Remove the supernatant and resuspend the pellet in 800 μL of FACS buffer.
7. Filter the sample through a 40 μm cell strainer into a 5 mL round-bottom polypropylene FACS tube. Rinse the tube that contained the sample with 700 μL of fresh FACS buffer and pass the liquid through the same cell strainer into the FACS tube (*see* **Note 5**).

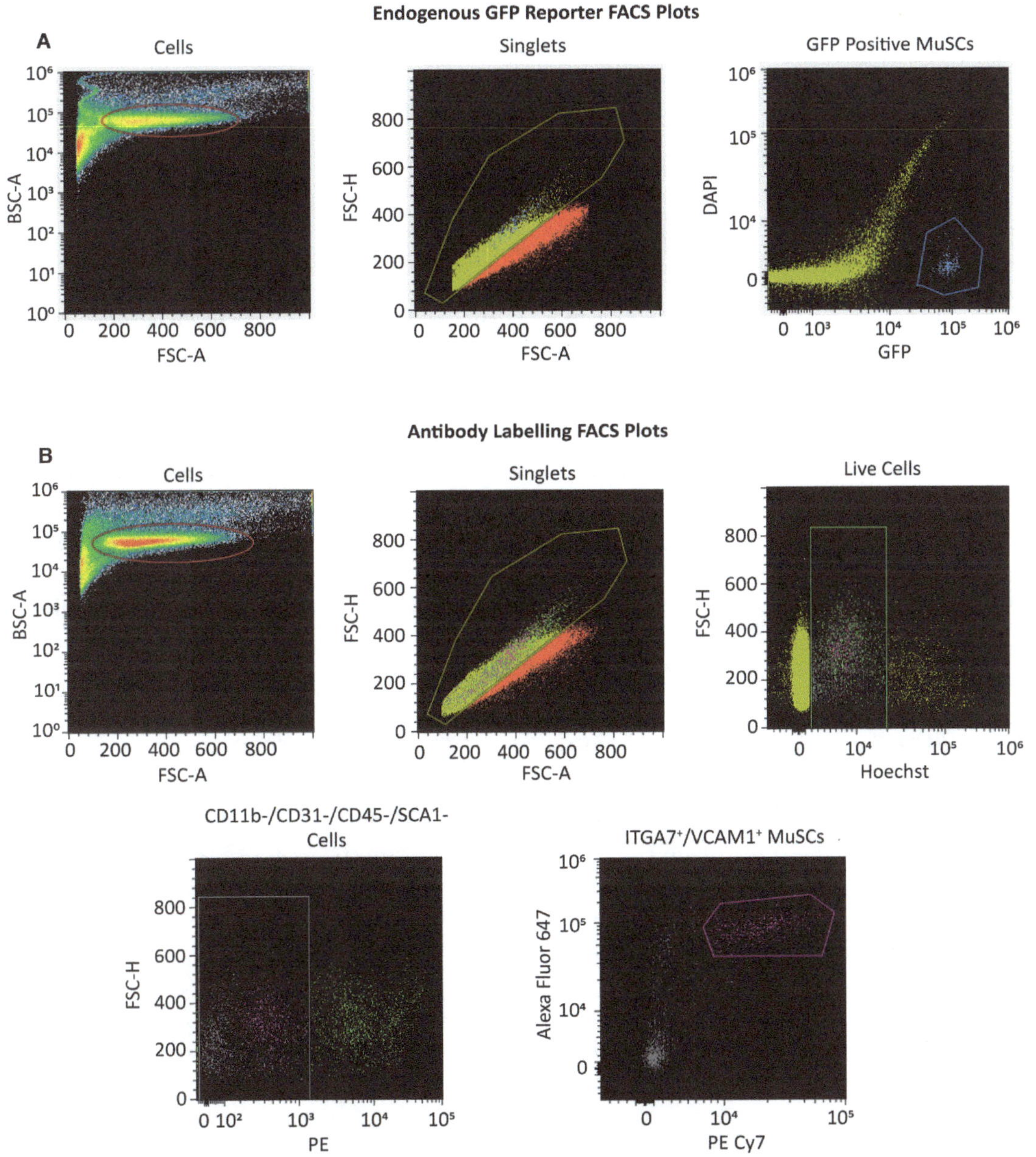

Fig. 1 FACS plots of muscle stem cell isolation. (**A**) FACS plots for sorting MuSCs from an endogenous GFP reporter mouse. The gating strategy includes selecting for cells, then for singlets and finally sorting the DAPI$^-$ and GFP$^+$ population. (**B**) FACS plots for sorting MuSCs using antibody labeling. The gating strategy includes selecting for cells, then for singlets, then live cells by selecting for Hoechst$^+$ cells. Nonlineage cells are negatively selected against by sorting for CD11b$^-$, CD31$^-$, CD45$^-$, and SCA1$^-$. Lastly, MuSCs are selected by sorting the double-positive ITGA7$^+$ and VCAM1$^+$ population

8. Sort for MuSCs by gating for the GFP$^+$ population if using an endogenous reporter. For antibody-labeled MuSCs, sort for the ITGA7$^+$/VCAM1$^+$/CD11b$^-$/CD31$^-$/CD45$^-$/SCA1$^-$ population (Fig. 1).

3.3 Lysing, Permeabilizing, and Tagmenting Muscle Stem Cells for ATAC-Seq

1. Sort the MuSC population into a 0.2 mL microtube containing 30 μL ATAC-Seq lysis buffer (*see* **Note 6**).
2. Incubate the cells on ice for 5 min and then for another 3 min at room temperature.
3. Wash the nuclei by adding 100 μL of wash buffer to the microtube.
4. Centrifuge at 800 g for 10 min and remove the supernatant.
5. Resuspend the nuclei pellet in 10 μL of transposition mixture.
6. Incubate the samples at 37°C for 25 min, periodically mixing every 5–7 min (*see* **Note 7**).
7. Purify the DNA by column purification using the Qiagen QIAquick PCR Purification Kit following the manufacturer's guidelines. Elute the DNA into a final volume of 20 μL (*see* **Note 8**).
8. DNA can be stored at −20°C until library prep.

3.4 ATAC-Seq Library Preparation

1. Prepare the Library Prep PCR master mix.
2. Add 30 μL of the Library Prep PCR master mix to the 20 μL of the purified DNA from Step 3.3.7.
3. Run the following PCR program (*see* **Note 9**): 72°C for 5 min; 98°C for 30 s; 98°C for 10 s, 63°C for 30 s 12 cycles; 72°C for 60 s; Hold at 4°C.
4. The libraries can be stored at 4°C overnight and at −20°C until library purification.

3.5 ATAC-Seq Library Purification

1. Vortex the AMPure XP beads and allow the beads to reach room temperature.
2. Add the AMPure XP beads to the libraries at a ratio of 0.85× (v/v) and mix. This ratio of beads will remove the DNA fragments that are less than 200 bp in size.
3. Incubate the samples at room temperature for 8 min.
4. Place the tubes on a magnetic rack and wait until the solution becomes clear.
5. Remove the supernatant and do not allow the beads to get dry.

6. Wash the beads twice with 200 μL of 80% ethanol.
7. After the last wash, remove all the ethanol and allow the beads to dry for 2 min (*see* **Note 10**).
8. To elute the libraries, remove the tubes from the magnetic rack and add 15 μL of TE buffer to the samples and resuspend the AMPure beads. Incubate at room temperature for at least 5 min.
9. Place the tubes on a magnetic rack and allow the solution to clear.
10. Keep the supernatant. Samples can be kept at −20°C until they are sequenced.
11. Proceed to library quantification and quality control (see Figs. 2 and 3 and **Note 11**).
12. Follow standard protocols for sequencing and downstream analysis.

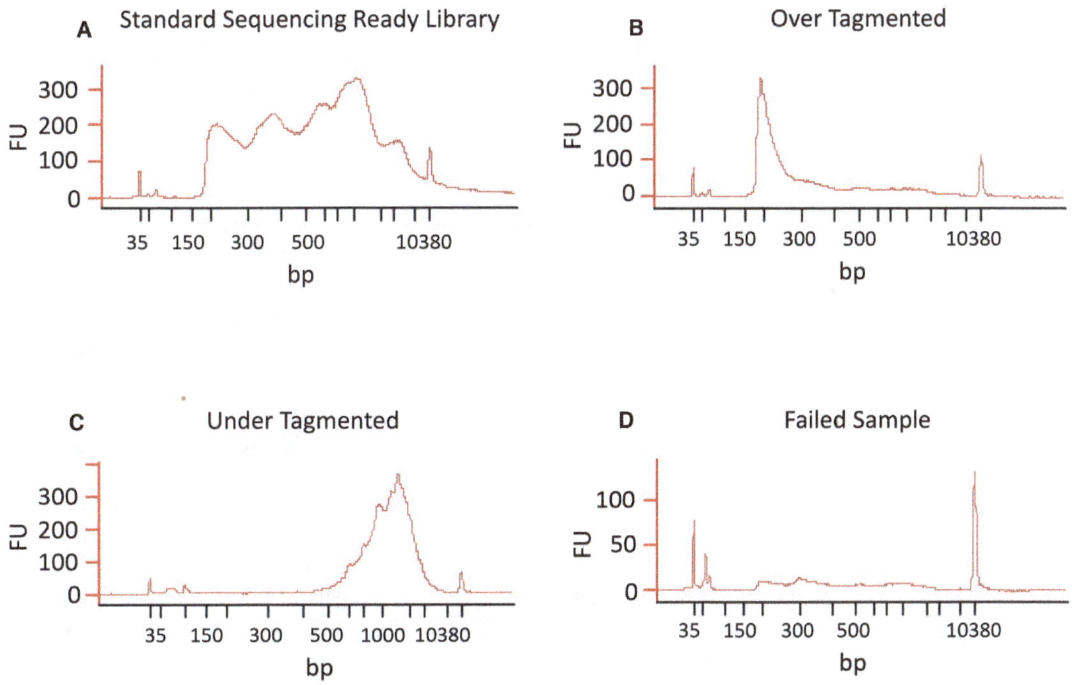

Fig. 2 Bioanalyzer profiles of ATAC-Seq libraries. (**A**) Bioanalyzer profile of a high-quality sequencing-ready library. (**B**) Bioanalyzer profile of an ATAC-Seq library where the DNA was over tagmented. (**C**) Bioanalyzer profile of an ATAC-Seq library where the DNA was under tagmented. (**D**) Bioanalyzer profile of a failed ATAC-Seq library

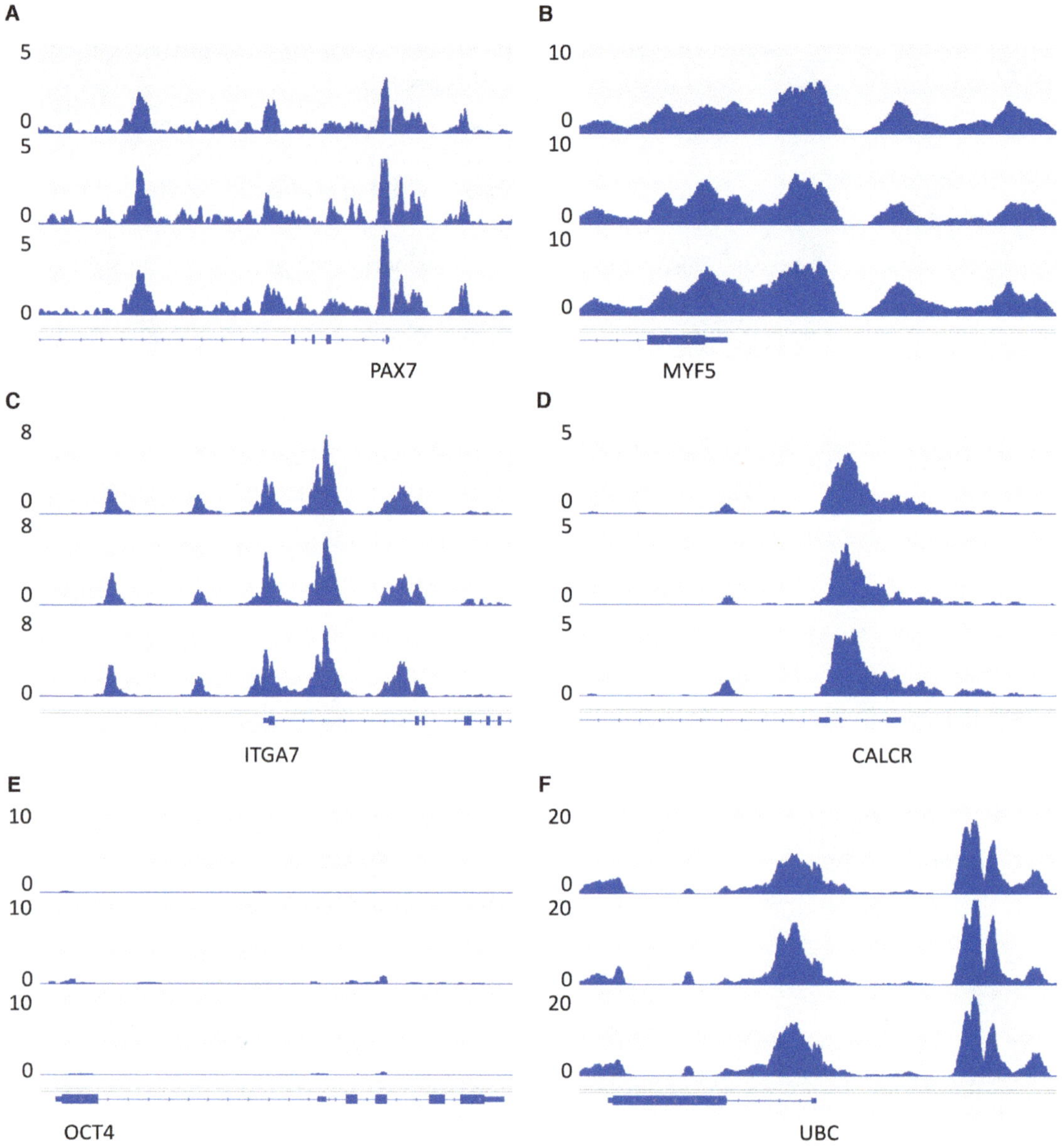

Fig. 3 Representative IGV tracks of sequenced freshly sorted MuSC ATAC-Seq libraries. MuSC ATAC-Seq IGV tracks of selected positive control genes (**A**) PAX7, (**B**) MYF5, (**C**) ITGA7, and (**D**) CALCR. (**E**) MuSC ATAC-Seq IGV tracks of negative control gene OCT4. (**F**) MuSC ATAC-Seq IGV tracks of housekeeping gene UBC. ATAC-Seq data were retrieved from GSE230622 [12]

4 Notes

1. It is important to properly mince the muscle to promote enzymatic digestion. Ensure there are no chunks of muscle visible after mincing.
2. It is important to coat pipettes with FBS to prevent the muscle stem cells from adhering to the pipette.

3. The muscle pellet is thick; therefore, it is best to cut the tip of the P1000.
4. It is important to wet the cell strainer with PBS before passing the muscle digest through. Due to the debris from the muscle, the cell strainers will become clogged. A minimum of two cell strainers will be necessary to fully filter all the sample.
5. Polypropylene tubes are preferred as MuSCs can adhere to polyethylene tubes.
6. For ATAC-Seq on quiescent MuSCs we typically sort 5000 cells. However, more or less cells can be used. When changing the number of cells or cell type, it may be necessary to optimize the tagmentation time and concentration of the Tn5 enzyme.
7. The activity of the Tn5 enzyme can change between every batch. When using a new batch of Tn5, it may be necessary to reoptimize the incubation time and the concentration of Tn5 used.
8. Doing a double elution can increase the yield of DNA. After the first elution, take the 20 μL of eluted DNA and add it back to the same column and centrifuge again.
9. The number of PCR cycles will need to be adjusted if using a different number of cells.
10. It is important that the AMPure beads do not get too dry as this will result in a worse elution of the libraries. When the beads are too dry, this can be seen as cracking in the bead pellet, or in clumping when trying to resuspend the beads. If the beads are too dry, the elution can be improved by incubating the beads with the elution buffer at 37°C for 5–10 min.
11. The size of the library will be dependent on how much the DNA has been cut by the Tn5 enzyme. If the samples are incubated for too long the fragment sizes will be small. Not long enough and the fragments will be too large. If you find that you are not getting appropriate fragment sizes you can change the incubation time or the concentration of the Tn5.

Acknowledgments

The authors thank Christian Young at the Lady Davis Institute for Medical Research core facility for help with fluorescence-activated cell sorting. This work was supported by a project grant (PJT-156087) from the Canadian Institute of Health Research (CIHR) to VDS.

Conflict of Interest The authors declare no conflicts of interest.

References

1. Tedesco FS, Dellavalle A, Diaz-Manera J et al (2010) Repairing skeletal muscle: regenerative potential of skeletal muscle stem cells. J Clin Invest 120:11–19
2. Ancel S, Stuelsatz P, Feige JN (2021) Muscle stem cell quiescence: controlling stemness by staying asleep. Trends Cell Biol 31:556–568
3. van Velthoven CTJ, Rando TA (2019) Stem cell quiescence: dynamism, restraint, and cellular idling. Cell Stem Cell 24:213–225
4. Corces MR, Trevino AE, Hamilton EG et al (2017) An improved ATAC-seq protocol reduces background and enables interrogation of frozen tissues. Nat Methods 14:959–962
5. Buenrostro JD, Wu B, Chang HY et al (2015) ATAC-seq: a method for assaying chromatin accessibility genome-wide. Curr Protoc Mol Biol 109:21.29.21–21.29.29
6. Giordani L, He GJ, Negroni E et al (2019) High-dimensional single-cell cartography reveals novel skeletal muscle-resident cell populations. Mol Cell 74(609–621):e606
7. Machado L, Esteves de Lima J, Fabre O et al (2017) In situ fixation redefines quiescence and early activation of skeletal muscle stem cells. Cell Rep 21:1982–1993
8. De Micheli AJ, Laurilliard EJ, Heinke CL et al (2020) Single-cell analysis of the muscle stem cell hierarchy identifies heterotypic communication signals involved in skeletal muscle regeneration. Cell Rep 30(3583-3595):e3585
9. Elizalde G, Munoz AZ, Almada AE (2023) Protocol for the isolation of mouse muscle stem cells using fluorescence-activated cell sorting. STAR Protoc 4:102656
10. Sahinyan K, Blackburn DM, Simon MM et al (2025) REST/NRSF preserves muscle stem cell identity by repressing alternate cell fate. Nat Commun 16: 7487
11. Sambasivan R, Gayraud-Morel B, Dumas G et al (2009) Distinct regulatory cascades govern extraocular and pharyngeal arch muscle progenitor cell fates. Dev Cell 16:810–821
12. Blackburn DM, Sahinyan K, Hernandez-Corchado A et al (2024) The E3 ubiquitin ligase Nedd4L preserves skeletal muscle stem cell quiescence by inhibiting their activation. iScience 27:110241

Chapter 21

Targeted Chromatin Profiling in Muscle Stem Cells Using a CUT&RUN-qPCR Strategy

Alexandre Pintart, Delia Cicciarello, and Isabella Scionti

Abstract

Chromatin immunoprecipitation coupled with quantitative PCR (ChIP-qPCR), even after extensive optimization, is often limited by low signal-to-background ratio, particularly when cell numbers are restricted. This represents a major challenge for muscle stem cells (MuSCs), a rare population that cannot be obtained in large quantities. To overcome these limitations, we adapted Cleavage Under Targets and Release Using Nuclease (CUT&RUN) into a targeted CUT&RUN-qPCR strategy. This approach enables sensitive analysis of chromatin–protein interactions at defined loci from low-input samples, without the need for genome-wide sequencing. Using limited numbers of cells (150,000 MuSCs), we provide a robust and scalable solution for targeted chromatin profiling in rare stem cell populations.

Keywords CUT&RUN-qPCR, Low-input samples, Muscle stem cells, Chromatin–protein interactions, Targeted chromatin profiling

1 Introduction

Skeletal muscle has a remarkable regenerative capacity that depends on a small pool of resident stem cells known as muscle stem cells (MuSCs). These cells are essential for muscle growth, repair, and long-term tissue maintenance [1]. Although a rare population within adult skeletal muscle, their ablation completely blocks skeletal muscle regeneration after acute injury, demonstrating that these cells are essential for muscle repair [2, 3]. Skeletal muscle regeneration is a highly coordinated biological process that relies on the activation of MuSCs and their interaction with the surrounding tissue environment [1, 4]. Following injury, MuSCs exit quiescence and rapidly activate a transcriptional program that enables them to proliferate and generate myogenic progenitors. These progenitor cells then expand, migrate toward the damaged region, and differentiate into myocytes, which fuse with each other or with newly regenerated muscle fibers to restore muscle integrity [5, 6]. This

Nathalie Didier and Caroline E. Brun (eds.), *Muscle Stem Cells*, Methods in Molecular Biology, vol. 3093,
https://doi.org/10.1007/978-1-0716-5643-3_21,

process is regulated by a dynamic interplay between intrinsic factors, such as myogenic transcription factors and epigenetic regulators, and extrinsic signals derived from the niche, including inflammatory cells, fibroblasts, and extracellular matrix components [4, 7–9]. As regeneration proceeds, MuSCs undergo differentiation under the control of key myogenic regulators such as MyoD, Myf5, Myogenin, and MRF4 [10]. These factors drive the establishment of the myogenic lineage and coordinate the expression of structural and contractile proteins required for new fiber formation. At the same time, a fraction of activated MuSCs returns to quiescence through self-renewal definition, thereby replenishing the stem cell reservoir and preserving the long-term regenerative capacity of the tissue [11]. The balance between differentiation and self-renewal, governed by epigenetic and transcriptional regulation and shaped by signals from the stem cell niche, is crucial for ensuring effective muscle regeneration while preserving the long-term stem cell pool. As such, MuSC fate is associated with extensive chromatin remodeling. Quiescent MuSCs maintain a largely repressive but poised chromatin state, whereas activation induces changes in chromatin accessibility that permit rapid transcription of myogenic genes [12–15]. These epigenetic transitions are necessary for the precise temporal control of muscle repair and help explain why chromatin profiling has become so important for understanding MuSC biology.

However, despite their biological importance, the rarity of MuSCs has long prevented comprehensive molecular characterization, since conventional chromatin profiling approaches, such as ChIP-seq, typically require large quantities of input material, often far more than can be obtained from freshly isolated MuSCs from the muscles of a single mouse. As a result, it was historically difficult to perform high-resolution chromatin profiling directly on these cells. This limitation meant that many early insights into MuSC biology were inferred indirectly or obtained from cultured myoblasts, which do not fully reflect the native stem cell state [16–18]. Moreover, understanding MuSC chromatin architecture is more than a technical achievement; it will open new frontiers in regenerative biology. Indeed, muscle degeneration, whether due to aging, injury, or dystrophies, is often associated with impaired MuSC function. By mapping the chromatin and epigenetic signatures that underlie self-renewal, key regulatory elements, and potential therapeutic targets can be identified. Additionally, these studies will provide a molecular framework for understanding how environmental signals, such as exercise, nutrition, or inflammation, affect muscle regeneration at the epigenetic level.

Over the last decade advances in low-input technologies such as CUT&RUN have made possible to profile chromatin in rare cell populations with far fewer cells than was previously required

[19, 20]. This approach allows researchers to directly map the epigenetic landscape of MuSCs and to uncover regulatory programs that govern self-renewal and differentiation. Here, we proposed an adapted CUT & RUN-qPCR method, that allows the study of chromatin–protein interactions at defined loci using a limited number of cells.

2 Materials

2.1 Mouse Model

For the purpose of this study, we used Pax7<tm2.1(cre/ERT2)Fan>/J strain crossed with $R26R^{EYFP/EYFP}$ to generate a strain in which all PAX-7^{pos} MuSCs express EYFP reporter protein after tamoxifen induction.

2.2 Dissection Tools

1. Iris scissors
2. Straight Cohan–Vannas Spring scissors
3. Half-curved Iris forceps
4. Straight Dumont forceps
5. 70% Ethanol

2.3 MuSC Isolation from Hindlimb Muscles

1. 37°C shaking water bath.
2. Cell culture incubator with a humidified atmosphere containing 5% CO_2 at 37°C.
3. Swinging bucket rotor centrifuge.
4. 50 mL centrifuge tube.
5. 15 mL centrifuge tube.
6. 40-μm and 30-μm Nylon cell strainers.
7. Sterile 60 mm culture dishes (one per mouse).
8. Wash Medium (WM): Ham's F-10 supplemented with 10% Horse Serum and 1× penicillin/streptomycin. Keep cold (4°C–8°C) throughout the protocol.
9. Muscle Dissociation Buffer (MDB): 800 U/mL Collagenase II solution prepared in WM. Prepare 10 mL for each mouse (*see* **Note 1**).
10. Stock Collagenase II solution: This solution is prepared by dissolving Collagenase II powder in 1× PBS so that the final concentration is 1000 U/mL. Thaw before use.
11. Stock Dispase solution This solution is prepared by dissolving Dispase powder in 1× PBS so that the final concentration is 11 U/mL.
12. 20 G needles.
13. ACK Buffer.
14. 1× PBS.

2.4 CUT&RUN (see Note 2)

1. Washing Buffer: HEPES 20 mM, NaCl 150 mM, Spermidine 0.5 mM, EDTA-Free Roche protease inhibitor.
2. Concanavalin A (ConA)-coupled magnetic beads.
3. Binding Buffer: HEPES 20 mM, KCl 20 mM, $CaCl_2$ 1 mM, $MnCl_2$ 1 mM.
4. Digitonin Buffer: washing buffer, digitonin 0.05% (*see* **Note 3**).
5. Antibody Binding Buffer: Digitonin Buffer, EDTA 2 mM.
6. pAG-MNase.
7. Digestion Buffer: Digitonin Buffer, $CaCl_2$ 1 mM.
8. Stop buffer: NaCl 340 mM, EDTA 20 mM, EGTA 4 mM, RNAase A 50 μg/mL, and Glycogen 50 μg/mL.

3 Methods

3.1 Before Beginning: Design and Check the Quality of Primers

1. Design primer pairs for CUT&RUN—qPCR targeting the genomic regions of interest, ensuring an annealing temperature close to 60°C and an amplicon length between 80 and 140 bp.
2. Include primer pairs for a negative control region, selected as a genomic locus distant from the target sites or not bound by the protein of interest. Quantify enrichment using the $2^{-\Delta\Delta CT}$ method.
3. Assess primer specificity by amplifying the target regions and performing melting curve analysis at the end of the qPCR run using the instrument software (*see* **Note 4**).

3.2 Muscle Dissection and Digestion to Isolate MuSCs

1. Prewarm at 37°C the MDB, the Stock dispase and the Stock collagenase II.
2. Prepare 60-mm petri dishes with 5 mL WM (one dish per mouse).
3. Anesthetize the mouse with isoflurane in a specific induction chamber prior to cervical dislocation euthanasia.
4. Place the mouse in supine position on a clean polystyrene board and spray hindlimbs with 70% Ethanol.
5. Collect the hindlimb muscles in the Petri dish and trim away any visible fat and tendon.
6. Transfer all muscles to the lid of the initial dish and mince the muscles enough to obtain a pulp first with scissors then with a scalpel.
7. Transfer the minced muscle from each mouse to an individual 50-mL conical tube with 10 mL MDB.

8. Vortex, seal the tube(s) well with Parafilm and incubate in a 37°C water bath with agitation (70 rpm) for 1 h. Tubes should completely submerged.
9. After 1 h take out the tube pipet up and down ten times with a 10 mL pipet then fill each tube up to 50 mL with cold WM.
10. Centrifuge in a swinging bucket rotor at 500*g* for 5 min. Aspirate supernatant down to about 8 mL.
11. Add 1 mL stock Collagenase II solution and 1 mL stock Dispase II solution.
12. Vortex and resuspend the pellet with a 10 mL serological pipet until the suspension travels up and down the pipet smoothly without clogging.
13. Seal the tube with Parafilm and incubate in a 37°C water bath with agitation (60–70 rpm) for 30 min.
14. Using a 10 mL syringe and 20-gauge needle, aspirate and eject the muscle suspension in and out of the syringe ten times, avoiding foaming.
15. Place a 40-μm Nylon cell strainer on a fresh 50 mL conical tube.
16. Add 30 mL WM to the original tube and resuspend the pellet with a 10 mL pipet. Transfer to the cell strainer and allow filtration by gravity.
17. Add 10 mL WM to the original tube. Swirl to rinse, and transfer to the same cell strainer.
18. Use a 20–200 μL pipet to collect any remaining liquid on the underside of the strainer.
19. Centrifuge at 500*g* for 5 min at 4°C. Remove all supernatant by aspiration immediately after centrifugation.
20. Lyse red cells with 1 mL ACK Buffer. Vortex until the solution tunrs a pink/red (10–50 s).
21. Fill all tubes to 50 mL with WM to the original tube. Centrifuge at 500*g* for 5 min. Remove all supernatant by aspiration immediately after centrifugation.
22. Resuspend cell pellet in 500 μL WM.
23. Pass the sample through a 30 μm filter before sorting for EYFP fluorescent reporter by FACS (Fig. 1). All hindlimb muscles from one male mouse can readily generate 150,000–200,000 EYFP-positive MuSCs.

3.3 CUT&RUN on Isolated MuSCs

For efficient binding, we recommend incubating 150,000 MuSCs with 10 μL of Concanavalin A bead suspension in a microcentrifuge tube for each CUT&RUN reaction. Given the limited availability of MuSCs, cells from multiple mice can be pooled prior to FACS

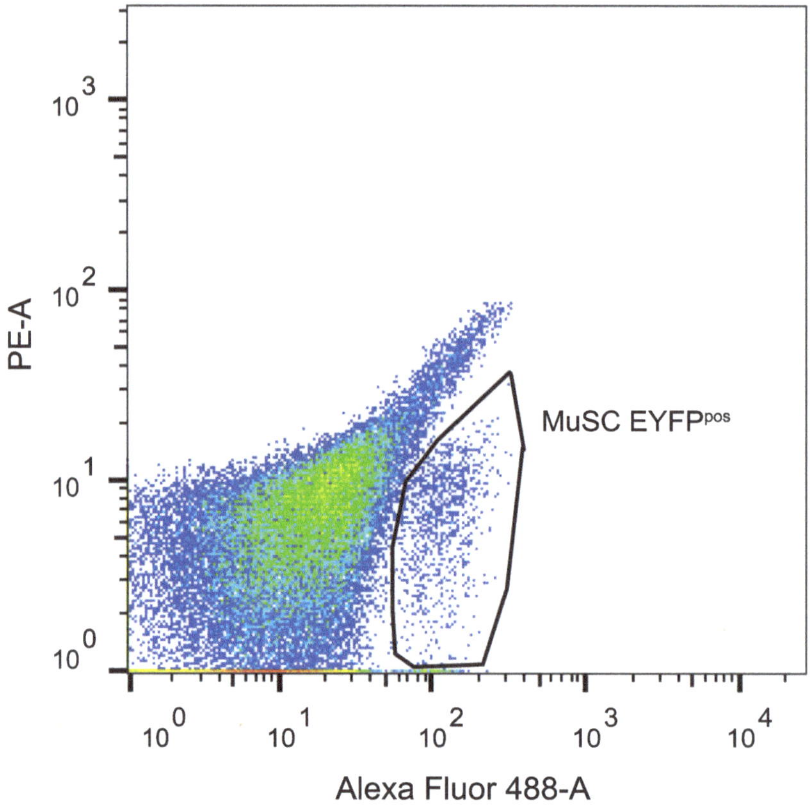

Fig. 1 Cytometry profile of isolated MuSC EYFP^{pos}

to obtain sufficient numbers. The sorted cells can then be divided into separate aliquots of 150,000 cells for individual CUT&RUN reactions and input sample, enabling the analysis of multiple histone marks and/or transcription factor binding profiles.

1. Centrifuge FACS-sorted cell suspension for 3 min at 350 × *g* at room temperature. For each reaction or input sample (150,000 cells), add 100 μL of Washing Buffer and resuspend the cell pellet by gently pipetting up and down.
2. Gently resuspend the ConA-coupled magnetic beads in the stock bottle by slow inversion.
3. Calculate volume of beads needed (10 μL per sample) and transfer this volume of beads to a 2.0 mL tube containing 100 μL chilled Binding Buffer (*see* **Note 5**).
4. Mix tube contents by inversion and place on magnetic separator. When liquid clarifies, remove the supernatant with p1000 pipet.
5. Repeat Binding Buffer wash twice. After second wash, resuspend the beads in a volume of Binding Buffer equal to the

initial volume of bead suspension (10 μL per final sample) and place cell-bead suspension on ice.

6. Add to each sample 10 μL of prepared ConA bead slurry. Resuspend cells thoroughly by pipetting to prevent ConA bead clumping. Incubate for 5 min at room temperature.
7. Place the tube on the magnetic rack until the solution turns clear, then remove and discard the liquid.
8. Remove tube from the magnetic rack. Add 450 μL of Antibody Binding Buffer per reaction, gently mix by pipetting up and down and place on ice.
9. Add the appropriate amount of antibody to each reaction and mix gently by pipetting up and down. Incubate tubes overnight at 4°C (*see* **Note 6**).
10. Wash cells twice in 500 μL chilled Digitonin Buffer and mix beads thoroughly by inversion (*see* **Note 7**). Place samples on magnetic separator until the solution turns clear and then remove supernatant using a p1000 pipet.
11. Add 1.5 μL of pAG-MNase to 50 μL of chilled Digitonin buffer and mix each sample gently by pipetting up and down.
12. Incubate tubes at 4°C for 1 h to allow pA/G-MNase-antibody chromatin binding.
13. Place the tube on the magnetic rack until the solution turns clear, then remove and discard the liquid.
14. Wash cells twice in 1 mL chilled Digitonin Buffer and mix beads thoroughly by inversion. Place samples on a magnetic separator until the solution turns clear and then remove supernatant using a p1000 pipet (*see* **Note 8**).
15. Remove tubes from the magnetic rack. Activate the pAG-MNase by adding 150 μL of chilled Digestion Buffer to each tube and mix by pipetting up and down. Incubate samples at 4°C for 30 min.
16. Add 100 mL chilled Stop buffer to each tube and gently mix using a p1000 pipet. Place samples at 37°C for 10 min.
17. Centrifuge samples at 4°C for 2 min at 16,000*g*. Place the tube on a magnet rack until the solution is clear.
18. Transfer the supernatant to a 2 mL tube These are the enriched chromatin sample.
19. Extract the DNA from the supernatant by using DNA spin columns (Qiagen MinElute PCR purification Kit) or phenol/chloroform extraction followed by ethanol precipitation.
20. Elute the purified DNA in 50 μL of TE buffer.
21. Assess the concentration of the eluted DNA sample using a Nanodrop Spectrophotometer.

22. Run qPCR of all the samples using three technical replicates per experimental replicate.

4 Notes

1. Muscle dissociation buffer should be prepared freshly by dissolving Collagenase II powder in WM before the dissection and kept on ice until needed.
2. All buffers are prepared extemporaneously.
3. The amount of digitonin varies depending on cell type and should be determined by the user.
4. A single, well-defined peak in the melting curve indicates specific amplification of a homogeneous product, whereas multiple peaks are indicative of nonspecific amplification or potential contamination.
5. It is very important to use equal amounts of beads in all samples, to avoid sample-to-sample variation in epitope binding and in the release of digested chromatin.
6. The amount of antibody required for CUT&RUN varies and should be determined by the user.
7. After second wash, use a p200 pipet to remove every last mL of liquid. This step is important, since efficient MNAse activity requires efficient removal of EDTA from the buffer. A dead volume of wash buffer may allow excess EDTA to be carried over.
8. After second wash, use a p200 pipet to remove every last mL of liquid. This step is important to completely remove nonspecific binding of pA/G-MNase.

Acknowledgments

I.S. receives support of grants from the Agence Nationale de la Recherche (ANR-24-CE14-1484 and ANR-24-CE14-5834) and the Association Française contre les Myopathies (AFM)-Téléthon.

Competing Interests The authors declare no competing interests.

References

1. Relaix F, Bencze M, Borok MJ et al (2021) Perspectives on skeletal muscle stem cells. Nat Commun 12:692
2. Lepper C, Partridge TA, Fan C-M (2011) An absolute requirement for Pax7-positive satellite cells in acute injury-induced skeletal muscle regeneration. Development 138:3639–3646
3. Sambasivan R, Yao R, Kissenpfennig A et al (2011) Pax7-expressing satellite cells are indispensable for adult skeletal muscle regeneration. Development 138:3647–3656
4. Saclier M, Yacoub-Youssef H, Mackey AL et al (2013) Differentially activated macrophages orchestrate myogenic precursor cell fate during human skeletal muscle regeneration. Stem Cells 31:384–396
5. Robinson DCL, Dilworth FJ (2018) Epigenetic regulation of adult myogenesis. Curr Top Dev Biol 126:235–284
6. Yoshida N, Yoshida S, Koishi K et al (1998) Cell heterogeneity upon myogenic differentiation: down-regulation of MyoD and Myf-5 generates "reserve cells". J Cell Sci 111(Pt 6):769–779
7. Mouradian S, Cicciarello D, Lacoste N et al (2024) LSD1 controls a nuclear checkpoint in Wnt/β-Catenin signaling to regulate muscle stem cell self-renewal. Nucleic Acids Res 52(7):3667–3681
8. Nakka K, Hachmer S, Mokhtari Z et al (2022) JMJD3 activated hyaluronan synthesis drives muscle regeneration in an inflammatory environment. Science 377:666–669
9. Cicciarello D, Schaeffer L, Scionti I (2022) Epigenetic control of muscle stem cells: focus on histone lysine demethylases. Front Cell Dev Biol 10:917771
10. Zammit PS (2017) Function of the myogenic regulatory factors Myf5, MyoD, Myogenin and MRF4 in skeletal muscle, satellite cells and regenerative myogenesis. Semin Cell Dev Biol 72:19–32
11. Zammit PS, Golding JP, Nagata Y et al (2004) Muscle satellite cells adopt divergent fates: a mechanism for self-renewal?. J Cell Biol 166:347–357
12. Okafor AE, Lin X, Situ C et al (2023) Single-cell chromatin accessibility profiling reveals a self-renewing muscle satellite cell state. J Cell Biol 222:e202211073
13. Aguilar CA, Pop R, Shcherbina A et al (2016) Transcriptional and chromatin dynamics of muscle regeneration after severe trauma. Stem Cell Rep 7:983–997
14. Liu L, Cheung TH, Charville GW et al (2013) Chromatin modifications as determinants of muscle stem cell quiescence and chronological aging. Cell Rep 4:189–204
15. Massenet J, Gardner E, Chazaud B, Dilworth FJ (2021) Epigenetic regulation of satellite cell fate during skeletal muscle regeneration. Skelet Muscle 11:4
16. Magli A, Incitti T, Kiley J et al (2017) PAX7 targets, CD54, Integrin α9β1, and SDC2, Allow Isolation of Human ESC/iPSC-derived myogenic progenitors. Cell Rep 19:2867–2877
17. Machado L, Esteves de Lima J, Fabre O et al (2017) In situ fixation redefines quiescence and early activation of skeletal muscle stem cells. Cell Rep 21:1982–1993
18. Rao VK, Shankar SR, Taneja R (2019) Chromatin Immunoprecipitation in skeletal myoblasts. In: Rønning SB (ed) Myogenesis: methods and protocols. Springer, New York, NY, pp 43–54
19. Meers MP, Bryson TD, Henikoff JG, Henikoff S (2019) Improved CUT&RUN chromatin profiling tools. eLife 8:e46314
20. Robinson DCL, Ritso M, Nelson GM et al (2021) Negative elongation factor regulates muscle progenitor expansion for efficient myofiber repair and stem cell pool repopulation. Dev Cell 56(1014–1029):e7

Check for updates

Chapter 22

High-Throughput Chromosome Conformation Capture Immunoprecipitation in Skeletal Muscle Stem Cells

Jimmy Massenet, Luca Caputo, Chiara Nicoletti, Monica Nicolau, Jesus R. Barajas and Pier Lorenzo Puri

Abstract

Fate determination of muscle stem cells (MuSCs) is associated with the regulation of gene expression across multiple layers of chromatin organization, ranging from local chromatin accessibility to higher-order three-dimensional (3D) genome architecture driven by transcription factors, architectural proteins, and histone marks. Advancements in chromosome conformation capture (3C) technologies have facilitated the mapping of regulatory interactions with increasing precision. HiChIP-based genome-wide detection of chromatin interactions mediated by chromatin-bound proteins or histone modifications provides an optimal genomic reference for high-resolution detection of changes in 3D genome architecture. However, analyzing the regulation of chromatin conformation defined by specific proteins and integrating it robustly with chromatin accessibility and gene expression remains technically challenging in MuSCs. In this chapter, we present a reproducible workflow to generate a HiChIP dataset from a limited number of MuSCs. Finally, we provide integration tools/workflow to integrate HiChIP datasets with ATAC-seq and RNA-seq datasets.

Keywords Muscle stem cells, HiChIP, Epigenetic, Chromatin looping, 3D genome conformation, Multiomics integration

1 Introduction

The approximate length of our DNA is 2 m. This macromolecule must be compacted within the nucleus of a cell. To achieve this, the genome undergoes a hierarchical folding process, a strategy employed to reduce spatial requirements while preserving its fundamental function as the blueprint of the cell [1]. Nevertheless, the expression of genes, the repair and replication of DNA, and other processes occur within limited spatial domains. At the two-dimensional linear level, chromatin accessibility and histone modifications have been proven to play a critical role in the dynamic regulation of local gene expression [2–5]. Concurrently, over the past two decades, there has been a significant advancement in the understanding of tridimensional genome

Nathalie Didier and Caroline E. Brun (eds.), *Muscle Stem Cells*, Methods in Molecular Biology, vol. 3093,
https://doi.org/10.1007/978-1-0716-5643-3_22,

conformation, including its indispensable role in replication, mitotic segregation, genome stability, and the regulation of gene expression [6]. Chromosomes occupy discrete territories within the nucleus. Chromosome territories impose constraints on inter-chromosomal interactions while permitting the regulation of specific long-range interactions even between different chromosomes [7]. Within each chromosome, the organization of chromatin occurs in compartments that are classified as either transcriptionally active (A) or inactive (B) [8]. At a smaller scale, these compartments are organized in topologically associated domains (TADs), which correspond to regions with a higher frequency of interactions compared to outside regions [9]. TADs' boundaries are conserved not only between cell types but also, to a lesser extent, between different species [10]. These elements facilitate compartmentalized gene regulation, a critical aspect of cellular function. Furthermore, analysis of the TAD boundaries reveals an enrichment of factors, including architectural proteins such as CTCF, which serve to constrain regulation [9]. The structural organization of TADs facilitates the frequency of interactions between promoters and their distal regulatory elements (enhancers or silencers) while preserving their interaction within the TADs [11]. This hierarchical organization is imperative for timely and cell-type-specific regulation of gene expression, as demonstrated by the ectopic expression of genes when a TAD boundary is lost [11].

The regulation of gene expression plays a pivotal role in lineage selection [12]. In the context of muscle injury, MuSCs undergo a process known as adult myogenesis. This process involves the exit of quiescence of the MuSCs, followed by activation, proliferation, differentiation, and fusion to form new myofibers [13]. At each stage, modulation of gene expression is imperative for the progression of myogenesis. One layer of gene regulation is conferred by the stepwise expression of transcription factors, such as the paired box 7 Pax7, which is associated with MuSC stemness, and the myogenic regulatory factors (MRFs), including Myf5 and MyoD, which promote MuSC commitment to the myogenic lineage and in cooperation with later expressed MRFs (Myogenin and MRF4) and MEF2 family member induce formation of differentiated myofibers [14]. The stepwise expression of transcription factors regulating MuSC transition toward myofibers is influenced by a further regulatory layer consisting of combinatorial patterns of repressive (i.e., H3K27me3, H3K9me3) and permissive (i.e., H3K4me1/3, H3K27ac, H4acme) histone marks across the genome, as they instruct transcription factor and cofactor accessibility to nucleosomes at regulatory elements [15–17]. However, their mere presence, while informative on chromatin accessibility and potential presence of regulatory sequences, is insufficient to predict the contacts between target muscle gene promoters and their enhancers. To facilitate the folding of chromatin and the increase of the frequency of contacts between promoters and regulatory sequences, additional mechanisms are in place. These mechanisms encompass the presence of additional factors, including proteins and noncoding RNAs (ncRNAs), such as, for example, CTCF. CTCF can modulate the chromatin 3D structure [18] and the mediator complex, a large, protein complex comprised of a 26 protein core. This complex has been shown to facilitate interaction between promoter and enhancers [19]. A comprehensive understanding of those higher hierarchical structures in MuSCs is essential to fully comprehend how transcription factors in combination with architectural proteins and chromatin structure can modulate chromatin folding to regulatory sequences and regulate muscle gene expression to follow myogenic lineage [20]. This control is imperative for the acquisition and maintenance of muscle lineage, as well as for the repression of alternative lineages, during skeletal myogenesis [21]. A number of those mechanisms have been observed to be altered in muscle pathologies, such as dystrophies or sarcopenia [22]. However, the complete comprehension of the chromatin 3D structure alteration in such conditions remains to be elucidated.

To elucidate 3D chromatin architecture in cells, several techniques have emerged. Since the initial approach was the chromatin conformation capture (3C) method [23], a one-versus-one approach, multiple variation have been developed including circular chromosome conformation capture (4C or 3C-seq), Hi-C uses reverse PCR and sequencing to allow the identification of all the interactions of one specific locus [24–26], one-versus-many approach, and carbon-copy chromosome conformation capture (5C) requires a series of primers set to amplify multiple loci, a many-versus-many approach [27]. 3C-based techniques are based on the preservation of DNA-to-DNA contact by formaldehyde crosslinking, followed by enzymatic digestion, proximity ligation and subsequent detection of the resulting chimeric DNA molecules via qPCR or sequencing. These techniques lack an unbiased analysis of interactions occurring in the nuclei and require a priori knowledge of the specific locus of interest. Hi-C represents a major conceptual advance compared to previous techniques. This method combines the purification of proximity ligated DNA molecules, obtained in the 3C protocol, to high-throughput sequencing, allowing the unbiased identification of chromatin interactions genome-wide [8]. The progressive reduction of cost and enhancement of high-throughput sequencing have progressively permitted an increase in sequencing depth and the identification of smaller structures, such as the TADs, using Hi-C-based technologies. However, these achievements were accompanied by two significant expenses. Firstly, a substantial quantity of biological material, in the tens of millions of cells, was required. Secondly, the number of sequencing reads was considerable. The analysis of compartments and TADs, which needed 100 million valid pairs of reads to obtain respective resolutions of 100kb and 40kb was possible.

However, to reach the resolution that allows detection of intra-TAD loops between promoter and regulatory sequences or CTCF sites, the number of valid pairs of reads must reach one billion. To address these challenges, various enhancements to the technique have been developed, employing biotinylated enrichment-based methodologies and in situ digestion. Specifically, the preservation of nuclear structure has reduced the generation of random ligation between samples and allowed the generation of much denser maps recording more than 15 billion distinct contacts [28, 29].

While HiC enables the identification of chromatin interactions genome-wide, its depth may not be optimal for full detection of chromatin interactions mediated by specific proteins (i.e., transcription factors, or architectural proteins) or by histone modifications and involving specific genomic features (i.e., gene promoters or distal regulatory elements). To address this issue, an additional level of enrichment has been introduced into the standard Hi-C protocol by combining the strength of chromatin immunoprecipitation (ChIP) of chromatin-bound proteins or histone modification with the Hi-C-mediated capture of histone modifications, leading to the HiChIP [30]. In this approach, before the Hi-C library is amplified and sequenced, an immunoprecipitation for a specific chromatin-bound protein or histone modification (e.g., H3K27ac, CTCF, etc.) is performed to enrich for chromatin interactions mediated by the protein or histone modification of interest. For instance, H3K27ac HiChIP and CTCF HiChIP were recently used by our lab to map MYOD-mediated reconfiguration of chromatin interactions during somatic cell transdifferentiation [21]. Capture Hi-C involves the use of biotinylated probes for specific regions of interest, as, for example, promoters, to enrich for 3D interactions in which specific genomic regions (i.e., promoters) participate. A common variation of capture Hi-C is promoter capture Hi-C (pcHiC) [31]. Capture Hi-C has effectively mitigated the bias that is often associated with antibody-based methods. For a more in-depth review of Hi-C-based methods, please see [32].

While deciphering chromatin interactions provides essential information on genome architecture, changes or alterations in genome architecture may not necessarily result in corresponding changes in gene expression. Alteration of chromatin architecture often precedes detectable changes in gene expression, as we have shown during MYOD-mediated myogenic conversion of fibroblast [20]. To this point, integration of 3C-based techniques with chromatin accessibility datasets (i.e., ATACseq) and transcriptomic datasets (i.e., RNAseq) is an essential task [21]. Chromatin accessibility provides potential regulatory regions and helps to identify active enhancers and active promoters. Meanwhile, transcriptomic data directly provides a quantitative measure of gene transcription. In the specific case of MuSCs myogenesis, a multiomic integrative study is highly suggested.

In this chapter, we present a comprehensive optimized protocol for studying chromatin architecture by performing H3K27ac and CTCF HiChIP using muscle stem cells or myoblast-like cells using Arima Genomics HiC+ kit and xGen™ 2 S™ Plus DNA Library Prep Kits. Next, we describe the analysis of the dataset and how to define DNA loops mediated by H3K27ac and CTCF. Finally, we provide guidance on how to integrate such data with RNAseq and ATACseq datasets to obtain a multiomic picture of transcriptional regulation defined by CTCF and active enhancer loops in MuSCs.

2 Materials

2.1 Laboratory Equipment

1. Thermocycler 96-wells.
2. Qubit measurement fluorometer.
3. Tapestation system.
4. Sonicator Covaris S220 (Fisher Scientific, 500217).
5. Magnetic separator 0.2 mL (Permagen, MSR812).
6. Magnetic separator 1.5 mL (Fisher Scientific, 12321D).
7. 1.5 mL tubes with low DNA binding.
8. Covaris microTUBE AFA Fiber Pre-Slit Snap-Cap 6 × 16 mm (Covaris, 520045).
9. Screentape for Tapestation.

2.2 Reagents and Enzymes

1. Arima-HiC+ Kit (Arima genomics, A101020).
2. xGen™ 2 S™ Plus DNA Library Prep Kit (IDT, 10009877).
3. xGen™ UDI Primers 16 rxn (IDT, 10005975).
4. xGen™ 2S adapters S1 96rxn (IDT, 10009903).
5. AbFlex Histone H3K27ac (Active motif, 91193).
6. AbFlex CTCF (Active motif, 91285).
7. AMPure XP beads (Beckman, A63881).
8. cOmplete Mini, Protease inhibitor cocktail (Sigma, 11836153001).
9. Qubit™ dsDNA HS assay kit (Invitrogen, Q32854).

2.3 Buffers (see Note 1)

1. Day 1—Chromatin shearing buffer (130 μL/sample): 10 mM Tris-HCl (pH 8) and 0.1% SDS in deionized water.
2. Day 1—Beads wash buffer (880 μL/sample): 10 mM Tris-HCl (pH 8), 159 mM NaCl, 1.14 mM EDTA, 1.14% Triton X-100, 0.1% SDS, 0.114% sodium deoxycholate, 1× cOmplete Mini Protease inhibitor cocktail in deionized water (*see* **Note 2**).

3. Day 1—HiChIP blocking buffer (500 μL/sample): 10 mM Tris-HCl (pH 8), 140 mM NaCl, 1 mM EDTA, 1 % Triton X-100, 0.1% SDS, 0.1% sodium deoxycholate, 5 mg/mL BSA, 1× cOmplete Mini Protease inhibitor cocktail in deionized water.
4. Day 2—HiChIP low-salt wash buffer (3 mL/sample): 10 mM Tris-HCl pH 8, 140 mM NaCl, 1 mM EDTA, 1% Triton X-100, 0.1 % SDS, 0.1% sodium deoxycholate, 1× cOmplete Mini Protease inhibitor cocktail in deionized water (*see* **Note 2**).
5. Day 2—HiChIP high-salt wash buffer (2 mL/sample): 10 mM Tris-HCl (pH 8), 300 mM NaCl, 1 mM EDTA, 1% Triton X-100, 0.1% SDS, and 0.1% sodium deoxycholate in deionized water.
6. Day 2—LiCl wash buffer (1 mL/sample): 10 mM Tris-HCl (pH 8), 150 mM lithium chloride, 1 mM EDTA, 0.5% IGEPAL CA-630, 0.1% sodium deoxycholate in deionized water.
7. Day 2—Low EDTA TE buffer (1 mL/sample): 10 mM Tris-HCl (pH 8), 0.1 mM EDTA in deionized water.
8. Day 2, 3, and 4—80% Ethanol (400 μL/sample for day 2 and 3; 800 μL/sample for day 4): Prepare fresh 80% ethanol by mixing molecular biology grade ethanol with deionized water.
9. 0.5% Formaldehyde.
10. Phosphate-Buffered Saline (PBS) 1×.

3 Methods

The protocol can be performed with isolated quiescent MuSCs using the in situ prefixation of the mice tissues before the FACS or with nonprefixed activated MuSCs. For prefixed MuSCs, in short, prefixation is performed by anesthetizing the mouse using 2,2,2-Tribromoethanol at 250 mg/kg by intraperitoneal injection. Then the mouse is perfused with a 25-gauge needle attached to a 10 mL syringe through the left ventricle. Through the heart, the mouse is washed with 30 mL of PBS 1× and then fixed with 30 mL of 0.5% Formaldehyde for 5 min before to be washed with 2 M glycine [15, 33].

HiChIP can be performed with three to five million MuSCs. Because such a number cannot be obtained freshly with FACS-isolated MuSCs, we recommend, after each FACS sorting, to perform the initial chromatin crosslinking step of the protocol and flash-freezing the cell pellet at −80°C until sufficient material is collected.

All the procedures needing animal tissues described here must be carried out in accordance with the guide for the care and use of laboratory animals as adopted by the National Institutes of Health and been approved by the Institutional Animal Care and Use Committee or in accordance with your country of residency regulations.

Time requirement for the HiChIP is 4 days of sample preparation, including two stop points before day 2 and after day 3.

3.1 Chromatin Crosslinking (Day 0)

1. Spin down five million MuSCs at 500 × *g* for 5 min at 4°C.
2. Resuspend the cells in 5 mL of PBS 1×.
3. Crosslink the cells by adding 286 μL of 37% formaldehyde for a final concentration of 2%, invert the tube ten times to assure properly mix of the formaldehyde, and incubate for 5 min at RT.
4. Quench the fixation by adding 460 μL of Arima Genomic Stop Solution 1, invert the tube ten times rapidly to stop fixation, and incubate for 5 min at RT.
5. Spin down the cells at 500 × *g* for 5 min at 4°C.
6. The pellet is resuspended in 1 mL of PBS 1× and spins down at 500 × *g* for 5 min at 4°C a second time (*see* **Note 3**).
7. Supernatant is removed, and the dry pellet can be processed or freeze at −80°C (*see* **Notes 4 and 5**).

3.2 Chromatin Digestion (Day 1)

Preheat two slots of thermocycler 96-wells at 37°C and 62°C with a lid temperature at 85°C for an infinite time.

1. If chromatin was estimated as in Sect. 3.1 (*see* **Note 5**), adjust the number of MuSCs used to obtain 15 μg of chromatin by diluting with PBS 1× and pelleting. If not, use all five million MuSCs.
2. Resuspend the pellet in 24 μL of Conditioning solution by pipetting up and down before transferring the sample in a PCR tube and incubate at 62°C for 10 min.
3. Add 20 μL of the stopping solution by pipetting up and down and incubate at 37°C for 15 min.
4. Prepare the following mix for enzymatic digestion of the chromatin for each sample:

Digestion mix	Volume (μL)
Arima Buffer A	7
Arima Enzyme A1	1
Arima Enzyme A2	4
Total (per sample)	**12**

If several samples are processed, a master mix with 10% excess can be prepared. Keep the mix on ice until needed.

5. Add 12 μL of digestion mix to each sample by pipetting up and down and digest at 37°C for 1 h. Alternatively, chromatin digestion can be extended to O/N at 37°C.
6. Transfer the sample to a 1.5 mL tube and spin down at 10,000 × *g* for 10 min at 4°C. Discard the supernatant without disturbing the pellet.
7. Add 1.5 mL of deionized water without disturbing the pellet and spin down at 10,000 × *g* for 10 min at 4°C. Discard the supernatant without disturbing the pellet.
8. Resuspend the pellet in 75 μL of deionized water by pipetting up and down and transfer the sample to a new PCR tube.

3.3 Chromatin Biotinylation and Ligation (Day 1)

1. Prepare the following mix for biotinylating the sticky end for each sample:

Biotinylation mix	Volume (μL)
Arima Buffer B	12
Arima Enzyme B	4
Total (per sample)	**16**

If several samples are processed, a master mix with 10% excess can be prepared. Keep the mix on ice until needed.

2. Add 16 μL of biotinylation mix to each sample by pipetting up and down and incubate at RT for 45 min.
3. During the incubation, prepare and cool down the shearing system (Sonicator Covaris S220 recommended; see parameters in Sect. 3.3).
4. Prepare the following mix for ligation of the biotinylated DNA for each sample:

Ligation mix	Volume (μL)
Arima Buffer C	70
Arima Enzyme C	12
Total (per sample)	**82**

If several samples are processed, a master mix with 10% excess can be prepared. Keep the mix on ice until needed.

5. Add 82 μL of ligation mix to each sample by pipetting up and down. Immediately store 10 μL of the mixture in ice; this aliquot will be used for unsheared quality control.
6. Incubate the ligation mixture at RT for 15 min.

7. Transfer the sample into a 1.5 mL tube and spin down at 10,000 × *g* for 10 min at 4°C.
8. Remove the supernatant and resuspend the sample in 130 μL of chromatin shearing buffer by pipetting up and down. Store the sample on ice before starting the next step.

3.4 Chromatin Shearing (Day 1)

Covaris S220 shearing program:

Parameters	
Temperature	4°C
Power	3
Duty factor	5%
Time	5 × 60 s (300 s total)

1. Transfer the sample to a Covaris microTUBE AFA Fiber Pre-Slit Snap-Cap 6 × 16 mm.
2. Shear the sample using the provided parameters and transfer the sheared chromatin to a 1.5 mL tube in ice. Immediately aliquot 10 μL of the mixture; this aliquot will be used for sheared quality control.

3.5 Quality Control Preparation Part 1: Decrosslinking of the Quality Controls (Day 1) (see Note 6)

At this point, you should have two quality control aliquots, one preshearing control and one postshearing control.

1. Add 90 μL of Elution buffer in the two quality control aliquots.
2. Prepare the following mix for decrosslinking solution for each sample:

De-Xlink mix	Volume (μL)
Arima Buffer D	6
Arima Enzyme D	14.3
Total (per sample)	**20.3**

If several samples are processed, a master mix with 10% excess can be prepared.
Arima Enzyme D should be thawed at RT to avoid precipitation.

3. Add 20.3 μL of De-Xlink mix to the quality controls and pipette up and down.

4. Add 11.4 μL of Arima Buffer E to the quality control and pipette up and down before incubating the samples in a thermocycler with the following protocol:

Temperature (lid at 85°C)	Time
55°C	30 min
68°C	Overnight

3.6 Antibody Binding (Day 1)

While the quality controls are decrosslinking, continue with the following:

1. Place 30 μL of Protein A beads (*see* **Note 7**) in a 1.5 mL tube attached to a magnet. Incubate for 5 min before removing the supernatant and removing the tube from the magnet.
2. Resuspend the Protein A beads with 880 μL of cold beads wash buffer.
3. To preclear chromatin, add 120 μL of the sheared sample to the Protein A/beads wash buffer mixture and incubate on a rotator for 1 h at 4°C.
4. Place the sample on magnet and collect the supernatant containing the sheared chromatin in a new 1.5 mL tube.
5. Add the antibody of interest to the sheared chromatin. For H3K27ac, 2.5 μg of AbFlex Histone H3K27ac, or 5 μg of AbFlex CTCF. Mix by pipetting up and down and incubate overnight at 4°C on the rotator.
6. To prepare 30 μL of Protain A beads per sample for the next day, place the beads in a 1.5 mL tube attached to a magnet. Incubate for 5 min before aspiring the supernatant and removing the tube from the magnet.
7. Resuspend the beads in 500 μL of cold HiChIP blocking buffer, mix by pipetting up and down, and incubate overnight on a rotator at 4°C.

3.7 Chromatin Immunoprecipitation: Part 1 (Day 2)

Cool down a thermo-shaker at 4°C or place a thermo-shaker in a cold room for it to reach 4°C. Warm up a thermocycler 96-well at 55°C to hold with lid temperature at 85°C.

1. Remove the blocked protein A beads from the rotator, spin down for a few seconds on a tabletop centrifuge to collect any beads that are collected on the lid of the tube, and place the tubes against a magnet. Incubate in the dark for 5 min at RT.
2. Remove the supernatant and resuspend the protein A beads with the previously antibody-bound chromatin, pipette up and down to mix.
3. Incubate on rotator at 4°C for 4 h.

3.8 Quality Control Preparation Part 2: Quantification of Unsheared and Sheared Chromatin (Day 2)

1. While protein A is binding to the chromatin, remove the quality control from the thermocycler and add 120 μL of DNA purification AMPure XP beads (*see* **Note 8**), and incubate in the dark for 10 min at RT.
2. Place the tubes in a 0.2 mL magnet separator and incubate in the dark for 5 min at RT.
3. Remove supernatant and clean the beads on the magnet with 200 μL of freshly prepared 80% ethanol. Incubate for 1 min before discarding supernatant.
4. Repeat Steps 2 and 3 one more time.
5. Dry the beads on the magnet by incubating for 3–5 min at RT (*see* **Note 9**). During this time with a 20 μL tip, remove any ethanol collected at the bottom of the tube.
6. Remove the tubes from the magnet and resuspend the beads in 30 μL of Elution buffer, incubate for 10 min in the dark at RT.
7. Place the tubes in the magnet separator and incubate in the dark for 10 min at RT.
8. Collect supernatant in a new tube and use 2 μL to quantify the amount of chromatin with Qubit (*see* **Note 10**). A 1% agarose electrophoresis gel should be run to assess proper shearing (Fig. 1).

3.9 Quality Control Preparation Part 3: Quantification of Biotinylated Chromatin (Day 2)

1. Dilute 75 ng of the unsheared quality control in 50 μL of Elution buffer in a PCR tube.
2. Add 50 μL of Arima QC beads for biotin enrichment, and incubate in the dark for 15 min at RT.
3. Place the tubes in a 0.2 mL magnet separator and incubate in the dark for 1 min at RT until the solution is clear.

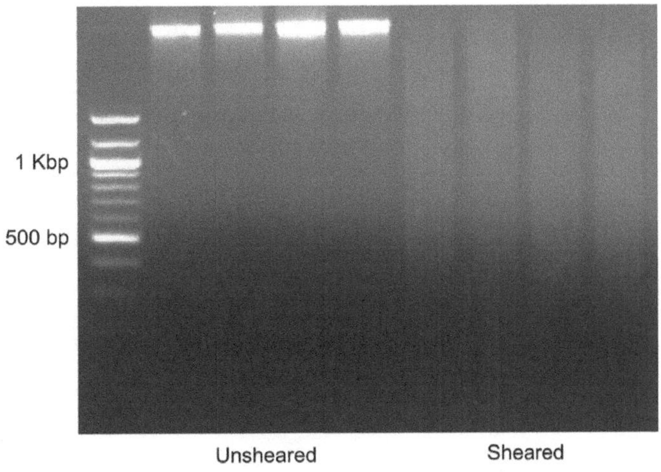

Fig. 1 Examples of H3K27Ac HiChIP quality controls and visualization of the results. Representative image of an electrophoresis gel of the expected unsheared and sheared DNA after sonication

4. Remove supernatant and clean the beads on the magnet with 200 μL of Wash buffer. Incubate for 2 min on the 55°C thermocycler.
5. Repeat Steps 3 and 4 one more time.
6. Place the tubes in a 0.2 mL magnet separator and incubate in the dark for 1 min at RT until the solution is clear.
7. Discard supernatant and clean the beads on the magnet with 100 μL of Elution buffer.
8. Place immediately the tubes in a 0.2 mL magnet separator and incubate in the dark for 1 min at RT until the solution is clear.
9. Discard supernatant and clean the beads on the magnet with 70 μL of Elution buffer.
10. Place immediately the tubes in a 0.2 mL magnet separator and incubate in the dark for 1 min at RT until the solution is clear.
11. Discard supernatant and resuspend the beads on the magnet with 30 μL of Elution buffer.
12. Use 2 μL of the resuspended beads to determine the amount of biotinylated chromatin with Qubit (*see* **Note 11**).

3.10 Chromatin Immunoprecipitation: Part 2 (Day 2)

1. When the incubation is over, remove the chromatin with protein A beads from the rotator, spin down for a few seconds on a tabletop centrifuge, and place the tubes against a magnet. Incubate in the dark for 5 min at RT.
2. Remove supernatant and add 1 mL of HiChIP low-salt wash buffer, resuspend by pipetting up and down before spinning down for a few seconds with a tabletop centrifuge, and incubate on the cold thermo-shaker at 4°C for 3 min at 1000 rpm.
3. Repeat the wash of Step 2 for two more times with HiChIP low-salt wash buffer.
4. Repeat the wash of Step 2 twice using buffer HiChIP high-salt wash buffer.
5. Repeat the wash of Step 2 once using buffer LiCl wash buffer.
6. Repeat the wash of Step 2 once using buffer low EDTA TE buffer.
7. Place the samples against the magnet and incubate in the dark for 5 min at RT. Discard the supernatant, detach the tubes from magnet, and resuspend the samples in 174 μL of Elution buffer and transfer it in a PCR tube (*see* **Note 12**).

3.11 Reverse Crosslinking Part 1 (Day 2)

1. Prepare the following mix for deXlinking for each sample:

DeXlink mix	Volume (μL)
Arima Buffer D	10.5
Arima Enzyme D	25
Total (per sample)	**35.5**

If several samples are processed, a master mix with 10% excess can be prepared.
Arima enzyme D should be thawed at RT to avoid precipitation.

2. Add 35.5 μL of deXlink mix to the samples and pipette up and down.
3. Add 20 μL of Arima Buffer E to the samples and pipette up and down before incubating the samples in a thermocycler with the following protocol:

Temperature (lid at 85°C)	Time
55°C	30 min
68°C	Overnight

3.12 Reverse Crosslinking: Part 2 (Day 3)

1. Transfer the sample into a 1.5 mL tube.
2. Add 240 μL of AMPure XP beads, and incubate in the dark for 10 min at RT (*see* **Note 8**).
3. Place the tubes in a 1.5 mL magnet separator and incubate in the dark for 5 min at RT.
4. Remove supernatant and clean the beads on the magnet with 200 μL of freshly prepared 80% ethanol. Incubate for 1 min before discarding the ethanol.
5. Repeat Steps 3 and 4 one more time.
6. Dry the beads on the magnet by incubating for 3–5 min at RT (*see* **Note 9**).
7. Remove the tubes from the magnet and resuspend the beads in 42 μL of Elution buffer, incubate for 10 min in the dark at RT.
8. Place the tubes in the magnet separator and incubate in the dark for 10 min at RT.
9. Collect supernatant in a new tube and use 2 μL to quantify the amount of DNA with Qubit (*see* **Note 13**).

Safe stop: Immunoprecipitated DNA can be stored at −20°C and the library preparation can be performed later.

3.13 HiChiP Library Preparation: Proximity Ligation Products Enrichment (Day 4)

Before start prepare a thermocycler at 55°C and prepare two aliquots of 200 μL Wash Buffer per sample and keep them at 55°C.

1. Add Elution buffer to the elute DNA from section 3-4-2 to bring the volume to 100 μL.
2. Add 100 μL of Enrichment beads to each sample and incubate for 15 min at room temperature.
3. Place the tubes in the magnet separator and incubate at RT until the solution is clear.
4. Remove the tubes from the magnet and add to the beads 200 μL Wash Buffer. Mix well by pipetting up and down ten times and incubate for 2 min at 55°C.
5. Repeat Steps 3 and 4 once more.
6. Wash beads with 100 μL Elution buffer.
7. Place the tubes in the magnet separator and incubate at RT until the solution is clear.
8. Remove supernatant and resuspend the beads in 40 μL of Elution Buffer.

The proximity-ligated DNA is now bound to the beads. Library amplification is performed on the bound DNA on beads. In the following steps you will have to use the Wash and Elution buffers provided by ARIMA in the HiC+ kit.

3.14 HiChiP Library Indexing (Day 4)

The protocol, while resembling the Swift library protocol, has been optimized to generate HiChIP library from streptavidin-bound DNA molecules. Before start prepare a thermocycler at 55°C and prepare 8 aliquots of 150 μL Wash Buffer per sample and keep them at 55°C.

1. Prepare the following dephosphorylation mix for each sample:

Dephosphorylation mix	Volume (μL)
Low EDTA TE	13
IDT Buffer W1	6
IDT Enzyme W2	1
Total (per sample)	**20**

   If several samples are processed, a master mix with 10% excess can be prepared.

2. Add 20 μL of the dephosphorylation mix to the sample and pipette up and down before incubating the samples in a thermocycler at 37°C for 10 min with the lid off. At the end of the incubation, immediately proceed to the next step. In these 10 min incubations, prepare the DNA end repair mix and keep it on ice until needed.
   If several samples are processed, a master mix with 10% excess can be prepared.

3. Place the tubes in the magnet separator and incubate at RT until the solution is clear.
4. Remove the tubes from the magnet and add to the beads 150 μL Wash Buffer. Mix well by pipetting up and down 10 times and incubate for 2 min at 55°C.
5. Repeat Steps 3 and 4 once more.

DNA end repair mix	Volume (μL)
Low EDTA TE	30
IDT Buffer G1	5
IDT Reagent G2	13
IDT Enzyme G3	1
IDT Enzyme G4	1
Total (per sample)	**50**

6. Wash beads with 100 μL Elution buffer.
7. Place the tubes in the magnet separator and incubate at RT until the solution is clear.
8. Discard the supernatant and resuspend the beads in 50 μL of DNA end repair mix.
9. Mix by pipetting up and down before incubating the samples in a thermocycler at 20°C for 20 min with lid off.
10. Meanwhile, prepare the following Primer ligation mix for each sample and keep it on ice:
11. Place the tubes in the magnet separator and incubate at RT until the solution is clear.
12. Remove the tubes from the magnet and add to the beads 150 μL Wash Buffer. Mix well by pipetting up and down ten times and incubate for 2 min at 55°C.
13. Repeat Steps 6 and 7 once more.

Primer ligation mix	Volume (μL)
Low EDTA TE	20
IDT Buffer Y1	3
IDT Enzyme Y3	2
Total (per sample)	**25**

14. Wash beads with 100 μL Elution buffer.

15. Place the tubes in the magnet separator and incubate at RT until the solution is clear.
16. Discard the supernatant and remove the tubes from the magnet and resuspend the beads in 25 μL of Primer ligation buffer.
17. For each sample, add 5 μL of a different premixed xGen™ UDI Primers pair to the ligation mix for a final volume of 30 μL per sample.
    WARNING: Premixed xGen™ UDI Primers correspond to the indexed i7 and i5 primers, and each sample should have a different primer pair to allow library pooling for sequencing.
18. Incubate the samples in a thermocycler at 25°C for 15 min with lid off.
19. Place the tubes in the magnet separator and incubate at RT until the solution is clear.
20. Remove the tubes from the magnet and add to the beads 150 μL Wash Buffer. Mix well by pipetting up and down ten times and incubate for 2 min at 55°C.
21. Repeat Steps 14 and 15 once more.
22. Wash beads with 100 μL Elution buffer.
23. Place the tubes in the magnet separator and incubate at RT until the solution is clear.
24. Discard the supernatant and remove the tube from the magnet.

Adaptors ligation mix	Volume (μL)
Low EDTA TE	30
IDT Buffer B1	5
IDT Reagent B2 - MID	2
IDT Reagent B3	9
IDT Enzyme B4	1
IDT Enzyme B5	2
IDT Enzyme B6	1
Total (per sample)	**50**

25. Meanwhile, prepare the following Adaptors ligation mix for each sample and keep it on ice:
    WARNING: IDT Reagent B2—MID corresponds to xGen™ MID Adapters S1 96rxn. Each adaptor must be used for one sample only.
    If several samples are processed, a master mix with 10% excess can be prepared.

26. Resuspend beads in 50 μL of the adaptors ligation mix by pipetting until the solution is homogenous.
27. Incubate the samples in a thermocycler at 40°C for 10 min with lid off.
28. Place the tubes in the magnet separator and incubate at RT until the solution is clear.
29. Remove the tubes from the magnet and add to the beads 150 μL Wash Buffer. Mix well by pipetting up and down ten times and incubate for 2 min at 55°C.
30. Repeat Steps 23 and 24 once more.
31. Wash beads with 100 μL Elution buffer.
32. Place the tubes in the magnet separator and incubate at RT until the solution is clear.
33. Discard the supernatant and remove the tube from the magnet.
34. Resuspends the beads in 22 μL of Elution buffer
35. Samples can now be stored at 4°C for up to 3 days or immediately quantified using the KAPA Library Quantification.

3.15 HiChIP Library Quantification (Day 4)

1. Prepare 1:1000 dilution of the enrichment-bound HiChIP libraries. It is important to vigorously mix the library, then immediately add 1 μL of library to 999 μL of deionized water.
2. Prepare a qPCR master mix according to the table below

qPCR master mix	Volume (μL)
qPCR Master Mix	5
Illumina Primer mix	1
Water	2
Total (per sample)	**8**

If several samples are processed, a master mix with 10% excess can be prepared.

3. Add 2 μL of each standard from the KAPA Kit, diluted libraries, to each well of a qPCR plate containing 8 μL of the master mix.
4. Run the following protocol:

Step	Cycles number	Temperature (lid at 105°C)	Time
Denaturation	1	95°C	5 min
Denaturation	35	95°C	30 s
Annealing-elongation		60°C	45 s
Melting curve		65 to 95°C	

5. At this step, use the C_q value obtained from this qPCR in the Arima-HiChIP QC Worksheet to calculate the required number of PCR cycles for library amplification.

3.16 Library Amplification (Day 4)

1. Prepare the following PCR mix for each sample and keep it on ice:
If several samples are processed, a master mix with 10% excess can be prepared.

PCR mix	Volume (μL)
Low EDTA TE	10
IDT Reagent R1	5
IDT Reagent R2	4
IDT Buffer R3	10
IDT Enzyme R4	1
Total (per sample)	**30**

2. Add 30 μL of the PCR mix to the sample and pipette up and down before running the PCR with the following program:

Step	Cycles number	Temperature (lid at 105°C)	Time
Denaturation	1	98°C	30 s
Denaturation	9–12 depending on starting material (*see* **Note 14**)	98°C	10 s
Annealing		60°C	30 s
Elongation		68°C	60 s
Hold	1	4°C	Hold

3. When the PCR reaction is over, add 35 μL of AMPure XP beads according to DNA quantity, and incubate in the dark for 10 min at RT (*see* **Note 8**).
4. Place the tubes in a 0.2 mL magnet separator and incubate in the dark for 5 min at RT.
5. Remove supernatant and clean the beads on the magnet with 200 μL of freshly prepared 80% ethanol. Incubate for 1 min before discarding the ethanol.
6. Repeat Steps 5 and 6.
7. Dry the beads on the magnet by incubating for 3–5 min at RT (*see* **Note 9**).

8. Remove the tubes from the magnet and resuspend the beads in 21 μL deionized water by pipetting up and down before incubating in the dark for 10 min at RT.
9. Collect supernatant in a new tube; use 1 μL to quantify the amount of chromatin with Qubit and 2 μL to analyze the size distribution with a Tapestation system (Fig. 2).

Safe stop: libraries can be stored at −20°C for several months before to be sequenced.

3.17 Paired-End Reads Sequencing Library Preparation

1. Calculate the molarity of each sample based on the DNA fragment analysis performed with Tapestation.
2. It is recommended to sequence 400 million reads per sample to obtain enough sequencing depth for differential analysis of the loops.
3. Equimolar libraries are pooled together in one mixture depending on the sequencing equipment requirement to obtain 400 million reads per sample. Ensure that the indexes used for each library pooled in the mixture are different to give a correct identity to each sample.

3.18 Data Analysis: HiChIP Alignment, Quality Control, and Interactions Calling

HiChIP reads quality is assessed with FASTQC, using default parameters. Alignment, filtering, normalization, and chromatin loop detection can be performed using the Arima implementation of MAPS (v2.0) bash script [34]. Reads, stored in paired-end fastq files, are aligned to the adequate reference genome, and only high-confidence alignments are kept by applying a minimum mapping quality threshold of 30 (-Q 30). Different filtering steps can be applied to remove technical artifacts and non-informative read pairs. The first

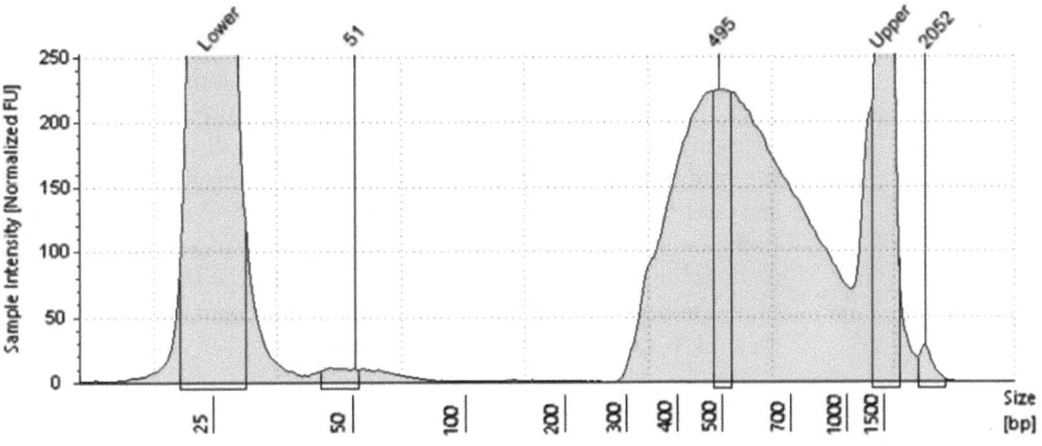

Fig. 2 Expected Tapestation size analysis result of an H3K27Ac HiChIP library pool

step in the pipeline is the read alignment and preprocessing of pair-end reads, performed with the tool feather (-F 1, input 0 to skip), which is based on the BWA aligner [35]. The second step, called MAPS (-M 1, input 0 to skip), performs read binning, normalization, and peak calling. Interactions separated by less than 1 kb are filtered out (-l 1000, length cutoff) to exclude self-ligation events or short-range contacts from the downstream analysis. Chromatin interactions are identified at a fixed genomic window (bin) resolution of 5 kb with parameter (-s 5000, bin_size), and interactions are considered up to a maximal genomic distance of 2 Mb with the parameter (-r 2000000, binning_range), allowing detection of long-range interactions. Distance-dependent interaction decay is explicitly modeled during normalization, and a statistical significance threshold for the identified chromatin interactions is set through the false discovery rate parameter (-d 2, which corresponds to fdr<0.01). The parameter (-f 1, where 1 is the value for deep-sequenced samples and 0 for shallow-sequenced ones) allows the user to collect and store in a tab-separated file all the information regarding the sample run, i.e., number and percentage of aligned reads, number and percentage of PCR duplicates, number and percentage of intra-chromosomal pairs (split in all and long-range ones, the latter defined as >15 kb genomic distance interactors), number of interactions detected, aa well as ChIP-specific quality parameters, like number of ChIP-seq peaks called with MACS2 from the data (if parameter -C 1, call-peak, was specified, input 0 to skip peak calling) and FRIP (Fraction of Reads in Peaks) [36]. This information will be stored in a file ending in "Arima_QC_deep.txt". Another parameter needed when performing peak calling is -p (peak_type), which can be either "broad," for most histone modifications, or "narrow," for Transcription Factors (TFs) HiChIP. Both options lead to MACS2 peak calls with a broad peak profile. When ChIP peaks are not called directly from HiChIP data and previous peak calls from independent ChIP-seq experiments can be used through the parameter -m (macs2_filepath), followed by the complete path to the BED/NARROWPEAK file with the peak calls. Finally, the pipeline will generate a plethora of intermediate and final output files, among which "sig3Dinteractions.bedpe" files will contain the significant interactions and can be directly visualized on the Integrative Genomic Viewer (IGV) [37], the arcplots, which can instead be visualized on Washington University (WashU) Epigenome Browser (Fig. 3) [38], the metaplots (i.e., tornado plots showing the read enrichment of the HiChIP data, generated if the parameter -P 1, plot, was specified), and the hic matrices, per chromosome (if parameter -H 1, generate_hic, was specified).

3.19 Data Analysis: Readout of Sample Quality

Different readouts can be found in the "Arima_QC_deep.txt" output file, which can be used to assess the quality of the data. Ideally, mapped reads should be higher than 85% to ensure no sample contamination. PCR duplicates are expected to be lower than 60%.

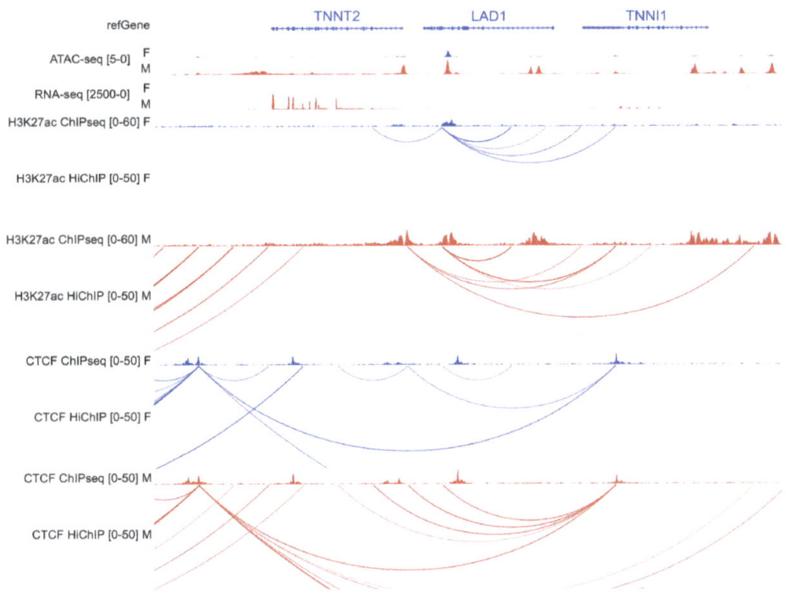

Fig. 3 Tracks of an integrated dataset from IMR90 in blue (F) and myoblast-like IMR90-MYOD cells in red (M) generated in our laboratory and visualized in Integrative Genomic Viewer (IGV) for the TNNT2 locus [21]. From top to bottom: Ruler, RefSeq gene, ATAC-seq for IMR90, ATAC-seq for IMR90-MYOD, RNA-seq for IMR90, RNA-seq for IMR90-MYOD, H3K27ac ChIPseq for IMR90, H3K27ac HiChIP loops for IMR90, H3K27ac ChIPseq for IMR90-MYOD, H3K27ac HiChIP loops for IMR90-MYOD, CTCF ChIPseq for IMR90, CTCF HiChIP loops for IMR90, CTCF ChIPseq for IMR90-MYOD, CTCF HiChIP loops for IMR90-MYOD

In our experience, the number of loops per sample for a histone HiChIP in human myoblast-derived cells is expected to be higher than 100,000 for good quality samples, while medium-quality ones may have between 50,000 and 100,000 loops. A lower number of chromatin interactions will impair the differential analysis between different experimental conditions. This number is variable depending on the species or the between histones and TF. If public datasets are available for your targets, it is recommended to estimate the expected number of interactions with them.

3.20 Data Analysis: Biological Replicates Concordance

The reproducibility of chromatin interactions across biological replicates can be quantified through the Irreproducible Discovery Rate (IDR) analysis, using IDR2D [39]. IDR2D provides Interaction scores that are transformed using inverse of log of the value (−log(x)). The reproducible interaction is identified using an overlap-based ambiguity resolution method. Non-standard chromosomes are excluded, and interaction pairs are merged using a maximum gap of 1 kb. IDR parameters are selected to balance sensitivity and stringency (max factor = 1.5, jitter factor = 0.0001, $\mu = 0.1$, $\sigma = 1$, $\rho = 0.2$, $p = 0.5$, $\varepsilon = 0.001$), ensuring robust detection of reproducible loops.

3.21 Data Analysis: Differential Chromatin Interactions

Differential chromatin interaction analysis between conditions is conducted using the DiffAnalysisHiChIP.r script implemented in FitHiChIP (v7.1) [40]. This approach models contact counts while accounting for distance-dependent interaction decay and technical variability across samples. Differential loops are filtered based on statistical significance, with a threshold of FDR < 0.05 (--DiffFDRThr), enabling the identification of condition-specific interactions. Interactions identified through the Arima pipeline need to be converted to the input format (BED) required by FitHiChIP before proceeding with the analysis. A required parameter for the DiffAnalysisHiChIP.r script is the chromosome size file (--ChrSizeFile), which can be easily retrieved from UCSC for different genomes releases, with filenames often ending in "chrom.sizes". Another required parameter is the BAM files generated by the Arima pipeline for each sample (--ChIPAlignFileList), ending in "shrt.bam". Other parameters are required to establish the classes of comparison and the number of replicates in each condition (--CategoryList, --ReplicaCount, --ReplicaLabels1, --ReplicaLabels2). The script generates a file containing the statistically significant differential interactions called "Loops_EdgeR_Default_SIG.bed". To apply different stringency parameters to filter significant interactions, it is advisable to use instead the "Loops_EdgeR_Default.bed" output file, which can be filtered by FDR, p-value, and log fold change (logFC).

3.22 Data Analysis: HiChIP Integration with ATAC-seq and RNA-seq

Integration of chromatin interaction data with transcriptomic and chromatin accessibility datasets can be performed using a combination of Bash command-line tools and downstream analyses in R. Genomic interval overlaps between HiChIP chromatin interactions, ATAC-seq peaks, RNA-seq–derived gene annotations, and additional regulatory features can be computed using BEDtools intersect [41]. Depending on the analysis objective, different intersection modes can be applied: parameters (-wa -u) are used to retain a unique list of overlapping genomic features; parameters (-wao) is used to report all overlaps, including nonoverlapping intervals, enabling downstream quantitative integration in R; and parameter (-v) is used to identify not-overlapping features. Furthermore, the user can tune the degree of overlap (defined as fraction of peak) of each set of features using the parameters -f, -F, -r, and -e. Integration of HiChIP chromatin interactions with RNA-seq data (where differential genes were converted into their promoter genomic coordinates, defined as −1000/+200 bp from the gene TSS) enables the association of distal regulatory elements with target genes based on physical chromatin contacts rather than linear genomic proximity. Gene expression changes can be subsequently interpreted in the context of chromatin accessibility changes, transcription factor occupancy (through direct overlap of the differential genes' promoters to ChIP-seq peaks of a specific TF or through motif analysis/foot-printing analysis of chromatin accessibility peaks), and three-dimensional chromatin interactions, facilitating the interpretation of the multilayered regulatory mechanisms governing transcriptional programs.

4 Notes

1. To maintain efficacy and consistency, the buffers must be prepared just before starting each indicated day and kept in ice. The buffers must not be prepared days in advance and cannot be stored for further usage. If several samples are processed, 10% excess can be prepared.
2. One cOmplete Mini Protease inhibitor cocktail tablet can be dissolved in 100 μL of deionized water to obtain a 100× solution to use for buffer preparation. The solution can be stored at −20°C.
3. From this step, the samples need to be maintained in ice except if stated otherwise. This reduces the risk of decrosslinking and will increase the quality of the Hi-C.
4. From this step, the samples can be stored as dry pellet at −80°C for months. If working with limited isolated MuSCs, they can be stored until the number is high enough to pool five million MuSCs for each experimental condition.
5. It is highly recommended to start part 3.2 with 15 μg of chromatin for each sample; starting material of five million MuSCs is sufficient for that. If working with more cells, it is possible to use a portion to quantify the amount of chromatin. In this case, 0.5 million cells need to be resuspended in 174 μL of Elution buffer and go through Sect. 3.5 and through Sect. 3.9 up to Qubit quantification to estimate the amount of chromatin.
6. The QC can be stored at −20°C if not performed directly.
7. When possible, place protein A beads in the dark as they are light sensitive.
8. AMPure XP beads need to be warmed up at RT and well homogenized before being used. They can be taken out of the fridge at the beginning of the day to be ready for DNA isolation of quality control. When possible, place AMPure XP beads in the dark as they are light sensitive. Cleaning steps are repetitive, and errors may happen; it is suggested to conserve first supernatant from beads and used beads after sample elution as backup at −20°C.
9. When drying the beads, they should not get over-dry; otherwise, solubilization of the DNA in the Elution buffer will be impaired. The beads reflect light when wet and should

remain slightly shiny before the addition of the elution buffer to ensure a proper drying method.

10. Calculate the quantity of chromatin present in the unsheared and sheared samples by multiplying by 520 and 390, respectively. This will inform you on your starting material and the amount recovered after shearing. The quantity of sheared chromatin will allow the calculation of ChIP efficiency in Sect. 3.4.2. Example: if you quantify 15 ng/μL in the unsheared control and 10 ng/μL in the sheared control. This means you have 7.8 μg of chromatin in the unsheared sample and 3.9 μg of chromatin in the sheared sample. Unsheared and sheared quality control can be run on a gel electrophoresis machine to verify the size of sheared chromatin. The size range should be between 200 and 600 bp, with a peak around 400 bp.
11. Calculate the quantity of biotinylated chromatin present in the diluted unsheared quality control by multiplying Qubit result by 7. Compare this result with the starting material of 75 ng used for this biotin enrichment. This informs you on the quality of the sample biotinylation; it should be greater than 15% to be of good quality. Example: if you have a Qubit result of 3 ng/μL, your total quantity if enriched biotinylated chromatin is 21 ng, which represents 28% of the 75 ng of starting material. Your efficiency is superior to 15% and the sample is of good quality.
12. The chromatin is now attached to the beads.
13. Calculate the quantity of chromatin present in the immunoprecipitated chromatin by multiplying the Qubit concentration by 42. It gives you information about the quantity of material you immunoprecipitated. Compare this result with the starting material of sheared chromatin you had estimated in Sect. 3.3.3. Expected results range between 0.5% and 2% depending on the abundance of the target protein. Example: if you have a Qubit result of 0.5 ng/μL, your total quantity of immunoprecipitated chromatin is 21 ng, if your quantity of sheared chromatin is 10 μg as in Sect. 3.3, your efficiency of immunoprecipitation is 0.54%, the quantity is enough for library preparation and the efficiency is in the expected range.
14. The number of PCR cycles used for library amplification depends on the result obtained in Sect. 3.5.4.

Competing Interests

The authors declare no competing interests.

References

1. Tolhuis B, Palstra RJ, Splinter E et al (2002) Looping and interaction between hypersensitive sites in the active beta-globin locus. Mol Cell 10:1453–1465
2. Crawford GE, Davis S, Scacheri PC et al (2006) DNase-chip: a high-resolution method to identify DNase I hypersensitive sites using tiled microarrays. Nat Methods 3:503–509
3. Kouzarides T (2007) Chromatin modifications and their function. Cell 128:693–705
4. Yekelchyk M, Guenther S, Braun T (2023) Assay for transposase-accessible chromatin using sequencing of freshly isolated muscle stem cells. In: Asakura A (ed) Skeletal Muscle Stem Cells. Springer, US, New York, NY, pp 397–412
5. Ding D, Braun T (2023) Efficient genome-wide chromatin profiling by CUT&RUN with low numbers of muscle stem cells. In: Asakura A (ed.) Skeletal muscle stem cells. Springer, US, New York, NY, pp 413–430
6. Bouwman BAM, Crosetto N, Bienko M (2022) The era of 3D and spatial genomics. TIG 38:1062–1075
7. Cremer T, Cremer M (2010) Chromosome territories. Cold Spring Harbor Perspect Biol 2:a003889
8. Lieberman-Aiden E, Van Berkum NL, Williams L et al (2009) Comprehensive mapping of long-range interactions reveals folding principles of the human genome. Science 326:289–293
9. Dixon JR, Selvaraj S, Yue F et al (2012) Topological domains in mammalian genomes identified by analysis of chromatin interactions. Nature 485:376–380
10. McArthur E, Capra JA (2021) Topologically associating domain boundaries that are stable across diverse cell types are evolutionarily constrained and enriched for heritability. Am J Hum Genet 108:269–283
11. Lupiáñez DG, Kraft K, Heinrich V et al (2015) Disruptions of topological chromatin domains cause pathogenic rewiring of gene-enhancer interactions. Cell 161:1012–1025
12. Sartorelli V, Puri PL (2018) Shaping gene expression by landscaping chromatin architecture: lessons from a master. Mol Cell 71:375–388
13. Bentzinger CF, Wang YX, Dumont NA, Rudnicki MA (2013) Cellular dynamics in the muscle satellite cell niche. EMBO Rep 14:1062–1072
14. Le Grand F, Rudnicki MA (2007) Skeletal muscle satellite cells and adult myogenesis. Curr Opin Cell Biol 19:628–633
15. Dong A, Liu J, Lin K et al (2022) Global chromatin accessibility profiling analysis reveals a chronic activation state in aged muscle stem cells. iScience 25:9
16. Dong A, Preusch CB, So W-K et al (2020) A long noncoding RNA, LncMyoD, modulates chromatin accessibility to regulate muscle stem cell myogenic lineage progression. Proc Natl Acad Sci U S A 117:32464–32475
17. Huang Z, Hu L, Liu Z, Wang S (2025) The functions and regulatory mechanisms of histone modifications in skeletal muscle development and disease. IJMS 26:3644
18. Narita T, Kilic S, Higashijima Y et al (2025) Disentangling the architectural and non-architectural functions of CTCF and cohesin in gene regulation. Nat Genet 57:3137–3151
19. Allen BL, Taatjes DJ (2015) The Mediator complex: a central integrator of transcription. Nat Rev Mol Cell Biol 16:155–166
20. Dall'Agnese A, Caputo L, Nicoletti C et al (2019) Transcription factor-directed re-wiring of chromatin architecture for somatic cell nuclear reprogramming toward trans-differentiation. Mol Cell 76:453–472.e8
21. Nicoletti C, Massenet J, Pintado-Urbanc AP et al (2025) MYOD represses gene expression from non-E-box motifs. Genes Dev 39:1299–1317
22. Massenet J, Weiss-Gayet M, Bandukwala H et al (2024) Epigenetic control of myogenic identity of human muscle stem cells in Duchenne muscular dystrophy. iScience 27:12
23. Dekker J, Rippe K, Dekker M, Kleckner N (2002) Capturing chromosome conformation. Science 295:1306–1311
24. Simonis M, Klous P, Splinter E et al (2006) Nuclear organization of active and inactive chromatin domains uncovered by chromosome conformation capture-on-chip (4C). Nat Genet 38:1348–1354
25. Van DeWerken HJG, DeVree PJP, Splinter E et al (2012) 4C technology: protocols and data analysis. Methods in enzymology. Elsevier, pp 89–112
26. Stadhouders R, Kolovos P, Brouwer R et al (2013) Multiplexed chromosome conformation capture sequencing for rapid genome-scale high-resolution detection of long-range chromatin interactions. Nat Protoc 8:509–524
27. Dostie J, Richmond TA, Arnaout RA et al (2006) Chromosome Conformation Capture Carbon Copy (5C): a massively parallel solution for mapping interactions between genomic elements. Genome Res 16:1299–1309
28. Rao SSP, Huntley MH, Durand NC et al (2014) A 3D map of the human genome at kilobase resolution reveals principles of chromatin looping. Cell 159:1665–1680
29. Belaghzal H, Dekker J, Gibcus JH (2017) Hi-C 2.0: an optimized Hi-C procedure for

high-resolution genome-wide mapping of chromosome conformation. Methods 123: 56–65
30. Mumbach MR, Rubin AJ, Flynn RA et al (2016) HiChIP: efficient and sensitive analysis of protein-directed genome architecture. Nat Methods 13:919–922
31. Mifsud B, Tavares-Cadete F, Young AN et al (2015) Mapping long-range promoter contacts in human cells with high-resolution capture Hi-C. Nat Genet 47:598–606
32. Jerkovic´ I, Cavalli G (2021) Understanding 3D genome organization by multidisciplinary methods. Nat Rev Mol Cell Biol 22:511–528
33. Machado L, Esteves de Lima J, Fabre O et al (2017) In situ fixation redefines quiescence and early activation of skeletal muscle stem cells. Cell Rep 21:1982–1993
34. Juric I, Yu M, Abnousi A et al (2019) MAPS: model-based analysis of long-range chromatin interactions from PLAC-seq and HiChIP experiments. PLoS Comput Biol 15:e1006982
35. Li H, Durbin R (2009) Fast and accurate short read alignment with Burrows–Wheeler transform. Bioinformatics 25:1754–1760
36. Zhang Y, Liu T, Meyer CA et al (2008) Model-based Analysis of ChIP-Seq (MACS). Genome Biol 9: R137
37. Robinson JT, Thorvaldsdóttir H, Winckler W et al (2011) Integrative genomics viewer. Nat Biotechnol 29:24–26
38. Zhou X, Maricque B, Xie M et al (2011) The human epigenome browser at Washington University. Nat Methods 8:989–990
39. Krismer K, Guo Y, Gifford DK (2020) IDR2D identifies reproducible genomic interactions. Nucleic Acids Res 48:e31
40. Bhattacharyya S, Chandra V, Vijayanand P, Ay F (2019) Identification of significant chromatin contacts from HiChIP data by FitHiChIP. Nat Commun 10:4221
41. Quinlan AR, Hall IM (2010) BEDTools: a flexible suite of utilities for comparing genomic features. Bioinformatics 26:841–842

Index

Nathalie Didier and Caroline E. Brun (eds.), *Muscle Stem Cells*, Methods in Molecular Biology, vol. 3093, https://doi.org/10.1007/978-1-0716-5643-3,

FSC
www.fsc.org
MIX
Papier | Fördert gute Waldnutzung
FSC® C083411

Kolibri 360 GmbH
Ferdinand-Jühlke-Straße 7
99095 Erfurt, Deutschland
produktsicherheit@kolibri360.de